Environmental Physics

Environmental Physics

Egbert Boeker

and

Rienk van Grondelle
Vrije Universiteit Amsterdam

JOHN WILEY & SONS
Chichester · New York · Brisbane · Toronto · Singapore

Other Wiley Editorial Offices

John Wiley & Sons, Inc., 605 Third Avenue,
New York, NY 10158-0012, USA

Jacaranda Wiley Ltd, 33 Park Road, Milton,
Queensland 4064, Australia

John Wiley & Sons (Canada) Ltd, 22 Worcester Road,
Rexdale, Ontario M9W 1L1, Canada

John Wiley & Sons (SEA) Pte Ltd, 37 Jalan Pemimpin #05-04,
Block B, Union Industrial Building, Singapore 2057

Library of Congress Cataloging-in-Publication Data

Boeker, Egbert.
    Environmental physics / Egbert Boeker, Rienk van Grondelle.
        p.   cm.
    Includes bibliographical references and index.
    ISBN 0-471-93931-5 (cloth):—ISBN 0-471-95110-2 (pbk.)
    1. Environmental sciences.   2. Physics.   3. Atmospheric physics.
I. Van Grondelle, Rienk.   II. Title.
GE105.B64   1994
628—dc20   94-9542
                                                          CIP

British Library Cataloguing in Publication Data

A catalogue record for this book is available from the British Library

ISBN 0 471 93931 5 (cloth)
ISBN 0 471 95110 2 (paper)

Typeset in 10/12pt TImes by Thomson Press (India) Ltd, New Delhi, India
Printed and bound in Great Britain by Biddles Ltd, Guildford and King's Lynn

Contents

Foreword

Environmental science: a great necessity and a great dilemma! All aspects of science meet and the environmental scientist is supposed to be knowledgeable in all of them. The future of human survival will depend heavily on environmental science being taken seriously and not being dominated by windbags. The problems to be faced in environmental science are real and are extremely difficult to understand and to solve.

Physics of all kinds is heavily involved in our surroundings, and a sound background in this discipline is indeed indispensable. This textbook demonstrates the utility of physics for understanding processes in the environment. It takes a middle course and combines basic principles with their application to important questions of environmental science. The book cannot stand alone; it requires on the one hand a textbook of basic physics for an in-depth understanding of the underlying principles. On the other hand it requires detailed treatises of different aspects of environmental science. *Environmental Physics* forms a bridge between the two poles and helps students to find their own way from the basic physics courses to the course room of practical environmental scientists.

I hope that this textbook helps to educate a new generation of knowledgeable and responsible environmental scientists who not only criticize but actively and creatively help to sustain life on earth for a long time to come.

Richard R. Ernst
Professor of Physical Chemistry
ETH-Zürich

Preface

The need to write this book originated in the Physics Department of a medium sized European university, where it was observed that physicists in their professional career become more and more confronted with environmental problems. Physical methods and physical scientists will be needed to analyse environmental data, to estimate environmental effects of production and consumption, to analyse the consequences of policy measures, to prepare government regulations with well-defined physical aspects and to handle vast amounts of data by computer methods. The last applies both for organizing and visualizing experimental data and for the design of models from a more theoretical point of view.

Environmental Physics is more than the study of the physical environment; it indeed comprises aspects of atmospheric physics, soil physics and also many aspects of applied physics and of engineering. Rather than giving a superficial discussion of all these aspects we wish to transmit to the reader an awareness of the physics which will help to analyse, prevent or mitigate environmental problems. Therefore, some major experimental and theoretical methods are discussed with this objective in mind.

We usually avoid the term 'solving' environmental problems, as the solution cannot come from physicists or from scientists alone and not even in the 'first place. A sustainable society can only be achieved by political decision making based upon a public awareness that certain perhaps unpopular measures should be taken now in order to prevent worse in the future. Physicists of course play a crucial role in defining the problems and in finding ways into the future. This book deals with the relevant physics; it is our experience, however, that students are motivated by putting the subject matter in a social context. Therefore we have added a final chapter on the context of society with some 'topics for class discussion' which should clarify the social relevance of the subject matter.

This book is written at an advanced undergraduate level, that is for readers who have a general knowledge of physics and a working knowledge of mathematical methods. The usual physics degree courses only scantily discuss subjects like diffusion or spectroscopic methods. Therefore each chapter starts in a rather elementary way, increasing in difficulty later on. The elementary topics may have been covered by other courses but in that case they are helpful to refresh the memory.

In Chapter 1 we present the organization of the book in a very elementary way. It should point out the cohesion between the many seemingly unrelated topics. Some exercises are added to each chapter which are essential in understanding the main text by looking at it for a second time. They do not in the first place aim at improving the student's problem solving ability.

It is well known that chemistry and biology have made substantial contributions to the identification and analysis of environmental problems. The boundary between these disciplines and physics is not always clear as physical methods are often used by the other natural sciences. Therefore, the methods to be discussed are widely used by the other disciplines as well.

A textbook is no encyclopaedia. Experts will notice that many scientific developments which seem promising such as magneto hydrodynamic energy conversion have been left out while others with the same chance of becoming important, such as nuclear fusion, are covered. These choices reflect the authors' subjective opinion of what subjects can be discussed on the level of this book with a chance that students will use them in their careers.

The authors and their colleagues are using the present text in three lecture courses of 20 to 25 hours each. We take the experimental chapters, 2 and 7, together and divide the rest into a course on transport problems (Section 4.1, Chapter 5 and sometimes 6) and one on climate and energy (Chapters 3, 4 and usually 8). The number of cross references to earlier chapters has been limited deliberately so one may choose the order of the chapters as one likes. When cross references do occur, the student will have no difficulty in reading the relevant part of the other chapter, wetting the appetite.

Finally, the authors will be grateful for any comments and suggestions from colleagues and students to improve the text and the exercises.

Amsterdam, 1 December 1993

Acknowledgements

The authors are grateful to the thesis student Matthieu Visser who wrote drafts of much of Section 7.6. They are also indebted to Paul van Kan who for a lecture course in environmental physics provided them with notes on physical methods of analysis. The computer group of the department was helpful in programming the greenhouse model discussed in Chapter 3.

The first author wrote part of the chapters while on sabbatical leave in the United States. He is grateful to Prof. Heymann for hospitality at Rice University and to Prof. Socolow for offering a position as visiting fellow at the Center of Energy and Environment Studies while at Princeton University.

The authors appreciate the discussions with Prof. Dunn of Reading on energy and thermodynamics. They acknowledge the contributions of Prof. Schuurmans to Chapter 3 and the discussions with Peter Siegmund on details of that text. The book profited from discussions with Prof. Ross on Section 4.2.2, Dr R. Meijer on the Stirling engine, Prof. Loman on Section 4.2.8, Dr P. Bosma and Dr D. G. de Groot on Section 4.4.1, Dr Al Cavallo on Section 4.4.2, from the comments of Professor Lidsky, Dr K. Abrahams and Dr A. J. Janssen on Section 4.5.1, Prof. van de Wiel on Section 4.5.2, Dr A. Bos on Section 4.5.3, Frans Berkhout and Prof. Hogervorst on Section 4.5.4, Prof. Aldama on Section 5.2, Prof. de Vries on Section 5.3, Ing. J. J. Erbrink on Section 5.6, Prof. Boersma and Prof. Plomp on Chapter 6, Dr H. van der Woerd on Section 7.6.1, D. P. J. Swart on Section 7.6.2, Dr R. D. Vis on Section 7.6.4, Dr W. Smit on Section 8.1, Dr H. Feiveson on Chapter 8, Prof. Tuininga and Dr J. Grin in the concept and contents of Chapter 8 and Prof. Andriesse on Section 8.2.

Many thanks are due to Annette Kik who so patiently typed and retyped the manuscript, which was hard work, even with editing facilities. Emiel Corten was very helpful in assisting with the use of FrameMaker on which the typing was done. Luigi Sanna has done a fine job in drawing and redrawing all figures.

1

Introduction: the Essentials of Environmental Physics

Environmental physics is defined in a broad sense as the physics concerning the identification and measurement of environmental problems; it is devoted to the prevention of problems and to the alleviation of remaining problems. It makes use of the instrumentation techniques that are being applied in various parts of physics and it needs the methods of mathematical physics. The problem field of environmental physics is concerned with society in general and its economic system in particular. This is illustrated in the following sections.

1.1 THE ECONOMIC SYSTEM

The economic system is sketched in Fig. 1.1, where it is very much simplified and reduced to its aspects of energy sources and raw materials on the one hand and the energy conversion and end uses of energy and materials on the other.

Let us look at the different parts of Fig. 1.1, most of which correspond to chapters in this book. Energy sources start to become polluting when they are converted into mechanical energy and electricity (Chapter 4) and put to end use in agriculture, industry, services, transport or in the household. Especially in the sector of the end uses, noise pollution may be experienced (Chapter 6). Air pollution may give rise to climate change (Chapter 3) and pollutants are transported by the air or the water to places far from its origin (Chapter 5). In all cases measurements are required to monitor pollution of all kinds (Chapters 2 and 7).

This book is aimed at physical scientists who want to apply their abilities to environmental problems. We therefore want to develop their specific abilities. Part of the strength of physical scientists lies in instrumentation, the development of measuring techniques and the application of existing techniques. Therefore in Chapter 2 a general introduction is given in spectroscopy while in Chapter 7 more methods are discussed more extensively. Another part of the abilities of

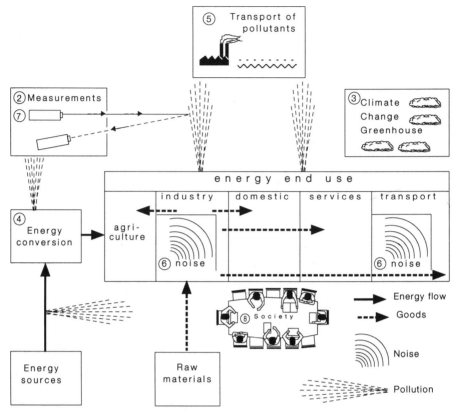

Figure 1.1 The economic system of production and consumption. Inputs are energy and raw materials, output goods and services. Pollution occurs at all stages. Indicated are the chapters where the emphasis is on certain parts of the system.

physical scientists lies in making simple models of complicated phenomena using mathematical techniques. Therefore the relevant portions of classical mathematical physics are discussed using environmental applications as a peg. Here the focus is on concepts and methods rather than on completeness.

As is clear from Fig. 1.1, many environmental problems originate in energy conversion and energy end use. Therefore Chapter 4 on energy takes a central part in the book. Here, the development of energy conservation methods in production and consumption and the design of alternative sources of energy is discussed.

Finally, environmental problems are defined as problems by human society and their solutions therefore can only be developed in a societal context. That is the reason why in Chapter 8 some attention is devoted to the societal framework in which environmental problems should be prevented, solved or mitigated.

In order to structure the book three areas of environmental physics have been selected to illustrate the physical context: the greenhouse effect that keeps the earth habitable, solar energy as a renewable energy source and transport physics that describes how pollutants move from one place to another. Examples of each of these subjects are given below.

1.2 LIVING IN THE GREENHOUSE

Life as we know it is intimately related to some physical properties of the sun–earth system. We mention:

(a) The surface temperature of the sun happens to be around 5800 K, which results in an emission spectrum with a maximum of emitted energy at wavelengths of 500 nm, corresponding to photon energies of 2.48 eV. These energies just correspond to excitations between molecular orbits or to specific ionization energies. Thus the solar surface temperature possesses exactly the right magnitude to induce photochemical reactions.

(b) The earth–sun distance (1.49×10^{11} m) and the earth radius (6.4×10^6 m) result in interception by the earth of 0.46×10^{-9} fraction of solar energy. With the assumptions that the sun emits black body radiation, has a radius of 6.96×10^8 m and a temperature of 5800 K one may calculate that the earth receives 1399 W m^{-2} from the sun. The actual value of the solar constant S is somewhat lower. Below we shall see that with the present value of the total emitted solar energy this—albeit with the help of some atmospheric gases—is precisely sufficient to maintain an average earth surface temperature of 288 K (15 °C).

(c) The mass of the earth ($m = 6 \times 10^{24}$ kg) and its radius are such that the earth can keep an atmosphere. The gravitation is sufficiently strong to tie essentially all the molecules.

A relatively small change in these parameters would have made life as we know it impossible. In this section we shall look a little closer at the second point: the temperature.

In a stationary situation the energy received from the sun equals the energy that the earth itself radiates into space. The first is given by the solar constant $S = 1353$ W m^{-2}, the solar energy intercepted by a square metre, perpendicular to the sun. For the latter it is assumed that in a first approximation the earth radiates as a black body of temperature T. Next one takes into account that a certain fraction of the sunlight, called the *albedo a*, is reflected back into space.*

---

*The albedo of the moon of about 0.06 is responsible for its bright appearance in the sky. The earth's albedo reflects back light that is illuminating the parts of the moon turned away from the sun—its grey part. Changes in the earth's albedo will change the brightness of this part of the moon.

The energy equation then becomes

$$(1 - a)\pi R^2 S = 4\pi R^2 \sigma T^4 \tag{1.1}$$

Here σ is the Stefan–Boltzmann constant and R the earth radius. With an estimated value of the albedo of about $a = 0.34$ one finds $T = 250$ K. One usually puts it at $T = 255$ K. This should be the radiation temperature of the earth, viewed from far away.

The surface temperature of the earth on the average is 288 K and of the atmosphere 255 K. The relatively high surface temperature is due to the presence of gases like water vapour, CO_2, O_3, N_2O and CH_4, to name the most crucial ones in the order of their contribution to the effect. These gases absorb most of the heat radiation of the earth and re-emit it back towards the earth's surface. To understand this one should realize that the differences in temperature of the sun and of the earth result in quite a different spectrum. This is shown in Fig. 1.2 where for temperatures of 5800 and 288 K the black body spectra are displayed.

The atmosphere is relatively transparent for short wavelengths but it absorbs the longer wavelengths because of the presence of the so-called greenhouse gases mentioned above. The atmosphere then operates as a kind of 'blanket', keeping the surface liveable. A similar process occurs in the agricultural greenhouses,

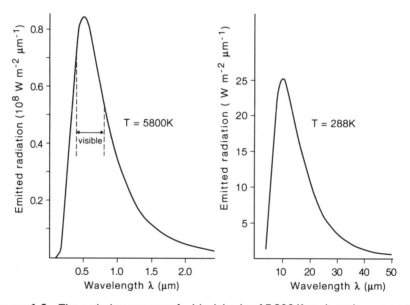

Figure 1.2   The emission spectra of a black body of 5 800 K such as the sun on the left and of a black body of 288 K such as the earth surface on the right. Note the differences in scale of both figures. Even at the peak of the earth's emission spectrum the solar emission is much higher.

although an important part of the effect there is that the cover prevents nightly convection and disappearance of air heated during the day.

At present the concentration of the greenhouse gases continuously increases due to natural and human causes. As we shall see in Chapter 2, the different gases absorb in different parts of the spectrum; therefore, when in a certain wavelength region the absorption would be 100% a further increase in concentration would not have any additional effect.

It is obvious that the increase in greenhouse gases will strongly influence the climate and on the average result in an increase in the surface temperature of the earth. It is not clear how fast the process develops and whether it would be detrimental for all regions on the earth. The mechanism of climate change involves an interplay of many physical and chemical processes that will be briefly discussed in Chapter 3. In this chapter the uncertainties of the various methods to calculate long term effects will also become apparent.

1.3 ENJOYING THE SUN

The sun not only maintains a liveable temperature on earth but part of its energy is stored by photosynthesis and another part acts as a driving mechanism behind the daily and annual change of the weather.

Photosynthesis is the process that plants, algae and a variety of (photosynthetic) bacteria use to store the light energy from the sun in a stable product. *All* photosynthetic organisms use the pigment chlorophyll and to a lesser extent carotene to absorb the incoming solar energy and to initiate the energy storage process.

The net reaction of photosynthesis is given by

$$6\,H_2O + 6\,CO_2 + 4.66 \times 10^{-18}\ \text{joule} \rightleftharpoons C_6H_{12}O_6 + 6\,O_2 \qquad (1.2)$$

During the night the reaction runs to the left, the respiration of the plant. During growth net energy is being stored. This may be retained by using the plant as food, by burning it as biomass or by letting nature follow its course, which sometimes results in dehydration of the plants and the formation of fossil fuels. The use of stored bioenergy is described in more detail in Chapter 4. At present it may suffice to notice that all uses of biomass let the process (1.2) run to the left, liberating bound CO_2, and thus increasing the greenhouse effect discussed above.

One way out of this dilemma is to grow new plants at the same rate that biomass is burned. We shall see that this is possible to some extent, but that at best it will produce only part of the energy required for a reasonable standard of living.

It is therefore useful to find ways of harnessing solar energy without the detour of biomass production. The easiest way is to use solar heat for heating homes or water for central heating and the taps. This will be discussed in Chapter 4, together with insulation to keep heat inside the homes, thus economizing on energy use. A more complicated way to harness solar energy is to convert solar radiation directly into electricity. The principles of solar cells will be discussed in Chapter 4 as well.

Finally one may use the fact that solar radiation is heating the earth more at the equator than at the poles. The result is that warm air rises at the equator and moves to the poles and cold air moves from the poles to the equator. A similar energy transport takes place in the great streams of the oceans. In both cases the actual flow is made complicated by the rotation of the earth, which relative to the earth is described by a Coriolis inertia force. The energy transport nevertheless gives rise to winds from which energy might be tapped. This process will be described in Chapter 4.

1.4 TRANSPORT OF MATTER, ENERGY AND MOMENTUM

Even without human interference many transport processes occur in nature. Under the influence of the sun as a driving force water is evaporated, forming vapour and clouds that move to another place where water precipitates again as rain and snow. These penetrate the earth's surface, forming ground water flows that may come to the surface again at another place. Here obviously not only matter but also energy is being transported. By evaporation a considerable amount of energy is stored in the water vapour as latent heat, which becomes free again on condensation. All this contributes to energy transport on a global scale, from the hot tropics to the colder regions of high latitudes in the order of $200\,\mathrm{MW\,m^{-1}}$.

Transport of matter without (much) energy transport occurs when dust particles are moved by the winds or when clay is moved by streams. Transport of energy without matter happens in the solar irradiation of the earth. Transport of momentum is usually coupled to that of matter or energy. However, in elastic collisions it is essentially momentum that is being transferred in the change of directions of the interacting particles.

In human society the transportation of matter, energy and momentum is being organized deliberately. Transport physics may be applied to the processes in factories where materials have to move and it describes the motion of fluids in pipes and pipelines. One of the reasons for studying transport processes in this book is that optimizing these may save energy and materials, thereby mitigating environmental problems. Another one is that environmental pollution is caused by movement of poisonous materials from one place, for example a waste disposal site, to ultimately the human body. The physical aspects here are transport by the ground and surface water or by the air. The other side of the coin is that by building high enough chimneys pollutants may be diluted to 'acceptable' concentrations when they reach the ground. These topics will be discussed in Chapter 5 with the appropriate differential equations.

One should be aware of the fact that biological processes which are outside the scope of physics play a crucial role here. Plants and animals may pick up pollutants; animals in particular may move to another place and plant debris may move on the winds and the waters; ultimately part of the pollutants will enter the food chain.

Finally, heat transport will be discussed in connection with the conversion and use of energy. Heat energy, for example, has to be kept inside human dwellings (in the winter) or outside (in the summer); in this context isolation and insulation will be studied.

1.5 THE SOCIAL AND POLITICAL CONTEXT

The natural environment has been changing even without human interference. In Section 3.3 it will be explained how the global temperature changes because of astronomical periodicities and in Fig. 3.13 it will be illustrated how the surface temperature varied even in prehistoric times. Animal life has been changing all the time, and species have appeared and disappeared; in earlier times this has happened not because of man but because of 'natural' causes.

In ancient times people changed the natural environment in order to supply food by agriculture; they improved the output by irrigation and built cities to optimize production even further. Since then dying a natural death no longer meant being devoured by wild animals, but was seen as the perspective of dying at home from old age.

Environmental protection therefore cannot mean keeping the natural environment as it is now but rather should mean environmental management or controlling environmental change. Its necessity is obvious. Air pollution occurs in big cities and near industrial concentrations and water pollution is readily observable in countries with a heavy use of fertilizers.

Some of these effects were already felt in ancient times. The old city of Rome, for example, was not a healthy place in which to live. The special aspect of modern times is the scale of the pollution and the fact that its effects are felt far away, often across country boundaries. One can no longer escape modern society in a forest, but will be confronted with dead trees from acid rain. One cannot look out of the window of an aeroplane without noticing the haze over industrial conglomerates; less visible but noted in the newspapers is the atmospheric deterioration resulting from chlorofluorocarbons, the CFCs, emitted by decaying refrigerators and aerosol cans.

The question then rises as to how environmental management should be arranged. This question falls outside the scope of physics, but physical scientists will participate in the discussions of the environmentalists and politicians.

Environmentalists distinguish nineteen components of the biosphere, displayed in Table 1.1, which one might consider as a checklist of discussion topics. All of these are changing in time, both by natural and by human causes. The point,

Table 1.1 Nineteen factors influencing the biosphere (Reproduced by permission of Cambridge University Press from B. L. Turner II (ed.), *The Earth as Transformed by Human Action, Global and Regional Changes in the Biosphere over the Past 300 Years*, Cambridge University Press, Cambridge, 1990, Tables 1–2, p. 6)

Land	Biota	Water	Chemicals and radiation	Oceans and atmosphere
Land transformation	Terrestrial fauna	Water management	Carbon	Atmospheric trace constituents
Forests	Marine biota	Water quality and flows	Sulphur	Marine environment
Soils	Flora	Coastal environment	Nitrogen and phosphorus	Climate
Sediment	Human population		Trace pollutants	
			Ionizing radiation	

however, is that the change from human causes in many cases has surpassed those from natural causes in recent years. Industrial emissions now multiply the annual natural releases of CO_2 by a factor of 1.2, of S by 2, of As by 3, of Cd by 7, of Hg by 10 and of Pb by 25. Moreover, new synthetic organic chemicals that did not exist before have been created and are being dispersed and their number is still increasing.

Looking at Table 1.1 one could, for example, note that forests are disappearing rapidly with the accompanying loss of species, both of animals and of plants. This results in a loss of biological diversity on earth that can be mitigated by conservation of selected species in zoos or natural reserves; only part of the diversity can be protected in this way. This seems to be serious, as natural plants are still sources of pharmaceuticals that are so complicated that they cannot be synthesized (as yet—one should add). Moreover, a stock of diversity is required to act as a 'pool' of properties from which new forms of plants can be bred if the existing crops become too vulnerable to diseases or to environmental change.

The social and political context in which environmental change takes place will be discussed in Chapter 8. Most important is the growth of the world population, which is increasing rapidly from 5 billion around 1990 to an estimated 10 billion in 2050. Since Malthus made his prediction people have asked whether nature could produce enough food and shelter to accommodate such an increase. Despite recurring pessimism this has been possible so far. The reason is that technological advances made much more efficient production possible and much of the poverty in the Third World may be ascribed to political and economic circumstances (military expenditures and a bad performance on the world market).

This book focuses on physical–technical aspects. In Chapter 8 they are put in a broader societal context, for it is clear to the authors that technical solutions alone, however necessary they are, will not be sufficient. As governments in various countries are beginning to realize, changes in the social system and ultimately a change in human behaviour, in the pattern of consumption and production, are unavoidable. The political question concerns who should pay for this change, the poor countries or the wealthy, and within each country again the question is who should pay the bill, the poor or the rich. Or will we try to pass on our problems to future generations?

It will be suggested that there is a wider philosophical context as well. A growing number of scientists believe that people should change their way of thinking about nature as a prerequisite for a change in living patterns and of human behaviour. It is pointed out that the scientific method consists of cutting a problem into smaller, solvable ones. Nature is then considered as a machine where every part can be studied separately.

There are two reasons why this world view should be questioned from a scientific point of view. In the first place it has become clear that even in classical physics the non-linearity of the differential equations governing, for example, weather and climate make accurate long time predictions questionable. The determinist aspect of the traditional world view is questioned. Moreover, recent experiments that have confirmed the basis of quantum mechanics and refuted the so-called Einstein–Podolski–Rosen paradox are quoted to support this view [1]. In the experiments it became clear that measurements performed several metres apart are correlated in such a way that one needs a single wave function to describe the system of particles.

It may well be that both arguments point to a more organic view of nature, even from a scientist's point of view. It should be considered whether besides the necessary political measures a new view of nature is also called for to preserve the environment and, ultimately, human life.

Reference

1. H. Primas, Umdenken in der Naturwissenschaft, GAIA, *Ecological Perspectives in Science, Humanities and Economics*, **1** (1992) 5–15.

2

Elementary Spectroscopy

Measurements in the environmental sciences often result in a spectrum. In this chapter we therefore show the type of information contained in the solar spectrum and the role of ozone in the upper atmosphere. In Chapter 7 we will discuss in more detail how to interpret spectral data and how to draw conclusions.

2.1 INTRODUCTION TO THE SOLAR SPECTRUM

Light reaching the earth from the sun is essential for life. The detailed balance between the inflow and outflow of solar energy establishes the temperature of the earth's surface. The absorption of solar light by the photosynthetic pigments drives a unique energy conversion process that enables plants, algae and a variety of photosynthetic bacteria to store solar energy into chemical free energy for later use.

The photosynthetic process has been the basis for all fossil fuels present on earth and is providing mankind with food and shelter. The development of artificial photosynthesis may be a way of coping with the energy crisis in the long run. Figure 2.1 shows the solar emission spectrum and how it overlaps with the spectra of a variety of pigments.

When the solar light arrives on the earth's surface, it is composed of a wide range of frequencies characteristic for (a) the emitter, the sun, (b) specific elements at the solar surface, and (c) the composition of the atmosphere, since the light is transmitted by the earth's atmosphere. Similarly, light reflected by the atmosphere or by the earth's surface and detected by a satellite contains information about their chemical composition.

It is well known that the presence of ozone (O_3) in the atmosphere protects the earth from the harmful solar UV radiation by absorption of all light with a wavelength less than 295 nm. Decrease of the O_3 layer will not only increase the amount of UV light at a particular wavelength but it will also allow the transmission of increasingly shorter wavelength radiation. This implies that

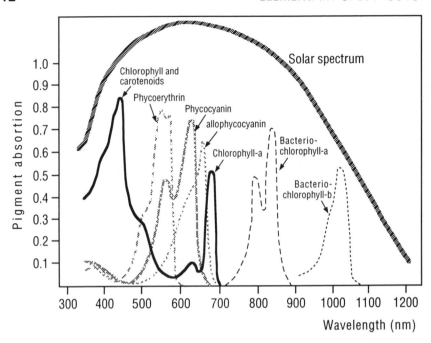

Figure 2.1 The solar spectrum as it reaches the surface of the earth. It overlaps with the important pigments of a variety of photosynthetic organisms. Chlorophyll *a*, the major pigment of higher plants, algae and cyanobacteria, absorbs red and blue light. In combination with carotenoids (such as *β*-carotene) they provide plants with their typical green colour. The major photosynthetic organisms of the world oceans, the cyanobacteria, absorb sunlight by a specialized pigment containing protein, the phycobilisome, which contains as major pigments phycoerythrin ($\pm$ 580 nm), phyco-cyanin ($\pm$ 620 nm) and allophycocyanin ($\pm$ 650 nm). All the light energy absorbed by the pigments can be used for photosynthesis. Bacteriochlorophyll *a* and bacterio-chlorophyll *b* are the major pigments of two classes of photosynthetic bacteria absorbing in the near infrared part of the spectrum

important biomolecules such as DNA and proteins, which were well protected against solar photons, may be damaged in the future.

Finally, the earth does not only absorb and reflect light, but it is also a light emitter. The emission is that of a black body at 288 K and thus far in the infrared. The detailed earth's energy balance is based on the emission and partial absorption of this infrared light and any change in any of these processes may disturb the balance. Since CO_2 shows a few strong absorption bands for infrared light, its presence in the atmosphere is of crucial importance in regulating the energy balance.

All the above-mentioned physical phenomena are intimately related to the field of atomic and molecular spectroscopy. Why are specific colours of the light emitted or absorbed by atoms and molecules? Can we quantify the amount of absorption? Is an absorption band narrow or wide and what does this imply? Can we quantify the changes in the atmospheric transmission spectrum due to changes in O_3 and CO_2 concentration (or the presence of other gases)? How do we quantitatively analyse the data accumulated by satellites or by detection from the earth's surface? Of a more fundamental nature are questions related to the solar energy conversion by photosynthesis and the design of artificial photosynthetic systems.

Since answering these questions requires a basic understanding of the interaction of light with matter and some knowledge of the terminology of spectroscopy we shall introduce the necessary concepts. For a more quantitative treatise of many of the theories discussed in this chapter, we refer to some general sources on spectroscopy [1–4] and to Chapter 7. We shall finish this chapter with a short discussion of the O_3 problem as viewed from spectroscopy.

2.1.1 BLACK BODY RADIATION

A body that absorbs and emits electromagnetic radiation without favouring particular frequencies is called a 'black body'. In first-order approximation the sun may be considered as a black body radiating at 5800 K. The intensity and spectral properties of the light emitted by a black body were originally described by the Stefan–Boltzmann law and Wien's law (1894). The Stefan–Boltzmann law formulates how the total energy density inside a black body relates to the temperature by

$$I(T) = \sigma T^4 \tag{2.1}$$

where I would be the density per unit area emitted by the black body through a pinhole in its surface, T is the absolute temperature and σ is the Stefan–Boltzmann constant, which is independent of the material and is equal to $\sigma = 5.67 \times 10^{-8}$ W m^{-2} K^{-4}. Therefore, 1 cm^2 of a black body at a temperature of 1000 K radiates about 5.7 W.

Wien's displacement law shows how the maximum of the energy distribution of black body radiation depends on the temperature by

$$\lambda_{max} T = \text{constant} \tag{2.2}$$

with the constant on the right-hand side equal to 2.989×10^{-3} m K. Therefore, a black body at 6000 K will emit maximally at 500 nm.

In the beginning of the twentieth century Rayleigh attempted to calculate the spectral distribution of a radiating black body from classical physical theory. His result, later corrected by Jeans, is given by

$$dU = \frac{8\pi \nu^2 kT}{c^3} \, d\nu \tag{2.3}$$

where dU is the energy density in the frequency interval between v and $v + dv$, k is the Boltzmann constant and c is the speed of light. The energy distribution that this law predicts is not realistic, as it leads to infinite energy density at high frequencies and is in total disagreement with Wien's displacement law.

Planck then solved the problem in one of the historic contributions to the emergence of quantum mechanics by making the assumption that the energy of an oscillator of the electromagnetic field is not continuously variable, but is restricted to integral multiples of the energy hv, where hv is called a quantum of energy and h is Planck's constant. Thus, Planck arrived at the Planck energy distribution:

$$dU = \frac{8\pi h v^3}{c^3}\left(\frac{e^{-hv/kT}}{1 - e^{-hv/kT}}\right)dv \qquad (2.4)$$

In the limit $v \to 0$, eq. (2.4) leads to the Rayleigh–Jeans law (2.3), while for $v \to \infty$ we obtain a form of Wien's law (2.2).

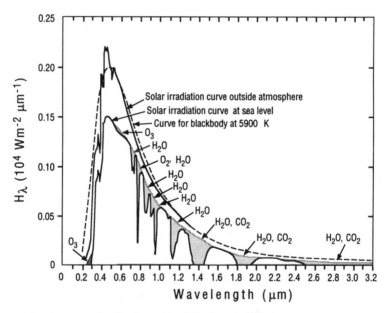

Figure 2.2 Spectral distribution of incident solar radiation outside the atmosphere and at sea level. Major absorption bands of some of the important atmospheric gases are indicated. (Reproduced by permission of McGraw-Hill from S. L. Valley (ed.), *Handbook of Geophysics and Space Environments*, McGraw-Hill, New York, 1965, Fig. 16.1, p. 16.2) The emission curve of a black body at 5900 K is shown for comparison

2.1.2 THE EMISSION SPECTRUM OF THE SUN

In Fig. 2.2 we show the emission spectrum of the sun, both measured outside the earth's atmosphere and measured at the earth's surface. In addition, the spectral distribution for black body radiation, as given by eq. (2.4) at a temperature of 5900 K is shown. The dotted areas in Fig. 2.2 indicate the specific contributions of the atmospheric gases to the absorption of solar radiation. We note the contribution of O_3 to the absorption of UV light between 200 and 300 nm, the strong absorption of H_2O in the near infrared and the strong infrared transitions of CO_2.

Figure 2.3a shows the absorption curve of the atmosphere measured over a horizontal distance of 1.8 km at sea level which is equivalent to the vertical atmosphere (it has the same optical density, to be introduced in eq. 2.29). Note the many and strong contributions of H_2O and O_2 extending far into the infrared with the complete cut-off of IR radiation by CO_2 for $\lambda > 12\,\mu m$. This latter aspect is further emphasized by the spectra of some of the individual atmospheric gases shown in Fig. 2.3b.

These spectra clearly demonstrate that the spectral distribution and intensity of sunlight reaching the earth's surface are to a large extent determined by the

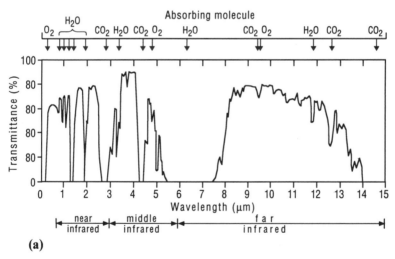

(a)

Figure 2.3 Absorption properties of the earth's atmosphere and of some of its major constitutions in the infrared region of the spectrum. (a) Horizontal absorption spectrum measured over a distance of 1.8 km at sea level. (Reproduced, with permission, from Hudson and Hudson (1975) in R. M. Measures, *Laser Remote Sensing*, Wiley-Interscience, New York, 1984, Fig. 4.6, p. 140) (b) Vertical absorption spectrum of the earth's atmosphere (bottom) and absorptions of individual compounds. (Reproduced, with permission, from M. Vergez–Delonde, Absorption des radiations infra-rouges par les gas atmosphériques, *J. Physique*, **25** (1964) 773)

(continued overleaf)

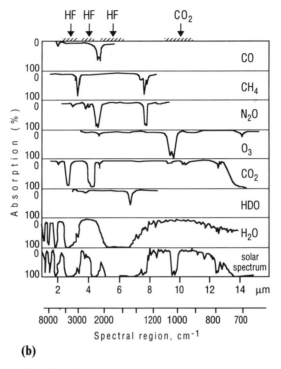

(b)

Figure 2.3 (*continued*)

absorption and light scattering properties of the earth's atmosphere. Similarly, the amount of light absorbed by the earth's surface in a certain frequency interval is determined by the distribution and spectral properties of the pigments of the vegetation and the scattering properties of the surface.

To obtain a basic understanding of these phenomena we shall proceed to discuss the interaction of light with matter and overview the spectral properties of atoms and molecules in Section 2.2.

2.2 INTERACTION OF LIGHT WITH MATTER

When light impinges on a medium containing a variety of atoms and/or molecules a number of processes may occur. First of all, due to the interaction between the electromagnetic wave and the matter energy may be absorbed or released. We will derive the equations governing the process which lead to the well-known Einstein coefficients. Then we introduce the concept of optical density of a medium in connection with Lambert–Beer's law. Finally, we apply these concepts to the ozone layer in the atmosphere.

2.2.1 THE TRANSITION ELECTRIC DIPOLE MOMENT

Consider an atomic or molecular system with discrete energy levels E_k and wave functions Ψ_k^0, which are the unperturbed solutions of the time-independent Schrödinger equation of the system

$$H_0\Psi_k^0 = E_k\Psi_k^0 \qquad (2.5)$$

with H_0 the Hamiltonian that describes the system. By switching on light with frequency ω we introduce a perturbation $H_1(t)$, which starts to mix the eigenfunctions Ψ_k^0. After some time t the system will be found in a state $\Psi(t)$, which is now a solution of the time-dependent Schrödinger equation

$$i\hbar\frac{\partial\Psi}{\partial t} = (H_0 + H_1(t))\Psi(t) \qquad (2.6)$$

For a complete set of eigenfunctions Ψ_k^0 the time-dependent wavefunction $\Psi(t)$ can be expressed as

$$\Psi(t) = \sum_k c_k(t)\Psi_k^0 e^{-iE_k t/\hbar} \qquad (2.7)$$

Insertion of this expression into eq. (2.6) yields a set of equations for the time derivatives of the coefficients $c_k(t)$, given by

$$\frac{dc_k(t)}{dt} = -\frac{i}{\hbar}\sum_n c_n(t)e^{i\omega_{kn}t}\langle\Psi_k^0|H_1|\Psi_n^0\rangle \qquad (2.8)$$

with $\omega_{kn} = (E_k - E_n)/\hbar$.

Taking Ψ_1^0 as the initial state, then at $t = 0$ we have $c_1(0) = 1$ and $c_k(0) = 0$ ($\forall k \neq 1$). This allows us to integrate eq. (2.8) for a time sufficiently short that $c_k(t) \approx 0$ ($\forall k \neq 1$) and $c_1(t) \approx 1$, which yields

$$c_k(t) = -\frac{i}{\hbar}\int_0^t e^{i\omega_{k1}t'}\langle k|H_1|1\rangle\,dt' \qquad (2.9)$$

The probability of finding the atomic or molecular system in the excited state k, thereby having absorbed a quantum $\hbar\omega_{k1}$ from the light beam, is given by $|c_k(t)|^2$. Equation (2.9) shows that $|c_k(t)|^2$ is directly related to the quantity $|\langle k|H_1|1\rangle|^2$. To obtain a useful expression for the transition probability we have to find the form of the perturbation $H_1(t)$.

An electromagnetic light wave is described by in-phase oscillating, perpendicular electric ($\mathbf{E}(t)$) and magnetic ($\mathbf{B}(t)$) fields, that both obey a scalar wave equation, provided that the velocity of the electromagnetic wave is $u = (\varepsilon\mu)^{-1/2}$. For a linearly polarized light wave the electric field is given by: $\mathbf{E}(t) = \mathbf{E}_0\cos\omega t$, where we have assumed that our atom/molecule is in the origin of the coordinate system and that the atomic/molecular dimensions are much smaller than λ. The response of the atom/molecule will be dominated by the interaction of the electric field with the electronic charge distribution. For electrically neutral molecules/atoms the leading term is the electric dipole and thus

$$H_1(t) = -\mathbf{\mu} \cdot \mathbf{E}(t) \tag{2.10}$$

where the electric dipole operator $\mathbf{\mu} = \sum_i q_i \mathbf{r}_i$, in which the sum is taken over all electronic charges q_i at positions $\mathbf{r}_i$.

Substitution of eq. (2.10) into eq. (2.9), performing the integration and calculating the probability $P_k(t) = |c_k(t)|^2$ that the system is in the state k at time t yields

$$P_k(t) = \frac{1}{\hbar^2} |\langle k|\mathbf{\mu}|1\rangle|^2 \frac{|\mathbf{E}_0|^2 \cos^2\theta \sin^2 \frac{1}{2}(\omega_{k1} - \omega)t}{(\omega_{k1} - \omega)^2} \tag{2.11}$$

where we have assumed that the frequency of the light is close to the resonant frequency of the transition $1 \to k$, and with θ the angle between the vectors $\mathbf{\mu}$ and $\mathbf{E}_0$.

In eq. (2.11) we see that the quantity

$$|\mathbf{\mu}_{k1}|^2 = |\langle k|\mathbf{\mu}|1\rangle|^2 \tag{2.12}$$

determines the probability $P_k(t)$: the larger $|\mathbf{\mu}_{k1}|^2$, the larger $P_k(t)$ and the stronger the transition $1 \to k$ in the absorption spectrum. The quantity $\mathbf{\mu}_{k1}$ is called the *transition dipole moment*. Note that the rate of light absorption due to the transition $1 \to k$ also depends on the orientation of $\mathbf{\mu}_{k1}$ relative to the electric field vector $\mathbf{E}_0$ by the term $\cos^2\theta$. For isotropic solutions or gases we will have to average over all possible orientations of $\mathbf{\mu}_{k1}$ relative to $\mathbf{E}_0$. For an ordered system eq. (2.11) leads to polarization effects.

The transition probability $P_k(t)$ as expressed in eq. (2.11) shows a marked frequency dependence, which in addition depends on time. Intuitively we would like to associate the absorption of light energy to a 'rate' and this requires a closer inspection of eq. (2.11). It is not difficult to show that if time proceeds the function $P_k(t)$ becomes sharply peaked around $\omega = \omega_{k1}$. If we further realize that in a normal spectroscopic experiment the incident light is far from monochromatic, it is more appropriate to average eq. (2.11) over a band of frequencies. This directly leads to the rate of population of level k:

$$\frac{dP_k}{dt} = \frac{\pi}{3\varepsilon_0\hbar} |\mathbf{\mu}_{k1}|^2 W(\omega) \equiv B_{k1} W(\omega) \tag{2.13}$$

where $W(\omega)$ represents the (time-averaged) energy density of the incident electromagnetic field at frequency ω. Note that we have averaged eq. (2.11) over all possible orientations of $\mathbf{E}_0$ relative to $\mathbf{\mu}_{k1}$.

The quantity $B_{k1} = \pi|\mathbf{\mu}_{k1}|^2/3\varepsilon_0\hbar^2$ is known as the Einstein coefficient for absorption and stimulated emission and is directly related to the extinction coefficient measured in an absorption experiment.

2.2.2 THE EINSTEIN COEFFICIENTS

In this section we will derive a simple relation between the rates of absorption, stimulated emission and spontaneous emission in an atomic/molecular system, schematically depicted in Fig. 2.4. There exist two energy levels, E_1 and E_2, populated by N_1 and N_2 atoms/molecules respectively. Three possible radiative processes which connect levels 1 and 2 are indicated in Fig. 2.4. Absorption and stimulated emission only occur with the light on and their rates (2.13) are summarized by $B_{12}W(\omega)$ and $B_{21}W(\omega)$ respectively, in which B_{12} is the Einstein coefficient for absorption and B_{21} the Einstein coefficient for stimulated emission. We define A_{21} as the Einstein coefficient for spontaneous emission from level 2 to level 1 which is a 'dark' process.

For the rate of population of level 1 we have

$$\frac{dN_1}{dt} = -B_{12}WN_1 + B_{21}WN_2 + A_{21}N_2 \tag{2.14}$$

Assuming a steady state, i.e. $dN_1/dt = 0$, it directly follows that

$$W(\omega) = \frac{A_{21}}{B_{12}(N_1/N_2) - B_{21}} \tag{2.15}$$

If we now take the situation that there is no external radiation field and that the system is in thermal equilibrium at a temperature T, then the ratio N_1/N_2 follows from the Boltzmann equation:

$$\frac{N_1}{N_2} = e^{\hbar\omega/kT} \tag{2.16}$$

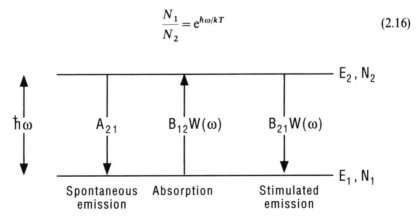

Figure 2.4 The three basic types of radiative processes. Indicated are the Einstein coefficients and the transition rates for two levels with energies E_1, respectively E_2, and occupation N_1, respectively N_2

Under these conditions the energy density frequency distribution is given by Planck's radiation law (2.4), taking into account eq. (2.16) and averaging over polarizations:

$$W(\omega) = \frac{\hbar\omega^3}{\pi^2 c^3} \frac{1}{e^{\hbar\omega/kT} - 1} \tag{2.17}$$

Since (2.17) should be identical to (2.15) it follows from eq. (2.16) that

$$B_{12} = B_{21} \tag{2.18}$$

$$\frac{A_{21}}{B_{21}} = \frac{\hbar\omega^3}{\pi^2 c^3} \tag{2.19}$$

From eq. (2.13) we deduce B_{12} and we obtain

$$A_{21} = \frac{\omega^3}{3\pi c^3 \varepsilon_0} |\mu_{21}|^2 \tag{2.20}$$

The importance of this result is that the same quantity $|\mu_{21}|^2$ determines the rates of all radiative processes and consequently common selection rules apply to them.

2.2.3 LAMBERT–BEER'S LAW

The microscopic quantities μ_{21}, B_{12}, B_{21} and A_{21} are connected to the macroscopic phenomenon of absorption by Lambert–Beer's law. When a beam of light passes through a sample of material, the light beam is usually absorbed, but sometimes amplified (in a laser). Absorption dominates when most of the atoms/molecules are in their ground states ($N_1 \gg N_2$) and when the intensity of the light beam is weak. In that case the rate equation for transitions between states 1 and 2 is given by

$$\frac{dN_1}{dt} = -B_{12} N_1 W(\omega) + A_{21} N_2 \tag{2.21}$$

In a steady state $dN_1/dt = 0$ and $A_{21} N_2 = B_{12} N_1 W$ is precisely the rate at which light energy is removed from the incident light beam. Let the light beam be propagated along the z axis. Since the beam is attenuated its energy W will be a function of z. Consider now a small slice of thickness dz and surface a as illustrated in Fig. 2.5. The amount of beam energy in this slice in the frequency interval ω and $\omega + d\omega$ is given by $W \, d\omega \, a \, dz$. It then follows that in the case of energy balance the rate of decrease of the beam energy equals the rate at which the light energy is removed from the beam by absorption. We find

$$-\frac{\partial W}{\partial t} d\omega \, a \, dz = N_1 B_{12} W(\omega) \hbar\omega \frac{a \, dz}{V} F(\omega) \, d\omega \tag{2.22}$$

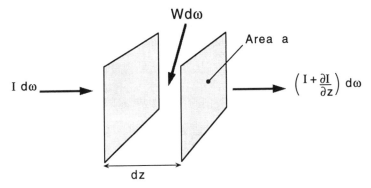

Figure 2.5 Passage of a light beam through a thin slice of sample, perpendicular to the direction of the beam

in which $a\,dz/V$ represents the fraction of atoms/molecules in the selected slice $a\,dz$ where V is the volume of the sample. Moreover, we have assumed that not all the atoms absorb precisely at the frequency ω_{k1}, but that some statistical spread in transition frequencies exists at which the atoms/molecules can absorb light. To account for this, we introduced $F(\omega)\,d\omega$, the fraction of transitions that occur in the frequency interval ω and $\omega + d\omega$ with

$$\int F(\omega)\,d\omega = 1 \tag{2.23}$$

Rearranging eq. (2.22) yields

$$\frac{\partial W}{\partial t} = - N_1 F(\omega) B W \frac{\hbar\omega}{V} \tag{2.24}$$

To obtain Lambert–Beer's law we rewrite eq. (2.24) in terms of the change in intensity I in $W\,m^{-2}$ of the light beam upon passage through the sample slice $a\,dz$. Inspection of Fig. 2.5 shows that the difference between the amount of energy entering and leaving the sample slice per unit of time precisely equals the rate of decrease of the beam energy. Thus we obtain $-\partial W/\partial t = -\partial I/\partial z$. Moreover, $W = I/cn$, with c the speed of light and n the index of refraction. Substitution into eq. (2.24) yields the required expression

$$\frac{\partial I}{\partial z} = -\frac{B_{12} N_1 \hbar\omega F(\omega)}{V nc} I \tag{2.25}$$

The solution of eq. (2.25) is straightforward and gives Lambert–Beer's law:

$$I(z) = I(0)e^{-Kz} \tag{2.26}$$

with K the absorption coefficient, given by

$$K = \frac{N_1 B_{12} \hbar \omega F(\omega)}{V cn} \tag{2.27}$$

Since B_{12} was calculated from eq. (2.13) in which an integration over a band was performed and since $F(\omega)$ is normalized, eq. (2.27) can be integrated to give

$$B_{12} = \frac{V cn}{N \hbar} \int_{\text{band}} \frac{K}{\omega} d\omega \tag{2.28}$$

Thus, from a simple integration of the measurable (!) absorption coefficient K over the absorption line, we obtain the Einstein coefficient for absorption (and consequently the two other Einstein coefficients). Moreover, since we have related B_{12} to $|\mu_{21}|^2$ (eq. (2.13)) the quantity $|\mu_{21}|^2$ is calculated directly.

It is common practice to use, instead of eq. (2.26), the following expression for the dependence of I on passage through a material with pathlength l:

$$I(l) = I(0) \times 10^{-\text{OD}} \tag{2.29}$$

where the optical density OD is defined as

$$\text{OD} = \varepsilon l C \tag{2.30}$$

Here, ε is the molar extinction coefficient (usually expressed in $\text{dm}^3 \, \text{mol}^{-1} \, \text{cm}^{-1}$), l the pathlength (usually in cm) and C the concentration of the sample (usually expressed in mol dm^{-3}). For example, in a leaf chlorophyll a has in its absorption maximum at 680 nm an extinction coefficient of $10^5 \, \text{dm}^3 \, \text{mol}^{-1} \, \text{cm}^{-1}$. The chlorophyll concentration in a leaf is about $10^{-3} \, \text{mol dm}^{-3}$ and the pathlength for light passing through a leaf is about 0.02 cm. Thus the OD of a single leaf at 680 nm is $\text{OD} = 10^5 \times 10^{-3} \times 2 \times 10^{-2} = 2$. Thus, the reduction in intensity of 680 nm light upon passage through a leaf is about a factor 100. As a consequence no red light is detected below the outer array of leaves of a tree.

It is not difficult to relate eqs. (2.29) and (2.30) to the expressions for K and B_{12} and we finally find for $|\mu_{21}|^2$:

$$|\mu_{12}|^2 = 1.01 \times 10^{-61} \int_{\text{band}} \frac{\varepsilon}{\omega} d\omega \tag{2.31}$$

where $|\mu_{12}|^2$ is obtained in $(\text{C m})^2$ and all the other constants have been replaced by their numerical values.

Returning to eqs. (2.29) and (2.30), we can relate the extinction coefficient ε to the absorption cross-section of a single molecule, σ, given by

$$\sigma = \pi r^2 P \tag{2.32}$$

with r the radius of the molecule and P the probability that a photon is absorbed upon the passage through the surface πr^2. Then

$$\varepsilon = \frac{\sigma N_A}{2.303} = \frac{\pi r^2 P N_A}{2.303} \qquad (2.33)$$

in which N_A is Avogadro's number. One may deduce this equation by using the technique of Fig. 4.39 with eqs. (4.240) and (4.241).

2.3 BIOMOLECULES, OZONE AND UV LIGHT

In this section we will briefly discuss the absorption of light in the UV region by biological molecules, such as proteins and nucleic acids. Since both the amount and the spectral composition of UV light incident on earth is totally determined by the ozone present in the atmosphere, we shall summarize processes that lead to the building up and destruction of the ozone layer and show that small variations in the ozone concentrations may lead to dramatic biological effects. These effects are largely related to the highly non-linear properties of Lambert–Beer's law.

2.3.1 THE SPECTROSCOPY OF BIOMOLECULES

Figure 2.6 shows the solar emission spectrum below 340 nm at the earth's surface in combination with the absorption spectrum of two important biomolecules: DNA, the carrier of the genetic code and α-crystallin, the major protein of the mammalian eyelens. The absorption of light by these biological molecules is essentially zero in the region $320 < \lambda < 400$ nm, which is called the near-UV or UV-A, intense in the region 200–290 nm, which is called the far-UV or UV-C and only overlaps with the solar spectrum in the wavelength region $290 < \lambda < 320$, the mid-UV or UV-B.

The absorption of DNA is due to the aromatic DNA bases guanine, thymine, cytosine and adenine and it peaks at about 260 nm, with a maximum extinction of $\varepsilon = 10^4$ dm^3 mol^{-1} cm^{-1} (expressed per mole of base). The electronic transitions which contribute to the absorption in the range 220–290 nm are predominantly oriented in the plane of the DNA base.

The absorption of proteins in the wavelength region 250–300 nm is mainly due to the aromatic amino acids tryptophan and tyrosine and it peaks at about 280 nm. For a protein with a number of n_{TRP} tryptophan residues and a number of n_{TYR} tyrosine residues the extinction at 280 nm is given by

$$\varepsilon_{280} = n_{TRP}5600 + n_{TYR}1100 \qquad \text{dm}^3 \text{ mol}^{-1} \text{ cm}^{-1} \qquad (2.34)$$

Around a wavelength of 190 nm the peptide bond by far dominates the absorption spectrum, but under normal solar UV conditions the peptide bond does not

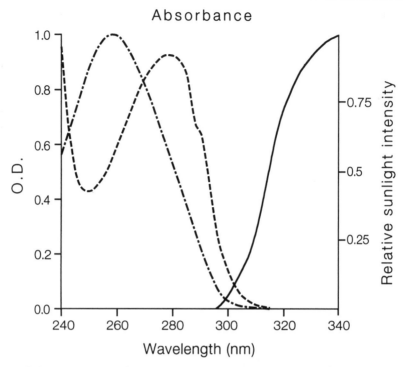

Figure 2.6   Absorption spectra of DNA ($-\bullet-\bullet$) and α-crystallin (----) in the wavelength region 240–340 nm. The solid line (———) indicates the solar emission spectrum in the same region, normalized at 340 nm

contribute to the absorption of far-UV light by proteins. For a typical biological cell the total absorption of solar UV due to protein would be about 10% of the absorption of nucleic acids.

Often proteins carry additional cofactors which allow proteins to carry out their specific tasks. For instance, in haemoglobin the haem group, a porphyrin molecule, is attached to the protein in a haem–protein complex which is active in oxygen binding. In those cases the absorption extends into the near-UV, visible and sometimes near-infrared region of the spectrum. Although in several cases the absorption of light in this wavelength range is crucial for the biological function (photosynthesis, vision) we shall not further discuss it here.

2.3.2 SOLAR UV AND LIFE

The interaction of solar UV with biological cells may lead to a severe damage of the genetic material (mutagenesis) or cell killing. In more complex, multicellular organisms exposure to solar UV may lead to damage of crucial parts of the

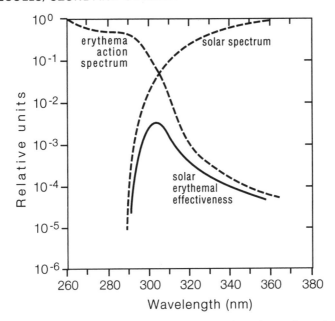

Figure 2.7 Erythema (sunburn) action spectrum, plotted together with the solar spectrum and the resultant 'solar effectiveness spectrum' for erythema (which is the product $E(\lambda)/I(\lambda)$ in eq. (2.35)). Note the logarithmic ordinate. (Reproduced by permission of Plenum Press from J. A. Parrish, R. R. Anderson, F. Urbach abd D. Pitts, *UV-A: Biological Effects of Ultraviolet Radiation with Emphasis on Human Responses to Longwave Ultraviolet*, Plenum Press, New York, 1978, Fig. 8.4, p. 119)

organism and, in the case of humans, to skin cancer. As the overlap between the solar spectrum at the earth's surface and the absorption spectra of nucleic acids and proteins in Fig. 2.7 shows, the mid-UV wavelength range will be of prime importance in these processes.

It is by now well-documented that in particular the chromophores of DNA form the prime target in the process that leads to photodamage. Figure 2.7 shows the action spectrum, which is the damage done by a unit of irradiation of a certain wavelength, for the production of erythyma (sunburn) in human skin. It also shows the UV tail of the solar spectrum. It may be noted that the efficiency of producing erythyma increases four to five orders of magnitude between 350 and 280 nm, precisely in the region where the solar UV spectrum collapses. The action spectrum must be largely ascribed to the direct absorption of solar UV by DNA. Although for $\lambda > 320$ nm the action spectrum tends to be higher than the DNA absorption, implying that there are more absorbers, the primary target is probably still DNA. Similarly, studies on cellular cultures where cell survival on cell mutagenesis was measured have identified the chromophores of DNA as the primary target molecules. The major photo product in DNA is the pyrimidine dimer, a covalent structure involving two thymine molecules or a cytosine–thymine pair. Its presence in DNA has major destructive effects on the reading and translation of the genetic code.

The damage D done to a biological system by solar UV can be calculated from the action spectrum $E(\lambda)$ by

$$D = \int_0^\infty E(\lambda)I(\lambda)\,d\lambda \qquad (2.35)$$

in which $I(\lambda)$ is the incident intensity. Since $E(\lambda)$ is a function with a large negative slope and $I(\lambda)$ a function with a large positive slope, even a minor change in $I(\lambda)$ may lead to large changes in D.

2.3.3 THE OZONE FILTER

Ozone (O_3) forms a thin layer in the stratosphere, with a maximum concentration between 20 and 26 km above the earth's surface (cf. Fig. 3.7). The atmospheric ozone absorbs essentially all the radiation below a wavelength of 295 nm, due to a strong optical transition at about 255 nm, which extends into the mid-UV region. Figure 2.8 shows the spectrum of ozone between wavelengths of 240 and

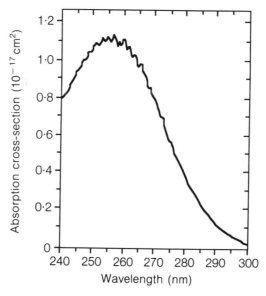

Figure 2.8 Absorption spectrum of ozone in the wavelength region 240–300 nm. (Reproduced by permission of Optical Society of America from E. C. Y. Inn and Tanaka, *J. Opt. Soc. Am.*, **43** (1953) 870)

300 nm. At the maximum (255 nm) the absorption cross-section is about 10^{-17} cm², it is about half-maximum at 275 nm and has decreased to about 10% of its maximum value at 290 nm. That ozone forms a very thin shield indeed is probably best demonstrated by the fact that the amount of O_3 in the atmosphere corresponds to a layer of 0.3 cm at standard temperature and pressure.

Ozone is constantly formed in the upper layer of the atmosphere through the combination of molecular oxygen (O_2) and atomic oxygen (O). The latter is formed through the photodissociation of O_2 in the 100 km region by light of wavelengths shorter than 175 nm. Sunlight excites the electronic transition between the triplet ground state of O_2 ($^3\Sigma_u^-$) and a triplet excited state ($^3\Sigma_g^-$). Once excited, the O_2 molecule may dissociate into two oxygen atoms: one in the ground state O (^3P) and one in the excited state $O(^1D)$. The latter is 2 eV above the ground state. As a consequence of these processes, the sunlight with $\lambda < 175$ nm is totally extinguished above the stratosphere. A much weaker absorption of O_2 occurs in the range $\lambda < 240$ nm. It involves a transition to the so-called Herzberg continuum ($^3\Sigma_u^- \leftarrow {}^3\Sigma_g^-$; maximum cross-section of about 10^{-23} cm²) and in that case the dissociation results in two ground state (^3P) O atoms.

Once formed, atomic oxygen attaches to O_2 to form O_3. The efficiency of O_3 formation by UV light is sensitive to a large number of factors, amongst which are the availability of O_2 and changes in stratospheric temperature and chemicals and on dust from volcanic eruptions.

Most of the ozone is produced above the equator, where the amount of incident solar UV light is maximal. Ozone formed at these latitudes then diffuses to the poles, where it is 'accumulated'. The effective thickness of the ozone layer may increase from $\leqslant 0.3$ cm at the equator to $\geqslant 0.4$ cm above the pole at the end of winter. The ozone concentration is sensitive to significant daily and seasonal fluctuations and tends to be highest in late winter and early spring.

Ozone is permanently being formed and broken down in the stratosphere and only a very small fraction of the formed ozone escapes down to the troposphere. There are basically two pathways for the destruction of ozone and the reformation of O_2:

$$O + O_3 \longrightarrow 2\,O_2 \tag{2.36}$$

$$O_3 + O_3 \longrightarrow 3\,O_2 \tag{2.37}$$

These reactions are the net result of a complex set of reactions catalysed by various gases and radicals. We mention explicitly atomic chlorine Cl, nitric oxide NO and hydroxyl radicals OH; the pathway along which they operate is shown in Fig. 2.9. Note that this scheme essentially corresponds to the net reaction (2.36).

How are the free radicals NO, Cl and OH produced? The OH radical is a product of the breakdown of H_2O vapour, for instance produced in the exhaust of supersonic aeroplanes. Although part of the Cl radical may be formed from HCl released by volcanoes, the major input of Cl into the stratosphere originates

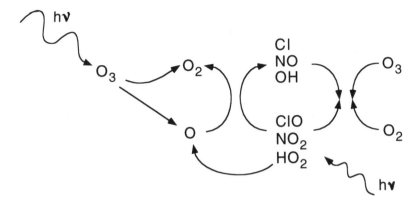

Figure 2.9 Breakdown of ozone to molecular oxygen, utilizing the free radicals Cl, NO and OH. Schematic representation. (Reproduced by permission of Greenwood Press from J. Jagger, *Solar UV Actions on Living Cells*, Praeger Special Studies, Praeger Publishing Division of Greenwood Press Inc., Westport, Connecticut, 1985, Fig. 9.2, p. 145)

from chlorofluorocarbons (CFCs), which are used as foam-blowing agents, refrigerants and propellants. CFCs are extremely stable in the troposphere. However, a small fraction may escape into the stratosphere, eventually reaching the upper stratosphere where they may be decomposed under the influence of UV light, thereby producing the Cl and ClO.

N_2O is also of anthropogenic origin and released from soils and waters where it has been formed as a fertilizer waste product. Similarly, the N_2O released at the earth's surface may eventually be photodecomposed and NO is formed. These radicals, together with the OH radical, remove some 99% of the stratospheric ozone. The processes that determine the amount and distribution of ozone in the atmosphere are illustrated in Fig. 2.10.*

The crux of the ozone problem is demonstrated by Figs. 2.6 and 2.8. Any small variation in the ozone concentration will lead to changes in both the amount of UV light at a particular wavelength and the transmission of shorter wavelength radiation. This is of course a direct consequence of Lambert–Beer's law. A certain percentage change in concentration of ozone yields the same percentage change in OD. A decrease of one OD unit at a particular wavelength implies a tenfold increase in the amount of light of that wavelength reaching the earth's surface. Since the action spectrum for damage to living cells or tissue is an exponentially increasing function (with decreasing λ; see Fig. 2.7) the amount of damage may

---

*For a more detailed discussion of the ozone problem and details of all the (photo-)chemical processes that occur refer to Refs. 5 to 9.

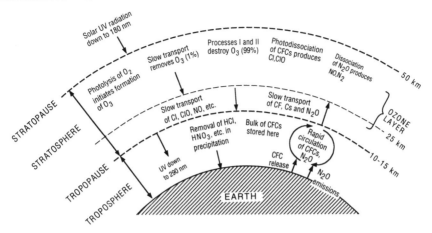

Figure 2.10   Processes that determine the concentration of ozone in the stratosphere. (Reproduced with permission from J. Jagger, NRC Report 1982)

dramatically increase, even with a relatively small decrease in the amount of ozone. That the amount and spectral distribution of the UV light is in fact a strong function of the ozone concentration is best illustrated by Fig. 2.11, where the irradiance at ground level has been measured at three different ozone concentrations.

Predictions from models for the atmospheric ozone production and breakdown indicate a stratospheric ozone depletion ranging from 5–20% due to CFC

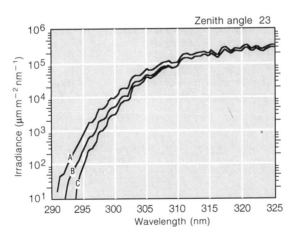

Figure 2.11   UV intensity at the earth's surface for various effective thicknesses of the ozone layer. $A = 273$ atm cm, $B = 319$ atm cm and $C = 388$ atm cm, all in the clear sky. (Reproduced with permission from T. Ito, UV-B observation network in the Japan Meteorological Agency, *Frontiers of Photobiology* (1993) 515)

and N_2O production. From results as presented in Figs. 2.6 and 2.8 one may predict that a 10% ozone depletion will result in a 45% increase in effective UV-B radiation. These alarming numbers illustrate the necessity to monitor the structure of the ozone layer accurately and to quantify the effects of increased UV on living organisms.

Exercises

2.1 Show that eq. (2.4) leads to eq. (2.3) for $v \to 0$ and to Wien's displacement law (2.2) for $v \to \infty$.

2.2 Calculate the frequency distributions given by eq. (2.4) for $T = 6000$, 2000, 800 and 300 K.

2.3 Calculate the total radiation energy per second emitted by the sun from eq. (2.1). Take the radius of the sun as 7×10^8 m.

References

1. P. W. Atkins, *Molecular Quantum Mechanics*, Oxford University Press, Oxford, 1983. An advanced text on spectroscopy.
2. P. W. Atkins, *Physical Chemistry*, Oxford University Press, Oxford, 1991. A general text with a large part devoted to spectroscopy and its applications.
3. R. Loudon, *The Quantum Theory of Light*, Oxford University Press, Oxford, 1983. A clear text, easy to read with some knowledge of quantum mechanics and electrodynamics.
4. S. Svanberg, *Atomic and Molecular Spectroscopy, Basic Aspects and Practical Applications*, Springer-Verlag, Berlin, 1992. A good text with many interesting applications.
5. J. Jagger, *Solar UV Actions on Living Cells*, Praeger Special Studies, Praeger Publishing Division of Greenwood Press Inc., Westport, Connecticut, 1985. A general text on the interaction of light with biomolecules in relation to the ozone problem.
6. J. H. Seinfeld, *Atmospheric Chemistry and Physics of Air Pollution*, John Wiley, New York, 1986.
7. H. S. Johnston, Atmospheric ozone, *Ann. Rev. Phys. Chem.*, **43** (1992) 1–32.
8. R. S. Stolanski, The Antarctic ozone hole, *Sci. Am.*, **258** (1988) 30.
9. R. P. Wayne, *Chemistry of the Atmosphere*, Clarendon, Oxford, 1985.

3

The Global Climate

The living conditions on earth are determined by the solar irradiation and the resulting mild temperatures. In this chapter a simple zero-dimensional model of the energy balance will first be presented and a computer representation of that model will be discussed. It will be shown how a change in the conditions on earth and in the atmosphere will affect the mean surface temperature of the earth. In the remaining part weather and climate as well as the climate change in the past and the predictions for the future will be discussed.

3.1 THE ENERGY BALANCE: A ZERO-DIMENSIONAL GREENHOUSE MODEL

Atmospheric scientists derive from their observations a very much simplified model for the radiation balance between the surface and atmosphere of the earth. One such model is displayed in Fig. 3.1. The numbers given are still the subject of discussion, but for a qualitative discussion any set will do.

On the left the incoming solar radiation averaged over the earth is given, taken as 100 units. The radiation is partly absorbed by the atmosphere and partly scattered back or reflected back into space. The remaining radiation at ground level is either absorbed or reflected back. The reflection coefficient, the albedo, of the earth's surface is usually taken as $a_s = 0.11$ as opposed to the range of $a_s = 0.34$ for the earth as viewed from space. Of course, the number $a_s = 0.11$ is an average over the earth's surface, where differences in vegetation and habitation give rise to rather differing albedos. Some representative numbers are given in Table 3.1 and it is clear that it will be difficult to find a mean value for the earth's surface. There are indications that the albedo could be in the order of 0.16. One should notice that the albedos for the planets in Table 3.1 refer them as viewed from space, including the effects of atmosphere and clouds.

On the right-hand side in Fig. 3.1 one finds the data for the long wavelength radiation. One observes the emission of the atmosphere, both upwards and downwards, the emission from the earth, most of which is absorbed by the

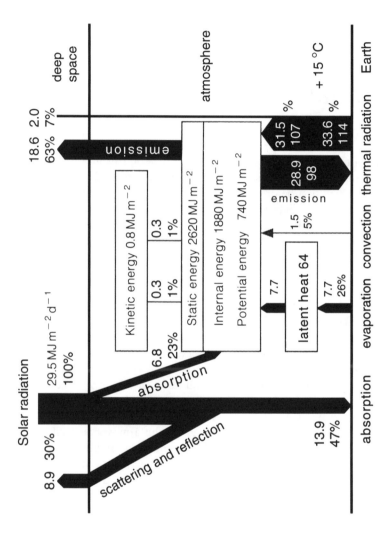

Figure 3.1 Mean global radiation balance model for the surface and the atmosphere of the earth. The mean solar irradiation is put as 100 units. Absolute numbers are given for energies; for fluxes both the absolute value per day is given and the value in % of the solar input. Note how both the bars for incoming solar radiation and for outgoing thermal radiation are split up. (Reproduced by permission of Department of Geography, ETH-Zürich, and A. Ohmura, *Climatic Changes, Ice Sheet Dynamics and Sea Level Variations*, ETH, Zürich, 1990, Fig. 12, p. 28)

Table 3.1 Albedo of various surfaces for visible light [1, 2]

Surfaces	%	Clouds	%	Planets	%
Horizontal water (low solar angle)	≈ 5	Cumulus	70–90	Earth	34–42
Fresh snow	≈ 85	Stratus	60–85	Moon	6–7
Sand desert	≈ 30	Altostratus	40–60	Mars	16
Green meadow	≈ 15	Cirrostratus	40–50	Venus	76
Deciduous forest	≈ 15			Jupiter	73
Coniferous forest	≈ 10				
Crops	≈ 10				
Dark soil	≈ 10				
Dry earth	≈ 20				

atmosphere, and finally the exchange between surface and atmosphere by evaporation and convection. In addition, the energy contents of the atmosphere are shown. It is interesting to note that the kinetic energy amounts to some 3% of the incoming solar radiation per day and is much smaller than the static energy content.

In order to obtain a qualitative understanding of the factors that influence the surface temperature of the earth, we have simplified Fig. 3.1 to an even simpler scheme, shown in Fig. 3.2. Albedos for the solar radiation are defined as a_a and a_s for atmosphere and surface respectively; the transmission of the atmosphere is defined as t_a. For the long wavelength radiation the albedo of the earth's surface is usually taken as zero, corresponding with the fact that it radiates as a black body at these wavelengths. (A quick introduction to black body radiation may be found in the text around eqs. (4.3) and (4.4).)

The transmission of the atmosphere for the infrared radiation is written as t_a'. It is more artificial to introduce an albedo a_a' for the atmosphere, but from Fig. 3.1 one notes that the emission upwards and downwards is different. Let us assume that the upwards radiation is just thermal, then the surplus downwards may be taken as radiation reflected back to earth, which is taken into account by a fictitious albedo a_a'. Finally, the interaction between the atmosphere and earth is written in first order as proportional to the temperature difference with a multiplication constant c. For the black body radiation one uses the Stefan–Boltzmann expression σT^4. The solar constant S should be divided by 4 as the energy entering the earth disk πR^2 is distributed over the total area $4\pi R^2$.

For a stationary situation one then finds *for the surface* in first order

$$(-t_a)(1 - a_s)\frac{S}{4} + c(T_s - T_a) + \sigma T_s^4(1 - a_a') - \sigma T_a^4 = 0 \qquad (3.1)$$

The first term describes the absorption, the second the non-radiative interaction

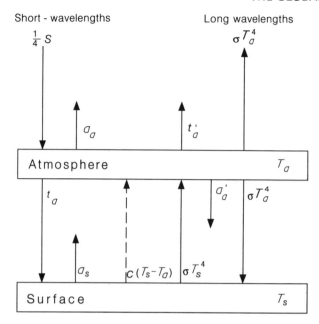

Figure 3.2 A zero-dimensional greenhouse model, easy to computerize and based on an energy balance. The incoming solar radiation is represented on the left as short wavelength with magnitude $S/4$ and the emitted long wavelength infrared radiation is represented on the right. The symbols a and t represent the fractions being reflected or transmitted at the location indicated. (Adapted and reproduced by permission of Academic Press Limited from Bent Sørensen *Renewable Energy*, Academic Press, London, 1979, Fig. 108, p. 206)

Table 3.2 Parameters of the zero-dimensional energy balance

Short wavelength	Long wavelength
$a_s = 0.11$	
$t_a = 0.53$	$t'_a = 0.06$
$a_a = 0.30$	$a'_a = 0.31$
$c = 2.5\,\mathrm{W\,m^{-2}\,K^{-1}}$[a]	

[a]From Fig. 3.1 one would deduce $c = 3.2$ $\mathrm{W\,m^{-2}\,K^{-1}}$; the value given here reproduces a reasonable surface temperature of $T_s = 288\,\mathrm{K}$.

between the atmosphere and surface, the third one the emitted radiation minus the backscattered and the last term the heat radiation from the atmosphere. Lost energy is taken as positive. *For the atmosphere* one finds similarly in first order

$$-(1 - a_a - t_a + a_s t_a)\frac{S}{4} - c(T_s - T_a) - \sigma T_s^4(1 - t_a' - a_a') + 2\sigma T_a^4 = 0 \quad (3.2)$$

Here the first term is the solar absorption, the second the non-radiative interaction, the third one the absorption of earth radiation by the atmosphere and the last one the atmospheric emission. Addition of the first terms of eqs. (3.1) and (3.2) shows that all solar energy is accounted for. Note that eqs. (3.1) and (3.2) contain parameters a_s and t_a only in the combination $t_a(1 - a_s)$. When the numbers given in Fig. 3.1 are used to deduce the albedo parameters one finds the values given in Table 3.2, assuming $a_s = 0.11$.

Examples

The examples shown below only represent trends due to a change of parameters. The resulting variations of global temperature should not be viewed in isolation since in reality many effects will occur at the same time, of which only a few are discussed.

1. The *white earth*. Assume that the earth's surface is covered with snow, both on land and on the oceans (ice and snow). The resulting albedo is high, let us say, consulting Table 3.1 $a_s = 0.75$. One then finds a surface temperature of 270 K, which is well below the freezing point of sea water. This means that a white earth is a stable solution of the energy balance equations. It is, however, so remote from the present situation that the transition to a white earth is hardly thinkable.

2. The *nuclear winter*. The climatic consequences of nuclear war have been seriously discussed in recent years. It was argued that the explosion of a few hundred nuclear warheads would lead to enormous fires of cities and of industrial installations which would bring great quantities of small particles of dust and smoke into the atmosphere, essentially cutting the earth off from sunlight. In the model of Fig. 3.2 this can be taken into account by changing the values of the parameters by 20% into the expected direction. With $a_a = 0.36$, $t_a = 0.43$, $t_a' = 0.05$ and $a_a' = 0.37$ one finds a surface temperature of 283 K, a cooling of 5 °C.

3. The *solar collector* world. The energy balance is influenced by many natural and human causes. From Table 3.1 it is clear that desertification, the change of green meadows and forests into desert increases the albedo. Similarly, the change of seas into polders increases the albedo. On the other hand, covering one-third of the surface area of the world with black solar collectors would

decrease the albedo a_s from 0.11 to 0.10. This by itself would change the surface temperature from 288.1 to 288.3 K, a negligible effect.

4. The *cool sun*. Astronomers believe that if the sun behaves like other stars of its type, some billions of years ago the solar constant S must have been considerably smaller than nowadays. Two billion years ago, for example, it would have been 85% of its present value and 4 billion years ago some 72% of its present value [3]. In our model this would lead to temperatures of 274.8 and 261.9 K respectively. Still life existed, requiring temperatures well above the freezing point of water. It is therefore believed that due to an elevated concentration of greenhouse gases in the atmosphere, simulated by a smaller transmission t'_a and higher albedo a'_a, these effects were largely compensated. The increase of the solar constant with time must to some extent have been compensated by binding of greenhouse gases, in particular CO_2.*

5. *Warming up*. The greenhouse effect is caused by a variety of gases, each contributing to a surface heating of some degrees. In Table 3.3 one finds the present situation.

It is well known that the concentration of CO_2 increases in time. Figure 3.3 shows the steady increase since the start of the industrial revolution (top) and precise recent measurements made in Hawaii, far from industries (bottom). The latter show, apart from the expected variation with the seasons, an average steady

Table 3.3 Present greenhouse warming of most important atmospheric trace gases and global warming potential after 100 years relative to the effect of CO_2. The last column is from Ref. 5, p. 60. The number of 3500 applies to CFC 11. As the trace gases have different asmospheric lifetimes their effects in time are different. Reproduced by permission of Deutsche Verlags-Anstalt from C. D. Schönwiese and B. Diekmann, *Der Treibhauseffekt*, Deutsche Verlags-Anstalt, Stuttgart, 1987, Table 9, p. 132 (originates from K. Y. Kondratyev and N. I. Moskalenko))

Trace gas	Present concentration (ppm)	Present warming effect (°C)	Global warming potential after after 100 yr
H_2O vapour	$2-3 \times 10^3$	20.6	
CO_2	345	7.2	1
O_3 (troposphere)	0.03	2.4	
N_2O	0.3	0.8	290
CH_4	1.7	0.8	21
Others (CFCs)		0.6	≈ 3500
Total		33.0	

*This fact and many similar effects suggest that the earth itself maintains an environment ready to support life. It is the basis of the hypothesis that earth itself behaves like a living being: GAIA (cf. Ref. 4).

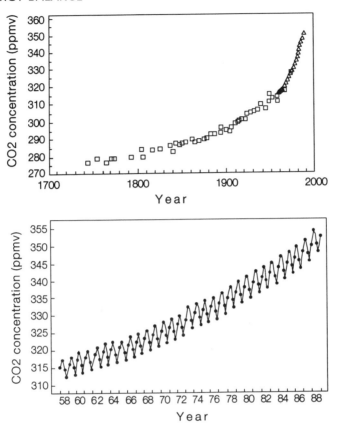

Figure 3.3 The increase of CO_2 concentration since 1750 (top) and since 1958 at Mauna Loa observatory, Hawaii (bottom). The top graph has been integrated into the bottom one and fits smoothly. (Reproduced from C. D. Keeling *et al.*, *Geophysical Monograph*, 55 (1989) 165–236 and other authors quoted in IPCC, p. 9, reproduced by permission of IPCC from J. T. Houghton, G. J. Jenkins and J. J. Ephraums (eds), *IPPC Scientific Assessment*, Cambridge University Press, 365 pp.)

growth, which is due to global burning of fossil fuels and the burning of wood in deforestation. In addition, methane and some other greenhouse gases increase in concentration and the question is what their influence will be on the temperature of the earth's surface. The estimated effect after 100 years of releasing 1 kg of trace gas relative to 1 kg of CO_2 is given in the last column of Table 3.3 as the *global warming potential*.

Below, we will discuss the effect of a doubling of the atmospheric CO_2 concentration. It should be mentioned that usually all greenhouse gases are taken into account and their consequences are evaluated in terms of an 'equivalent'

CO_2 concentration having the same effect as all greenhouse gases taken together. When one takes 1990 as the basis year and compares with the year 1765, which is the start of the industrial revolution, the man-made greenhouse gases double the total amount of these gases in about the year 2030, if no policy measures restricting emissions are enforced. In that year half the man-made temperature increase is due to CO_2 and the rest to the other greenhouse gases, while the contribution from water vapour is assumed to be constant ([5], business-as-usual scenario).

Radiative Forcing

Let us take the simplest possible atmospheric model as sketched in Fig. 3.4; it is even simpler than Fig. 3.2 as we now only consider total energy fluxes. The net radiation flux F_{TA} at the top of the atmosphere will vanish under equilibrium conditions. Assume a sudden doubling of effective CO_2 concentration. That would lead to an effective reduction of the earth's long wavelength radiation at the top of the atmosphere with a magnitude of ΔI and consequently cause a decrease in flux there of $\Delta F_{TA} = -\Delta I$. The energy balance at the top of the atmosphere requires a constant flux. Therefore the earth's surface temperature should rise by ΔT_s to compensate. This effect is called *radiative forcing*. The flux increase ΔI will be connected with the increase ΔT_s of the surface by the relation

$$\Delta I = \frac{\partial I}{\partial T_s} \Delta T_s \qquad (3.3)$$

The intensity I is the earth's radiation measured at the top of the atmosphere:

$$I = \varepsilon \sigma T_s^4 \qquad (3.4)$$

$$\Delta F_{TA} = -\Delta I$$

Top Atmosphere

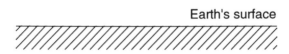

Earth's surface

Figure 3.4 A sudden decrease in flux F_{TA} at the top of the atmosphere (TA = top atmosphere)

from which it follows that

$$\frac{\partial I}{\partial T_s} = 4\varepsilon\sigma T_s^3 = \frac{4I}{T_s} = \frac{4}{T_s}(1-a)\frac{S}{4} \tag{3.5}$$

where $a = 0.34$ is the effective albedo of the surface and lower atmosphere. With numbers given before, we find $\partial I/\partial T_s = 3.1$ W m^{-2} K^{-1}. From eq. (3.3) it follows that

$$\Delta T_s = G\,\Delta I \tag{3.6}$$

in terms of cause ΔI and effect ΔT_s with $G = 0.3$ W^{-1} m^2 K a gain function. From more detailed models one usually estimates $\Delta I = 4.6$ W m^{-2} which leads to $\Delta T_s = 1.37$ K.*

Of course, there are uncertainties in all these numerical estimates, but, more importantly, a temperature rise will cause a lot of effects reinforcing it, and a few counteracting it. Let us mention some of them; in as much as they have to do with a change of albedo's one may check qualitatively these effects by studying eqs. (3.1) and (3.2) numerically.

Global Warming

Effects reinforcing global warming
(a) Melting of ice and snow will lower the albedo a_s.
(b) More water vapour in the air will lead to smaller transmission t'_a and higher backscattering a'_a.
(c) For the increase of the cloud cover one believes the same to be the dominant effect.
(d) Several processes will cause further increase in CO_2 concentration: a higher sea water temperature gives less CO_2 absorption in sea water; connected with this effect is that with higher polar temperatures there will be a smaller ocean circulation and decreasing CO_2 absorption; a faster decay of organic materials will give more CO_2 and CH_4.
(e) More CO_2 leads to an increased growth of plants which may lower the albedo a_s according to Table 3.1.

Effects mitigating global warming
(f) The so-called adiabatic lapse rate $-\partial T/\partial z$ for humid air decreases. This last effect is not obvious but is in practice considerable; therefore it will be discussed now as it illustrates the complexity of describing the reinforcing and mitigating mechanisms.

Consider a parcel of air rising in altitude in the lowest 10 km of the atmosphere.

---

*The number $\Delta I = 4.6$ W m^{-2} follows from Ref. 5 (p. 57, business-as-usual scenario, taken from 1765 to 2025).

As will be shown in Section 3.2 (eqs. (3.28) and (3.31)), its temperature will then decrease, resulting in a positive lapse rate $-\partial T/\partial z$. Assume that the CO_2 concentration doubles; then there are three possible effects on the lapse rate, as illustrated in Fig. 3.5. In case A the lapse rate does not change and there is just an overall increase in temperature. This situation is indicated with a dotted line in the middle and lower figures.

The middle figure, case B, indicates a positive feedback. That means that for a certain surface temperature T_s the atmosphere will be cooler than for a zero feedback. This implies that the total radiation leaving the earth would be too small, which must be corrected by a higher surface temperature, as is already

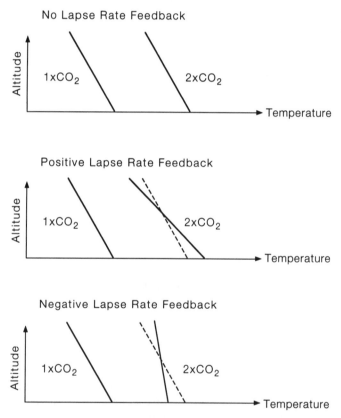

Figure 3.5 Schematic representation of CO_2 induced warming with (A) no lapse rate feedback, (B) a positive lapse rate feedback and (C) a negative lapse rate feedback. Calculations show that case C is realized. (Reproduced by permission of Kluwer Academic Publishers from Michael E. Schlesinger, Quantitative analysis of feedbacks in climate model simulations of CO_2 induced warming, in *Physically-Based Modelling and Simulation of Climate and Climate Change*, Part 2 (ed. M. E. Schlesinger) Kluwer, Dordrecht, 1988, Fig. 6, p. 687)

indicated in Fig. 3.5 (case B). For a negative lapse rate feedback, shown in Fig. 3.5 (case C), the reverse is true. Detailed calculations show that the magnitude of the lapse rate feedback depends on the latitude and the global effects depend on the climate model assumed, but it always turns out to be negative. Qualitatively this may be understood by realizing that with a higher temperature more sea water will evaporate. At higher altitudes it will condense, heating the upper layers more than without global warming.

The reinforcing and mitigating effects ascribed above may be described as a series of feedback mechanisms. A single feedback loop may be presented as in Fig. 3.6 as a system in which the output is partially fed back to the input. A signal V_s entering the loop receives an extra input V_F which leads to

$$V_1 = V_s + V_F \tag{3.7}$$

reinforced by the gain function G which leads to a generalization of eq. (3.6):

$$V_2 = GV_1 \tag{3.8}$$

This signal is picked up by H and produces

$$V_F = HV_2 \tag{3.9}$$

which was introduced in eq. (3.7). All taken together we have

$$V_2 = GV_1 = G(V_s + V_F) = G(V_s + HV_2) = GV_s + GHV_2 \tag{3.10}$$

or

$$V_2 = \frac{G}{1 - GH} V_s \tag{3.11}$$

When there are more loops, independent of each other, characterized by H_i, one could write $V_F = \Sigma H_i V_2$ and obtain

$$V_2 = \frac{G}{1 - \Sigma H_i G} V_s \tag{3.12}$$

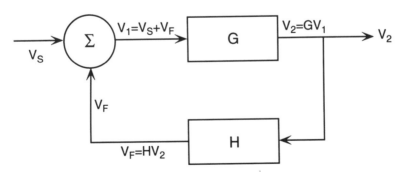

Figure 3.6 Feedback mechanism. At the output of the gain function the signal V_2 is picked up and reintroduced into G by means of H

Usually one writes $f_i = H_i G$ or

$$V_2 = \frac{G}{1 - \Sigma f_i} V_s \tag{3.13}$$

In the climate system the signal would be $-\Delta F_{TA} = \Delta I$ with the response ΔT_s which gives

$$\Delta T_s = \frac{G}{1 - \Sigma f_i} \Delta I = G_f \Delta I \tag{3.14}$$

where

$$G_f = \frac{G}{1 - \Sigma f_i} \tag{3.15}$$

In climate calculations the cumulative effects (b), (c) and (f) mentioned above yield $\Sigma f_i = 0.7$, which leads to $\Delta T_s = 4.2\,\mathrm{K}$ (see Ref. 6, which also discusses the ocean induced time lag of climate change) for an equivalent CO_2 doubling in the atmosphere.

It should be mentioned that more effects may influence the resulting temperature. One could imagine that continuing ozone destruction would destroy plankton in the upper oceans, thereby diminishing their capacity for CO_2 absorption and thus reinforcing global warming. One could believe that higher temperatures lead to more use of air conditioners, leading to more fossil fuel burning. Opposite to these effects, mitigating the problem, one could point to the aerosol particles in the atmosphere caused by the fossil fuels which will give more backscattering, increasing a_a and a_a'. Also, less need for home heating may cancel the effect of more need for cooling to some extent.

In conclusion, it is clear, even from our simplified discussion, that there are large numerical uncertainties in the predicted effects on global temperature due to CO_2 doubling, but there is little doubt that the effect is real. The best models of the moment, to be discussed in Section 3.3, indicate a temperature rise between 1.3 and 4.2 K.

Time Dependence

It should be realized that it will take some time to heat up the oceans because of their heat capacity, which originates in the fact that the top layer of the oceans circulates and distributes its heat energy. Let us concentrate on the oceans as they provide for 70% of the earth's surface. Let the heat capacity of the top layer per m^2 be c_s in $J\,m^{-2}\,K^{-1}$. The radiative forcing ΔI then has to supply the temperature increase both of the surface by $\Delta T_s / G_f$ and the increase in 'sensible' heat per second of $(d/dt)(c_s \Delta T_s)$, which gives

$$\Delta I = c_s \frac{d}{dt}(\Delta T_s) + \frac{\Delta T_s}{G_f} \tag{3.16}$$

A solution for a constant ΔI is easy to find as

$$\Delta T_s(t) = G_f(\Delta I)(1 - e^{-t/\tau_e}) \qquad (3.17)$$

with $\tau_e = c_s G_f$. The time lag τ_e is estimated to be between 50 and 100 years. For $t \to \infty$ one of course gets back to eq. (3.14).

3.2 ELEMENTS OF WEATHER AND CLIMATE

Weather is characterized by parameters such as the number of hours of sunshine, the rain (or more generally the precipitation), the clouds, the winds and the temperature. Climate is also characterized by these parameters, but then averaged over a period of time of some 30 years. Particularly for climate the variability of these parameters in the course of hours or days is included in the definitions. The climate therefore determines whether the natural surroundings are kind to human beings and are suitable for agriculture and industry.

Weather and climate are determined by the conditions of a film of gas around the earth. The horizontal scale is in the order of the circumference of 40 000 km. On a vertical scale the short term changes of the weather are taking place over the much smaller distance of some 10 km, in the so-called *troposphere*. This is the lowest part of the atmosphere, which together with its higher parts extends to a height of about 100 km. The vertical structure of the atmosphere is shown in Fig. 3.7 where it appears that the fall or rise of temperature defines subsequently the *troposphere*, *stratosphere*, *mesosphere* and *thermosphere*. The division lines which vary as a function of latitude are called *tropopause*, *stratopause* and *mesopause*.

One notices the increase in temperature above 80 km. This is due to the photodissociation of molecular oxygen O_2 into atomic O. These atoms strongly absorb light of wavelengths between 100 and 200 nm. Alternatively, dissociation into O^+ ions and electrons may occur, giving rise to an *ionosphere* which reflects radio waves. Similarly absorption of UV sunlight by O_2 between 20 and 40 km of height produces ozone O_3 which strongly absorbs between wavelengths of 200 and 300 nm and causes a general increase in temperature in these regions (cf. Chapter 2).

Motions of the atmosphere not only determine climate but the dispersion of pollutants as well. Therefore, we start with discussing the vertical structure and mobility in the atmosphere and next the horizontal motions.

The steady temperature decrease with increasing altitude in the lower troposphere, as shown in Fig. 3.7, can be understood by studying the adiabatic expansion of a rising parcel of air. Let us first look at the pressure. To a very good approximation a parcel of air is in hydrostatic equilibrium with the air above and below. Let the pressure at height z be $p(z)$; let the pressure at height $z + dz$ be $p + dp$. In vertical equilibrium the sum of the net pressure force dp and

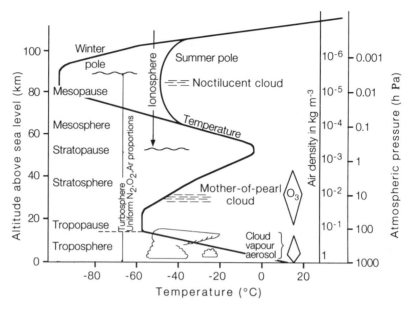

Figure 3.7 The vertical structure of the atmosphere. The altitude on the left scale corresponds with density and pressure on the right scale. Horizontally one finds the temperature with a significant dependence on the seasons around 80 km of altitude. The names of the atmospheric regions are indicated. (Adapted and reproduced by permission of Chapman and Hall, London, from Robin McIlveen, *Fundamentals of Weather and Climate*, Chapman and Hall, London, 1992, Fig. 3.1, p. 48)

the gravity $g\rho\,\mathrm{d}z$ vanish, leading to the equation

$$\mathrm{d}p = -g\rho\,\mathrm{d}z \qquad (3.18)$$

In the lower troposphere one may take the gravity acceleration g as a constant; the density ρ has a strong logarithmic decrease (as can be seen in Fig. 3.7) and it may not be taken as a constant. However, since the actual thermodynamic behaviour of the atmosphere can be approximated rather well by the equation of state of an ideal gas, the density ρ can be expressed in terms of the pressure p.

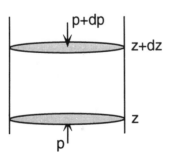

Figure 3.8 Derivation of the hydrostatic equation (3.18). An air parcel of unit area is in hydrostatic equilibrium with air

The basic equation of state for an ideal gas reads

$$pV = nR'T \tag{3.19}$$

Here n is the number of moles in the volume V (1 mole equals M gram when M is the molecular weight) and R' is the universal gas constant given in Appendix C without the prime. The mass m of the gas is given by

$$m = nM \times 10^{-3} \tag{3.20}$$

where the factor 10^{-3} takes care of the required kg units. By defining a specific gas constant

$$R = \frac{10^3}{M} R' \tag{3.21}$$

one may write

$$p = \frac{m}{V} RT = \rho RT \tag{3.22}$$

For a mixture of gases like air the specific gas constant of eq. (3.21) is corrected by taking a linear combination of the specific gas constants of the constituents as explained in Exercise 3.6. With eq. (3.18) it follows that

$$\frac{\partial p}{\partial z} = -g\rho = -\frac{g}{RT} p = -\frac{p}{H_e} \tag{3.23}$$

where for T some average value is taken in order to find an effective height $H_e = (RT)/g$. Therefore, for the pressure p one finds an exponential dependence on the altitude z:

$$p = p_0 e^{-z/H_e} \tag{3.24}$$

which explains the logarithmic scale on the right-hand side of Fig. 3.7. For a temperature $T = 250$ K one finds a decrease of a factor of 10 in pressure for a difference in altitude $\Delta z = 16.8$ km.

Let us now study the temperature change corresponding to the vertical motion of a parcel of air. It follows from the first law of thermodynamics

$$\delta Q = c_V \, dT + p \, dV \tag{3.25}$$

where c_V is the specific heat for constant volume V and δQ is the added heat. For a parcel of air which deforms and changes its volume but in principle retains its mass, it is appropriate to consider a unit of mass for which the volume can be expressed in the density by $V = 1/\rho$. One then finds

$$\delta Q = c_V \, dT + p \, d\left(\frac{1}{\rho}\right) = c_V \, dT + d\left(\frac{p}{\rho}\right) - \frac{1}{\rho} \, dp$$

$$= c_V \, dT + R \, dT - \frac{1}{\rho} \, dp = c_p \, dT - \frac{1}{\rho} \, dp \tag{3.26}$$

where $c_p = c_V + R$ is the specific heat at constant pressure. In passing it may be remarked that isobaric heating of a slab of air close to the ground during a sunny day will be found from $\delta Q = c_p \, dT$ (cf. Exercise 3.7). More important for transport of energy and matter are the rise and fall of a parcel of air. The reference process is the adiabatic case in which no exchange of heat with the surroundings occurs, so $\delta Q = 0$. The adiabat is the curve which describes the temperature T as a function of pressure p or altitude z. There are two extremes, the 'dry' case where the air contains no water vapour and the 'wet' case where the air is saturated with water. For the dry adiabatic process one finds from eq. (3.26)

$$c_p \, dT = \frac{1}{\rho} \, dp$$

$$dT = \frac{dp}{\rho c_p} = \frac{RT \, dp}{p c_p}$$

or

$$\frac{dT}{dp} = \frac{RT}{c_p p} \tag{3.27}$$

Using the second equality of eq. (3.23) one finds for the *dry adiabat*

$$\frac{\partial T}{\partial z} = \frac{\partial T}{\partial p} \times \frac{\partial p}{\partial z} = -\frac{g}{c_p} \tag{3.28}$$

The quantity $\Gamma_d = g/c_p$ is called the *dry adiabatic lapse rate*. Near the ground the magnitude is about $1\,°C$ per $100\,m$. The minus sign in eq. (3.28) is readily understood as rising air will expand and subsequently cool.

For wet, but not saturated, air the lapse rate is simply estimated by an extension of eq. (3.28). Call ω the mass fraction of water vapour; then one may use eq. (3.28) with the understanding that

$$c_p = (1 - \omega)c_{p,\text{air}} + \omega c_{p,\text{water vapour}} \tag{3.29}$$

When the air parcel rises it cools and part of the water vapour condenses, giving $d\omega < 0$. The positive heat of evaporation of a unit of mass of water vapour is denoted by ΔH_v. The positive amount of heat added to the parcel of air upon condensing of an amount $(-d\omega)$ becomes

$$\delta Q = (\Delta H_v)(-d\omega) = c_p \, dT - \frac{1}{\rho} \, dp \tag{3.30}$$

yielding by a similar derivation as before:

$$dT = \frac{1}{\rho c_p} \, dp - \frac{\Delta H_v}{c_p} \, d\omega \tag{3.31}$$

$$\frac{\partial T}{\partial z} = -\frac{g}{c_p} - \frac{\Delta H_v}{c_p} \times \frac{\partial \omega}{\partial z} = -\Gamma_d - \frac{\Delta H_v}{c_p} \times \frac{\partial \omega}{\partial z} = -\Gamma_s$$

The *saturated adiabatic lapse rate* Γ_s is smaller than the dry rate Γ_d. Cool air can contain little water vapour, so both ω and its derivative are small, resulting in lapse rates that are close. For warm air the saturated adiabatic lapse rate may be a factor 2 or 3 smaller than the dry one.

Stability and Vertical Motion of Air

The stability or instability of air becomes important when one studies the behaviour of a warm plume of smoke emitted from a chimney, or cloud formations and their influence on the energy balance of the earth. In both cases one has to consider the behaviour of a parcel of air with respect to its surroundings. Two extreme situations can be defined:

(a) an *unstable* atmosphere, where buoyancy forces enhance vertical motion, and
(b) a *stable* atmosphere, where buoyancy forces oppose vertical motion.

The vertical temperature gradient $\partial T / \partial z$ of the atmosphere is not only caused by vertical motion but also by horizontal motions, heating of the atmosphere by condensation and absorption of heat and radiation. When this gradient is considered as being given, one may distinguish two stability cases both for dry and for saturated air. They are illustrated in Fig. 3.9. On the left one observes that the lapse rate of the surrounding atmosphere is larger than the adiabatic lapse rate of a parcel of air. In such a case a parcel of air which rises because of

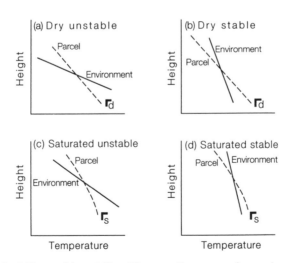

Figure 3.9 Stability and instability. The top diagrams refer to dry air, the bottom two to saturated air. The left diagrams refer to the situation that the lapse rate of the surroundings is larger than the adiabatic ones and the right diagrams refer to the inverse case. (Reproduced by permission of Chapman and Hall, London, from Robin McIlveen, *Fundamentals of Weather and Climate*, Chapman and Hall, London, Fig. 5.8, p. 128)

buoyancy will cool more slowly than its surroundings, therefore again experiencing buoyancy forces, and will continue to rise. Were it to experience negative buoyancy and move down, it would continue moving in that way. This is called *convective instability*.

In the cases on the right-hand side rising parcels cool more quickly than their surroundings, causing negative buoyancy forces that return the parcel to its original position. For similar reasons a descending parcel will also return. This is a *stable* atmospheric situation. One may note in passing that for lower temperatures the dry and wet lapse rates are almost identical. One also notes that warm wet air is very unstable, giving rise to strong convection and clouds during the day.

Horizontal Motion of Air and Water

Climate is partially determined by the energy flow from the equator to higher latitudes. It is estimated as some $200\,\mathrm{MW\,m^{-1}}$ integrated vertically. Three-quarters are transported by the atmosphere and one-quarter by the oceans. The transport equations will be discussed more thoroughly in Section 5.3. Some qualitative remarks, required for an understanding of climate modelling, may suffice now. Consider a volume $d\tau$ with mass $\rho\,d\tau$. Newton's equations of motion may be written as

$$\frac{d\mathbf{u}}{dt}\rho\,d\tau = \mathbf{F}_{press} + \mathbf{F}_{viscous} + \mathbf{F}_{Coriolis} + \mathbf{F}_{gravity} \tag{3.32}$$

The Coriolis term takes into account the fact that one is working in a rotating frame. Centrifugal terms are partly taken into account by the use of a local gravity constant g and the rest are neglected as the major motions are horizontal. Let us look at the individual terms of eq. (3.32).

Pressure Gradient Forces

The first term on the right-hand side is the pressure gradient force $\mathbf{F}_{press}$. It is easily found from Fig. 3.10 that the force in the x direction on the left of a volume element amounts to $p(x, y, z)\,dy\,dz$ and on the right to $-p(x + dx, y, z)\,dy\,dz$. That gives for the x component of the pressure gradient force $(-\partial p/\partial x)\,d\tau$ and for the force

$$\mathbf{F}_{press} = -\nabla p\,d\tau \tag{3.33}$$

Viscous Forces

Look at Fig. 3.11 which repeats Fig. 3.10 but with velocities added. Look at the x component $u_x(z)$ and assume that it is increasing in the z direction. The

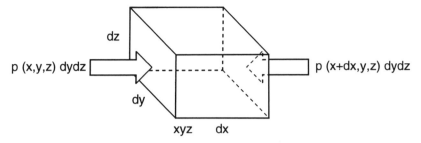

Figure 3.10   Derivation of the pressure gradient force

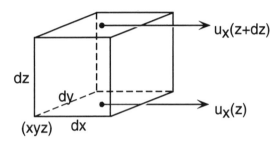

Figure 3.11   Derivation of the viscous force showing the increase of u_x

viscous forces at (x, y, z) will be proportional to $-\partial u_x/\partial z$ (Newton's assumption; cf. Section 5.4); the minus sign is added as it is the force exerted by the fluid on the element. As the proportionality constant μ, the *dynamic viscosity* is defined per unit area. The force on the bottom becomes

$$-\mu \frac{\partial u_x}{\partial z}\bigg|_{xyz} dx\, dy$$

Similarly the force on the top area becomes

$$+\mu \frac{\partial u_x}{\partial z}\bigg|_{xyz+dz} dx\, dy$$

The net force in the x direction therefore will become

$$F_x = -\mu \times \frac{\partial u_x}{\partial z}(z)\, dx\, dy + \mu \times \frac{\partial u_x}{\partial z}(z+dz)\, dx\, dy = \mu \times \frac{\partial^2 u_x}{\partial z^2}\, d\tau \quad (3.34)$$

and similarly in the y direction. In the meteorology the viscous force works in the horizontal directions as the z component of the velocity can be neglected.

Per unit of mass one finds for the viscous force in the x direction

$$\frac{\mu}{\rho} \times \frac{\partial^2 u_x}{\partial z^2} \tag{3.35}$$

The ratio μ/ρ is called the *kinematic coefficient of viscosity v.*

Inertial Forces

The Coriolis force written down in eq. (3.32) is a so-called inertial force, a fictitious force to be added to the physical forces in order to correct for the acceleration, here the rotation, of the coordinate system. It is written as

$$\mathbf{F}_{\text{Coriolis}} = -2\mathbf{\Omega} \times \mathbf{u}\rho\, d\tau \tag{3.36}$$

where vector $\mathbf{\Omega}$ is the angular velocity of the earth around its axis. Below, it is shown that the quantity $f = 2\Omega \sin \beta$, with β the geographical latitude, determines most of the physics of this force. Therefore f is called the *Coriolis parameter.*

Gravity

Finally, the gravity force, the last term in eq. (3.32) may be written as

$$\mathbf{F}_{\text{gravity}} = \mathbf{g}\rho\, d\tau \tag{3.37}$$

It is left as Exercise 3.9 to derive the hydrostatic equation (3.18) from eq. (3.32) and subsequent equations.

Example: Geostrophic Flow

Consider first the atmosphere. At higher altitudes (more than 500 m) the wind blows with a rather constant velocity, so the acceleration term on the left in eq. (3.32) vanishes to a good approximation. Viscous forces may be neglected as the earth's surface is far away and the vertical derivatives of the horizontal velocity components in eq. (3.34) are small anyway. Gravity is compensated by the hydrostatic decrease of pressure with altitude. Therefore the horizontal components of eq. (3.32) for a unit of volume $d\tau = 1$ reduce to

$$-\nabla p - 2\rho\mathbf{\Omega} \times \mathbf{u} = 0$$

or

$$-\frac{1}{\rho}\nabla p - 2\mathbf{\Omega} \times \mathbf{u} = 0 \tag{3.38}$$

On the equator the vector product $2\mathbf{\Omega} \times \mathbf{u}$ has vertical components only; therefore eq. (3.38) only makes sense at middle latitudes. In that region the vertical component of Ω which is $\Omega \sin \beta$ gives rise to a horizontal component in eq. (3.38). The resulting wind velocity is called the *geostrophic wind* u_G. By using eq. (3.38) one finds

$$u_G = \frac{|\nabla p|}{2\Omega \rho \sin \beta} = \frac{|\nabla p|}{f \rho} \qquad (3.39)$$

where f is the Coriolis parameter introduced above.

In eq. (3.38) the negative gradient of pressure p is perpendicular to the isobars in the direction of decreasing pressure. This vector has the same direction as $\Omega \times u$. In the Northern hemisphere it follows that the direction of the geostrophic velocity is perpendicular to the negative gradient to the right, while in the Southern hemisphere it is to the left. In both cases the velocity will be parallel to the isobars.

For the wind speeds an order of magnitude estimate can be made by using conservation of angular momentum. Consider a parcel of air with mass m that is in rest at the equator. In an inertial coordinate system it has an angular momentum $m\Omega R^2$ in the direction of Ω. The air parcel moves to the north, say, keeping its angular momentum. At a latitude β the velocity with respect to the earth in the east direction would be U. Conservation of its total angular momentum would then lead to

$$m(U + \Omega R \cos \beta)R \cos \beta = m\Omega R^2$$

which leads to

$$U = \Omega R \left(\frac{1}{\cos \beta} - \cos \beta \right) \qquad (3.40)$$

For a latitude β of 20° this would lead to a velocity of 55 m s^{-1} and for a latitude of 30° to a velocity of 130 m s^{-1}. In practice the geostrophic wind velocities are somewhat less than 60% of these values.

For the oceans eq. (3.39) holds as well. In fact, as a first-order approximation it is even better than in the atmosphere to ignore friction and gravity forces. In order to obtain an estimate of the geostrophic velocity in the oceans, one may take the gradients in the atmosphere and ocean to be equal. Differences in the atmospheric pressure will result in similar differences in the pressures at sea level. The essential difference between air and water then is the density, which is 1000 times higher for water than for air. Consequently, the geostrophic velocity should be a factor of 1000 smaller, i.e. 5–10 cm s^{-1}.

For the oceans one in practice measures average velocities up to 20 cm s^{-1}, whereas the wind velocities at the surface are up to 6 m s^{-1}. This of course has to do with the friction of the air, which plays an essential role at the earth's surface. In fact, it is the driving force behind the creation of waves.

Baroclinic Models

We will now show that models which describe horizontal motions essentially need to include a layered vertical structure of the atmosphere. We discuss *baroclinic* models, which means that the surfaces of constant pressure (isobaric surfaces) are inclined with respect to the surfaces of constant density. The latter are approximately the horizontal surfaces. According to the equation of state (3.22),

$$P = \rho RT \tag{3.41}$$

which implies that in a 'horizontal' plane of constant density the temperature varies with the pressure. It is useful to study its consequences. We recall the hydrostatic equation (3.18):

$$g = -\frac{1}{\rho}\frac{\partial p}{\partial z} \tag{3.42}$$

and the equations for a geostrophic flow (3.38)

$$-\frac{1}{\rho}\nabla p - 2\Omega \times \mathbf{u} = 0 \tag{3.43}$$

If **k** indicates the unit vector in the vertical direction this equation can also be written as

$$\mathbf{k} \times \mathbf{u} = -\frac{1}{\rho f}\nabla p \tag{3.44}$$

where the Coriolis parameter f was used. Substituting p from eq. (3.41) in the hydrostatic equation (3.42) gives

$$g = -\frac{1}{\rho}\frac{\partial \rho}{\partial z}RT - R\frac{\partial T}{\partial z} \tag{3.45}$$

Dividing on both sides by T leads to

$$\frac{g}{T} = -\frac{R\partial \rho}{\rho \partial z} - \frac{R\partial T}{T\partial z}$$

$$\frac{g}{T} = -R\left[\frac{\partial}{\partial z}(\ln \rho) + \frac{\partial}{\partial z}(\ln T) + \frac{\partial}{\partial z}(\ln R)\right] \tag{3.46}$$

$$\frac{g}{T} = -R\frac{\partial}{\partial z}(\ln p)$$

Similarly, the pressure p from eq. (3.41) may be substituted into eq. (3.44). Following precisely the same procedure but with the gradient operator instead of $\partial/\partial z$ gives

$$\frac{\mathbf{k} \times \mathbf{u}}{T} = -\frac{R}{f}\nabla \ln p \tag{3.47}$$

This equation is differentiated to z and using the last part of eq. (3.46) it gives

$$\frac{\partial}{\partial z}\left(\frac{\mathbf{k} \times \mathbf{u}}{T}\right) = -\frac{R}{f}\nabla\frac{\partial}{\partial z}\ln p = \frac{g}{f}\nabla\frac{1}{T} = -\frac{g}{fT^2}\nabla T \qquad (3.48)$$

Because of the vertical unit vector $\mathbf{k}$ this equation has horizontal components only. It follows that 'horizontal' temperature variations correspond with a change of $\mathbf{u}$ with z, a vertical wind shear. Realistic models for weather and climate in which a horizontal temperature variation is unavoidable should therefore describe at least two vertical layers with a different wind velocity. In his pioneering work Phillips [7] indeed used two vertical layers.

Clouds

In all realistic models cloud formation under the influence of the temperature distribution T as a function of the height z and of the humidity has to be taken into account. One therefore has to find the curves $T(z)$ on many places and construct curves like Fig. 3.12.

A moist parcel of air at the so-called lifting condensation level (LCL) will follow

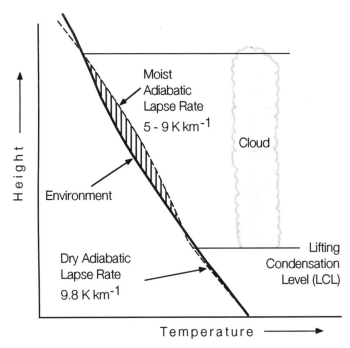

Figure 3.12 Cumulus cloud formation. From level LCL a parcel of air will rise, cool and condense, until the buoyant forces stop acting

the wet adiabat as indicated. By condensation it remains warmer than its surroundings until the moist adiabatic crosses the curve of the 'environment'. Of course, in practice an overshoot may increase the effect. On the other hand, for smaller clouds there will be some exchange of heat with the environment, mitigating the effects. The situation is therefore complex.

Reynolds Number *Re*

In practice, of course, for a realistic model all terms of eq. (3.32) need be taken into account. In Chapter 5 these more general considerations will be given. In deriving eq. (3.39) for the geostrophic wind or the hydrostatic equation (3.18) one in fact only considers two of the terms and neglects the remainder. One further step would be to estimate the relative importance of the terms in the equations of motion (3.32). An important concept which will be discussed more precisely later on (around eq. (5.185)) is to look at the dimensionless ratio of the magnitude of the inertia forces and of the viscous forces. This is expressed by the *Reynolds number Re*, defined as

$$Re = \frac{\text{magnitude inertia forces}}{\text{magnitude viscous forces}} = \frac{\left| \rho \times \dfrac{du_i}{dt} \right|}{\left| \mu \times \dfrac{\partial^2 u_k}{\partial z^2} \right|} \tag{3.49}$$

where u_i and u_k are components of the velocity **u**. In order to estimate the numerator and denominator for a specific problem one takes some characteristic velocity U and a length L leading to a time L/U. In eq. (3.49) one then substitutes the values corresponding to the right dimensions in the following way:

$$Re = \frac{\rho}{\mu} \times \frac{U \times U/L}{U/L^2} = \frac{\rho}{\mu} UL = \frac{UL}{\nu} \tag{3.50}$$

Instead of the dynamic viscosity μ one uses the *kinematic viscosity* $\nu = \mu/\rho$. Experience shows that for values of the Reynolds number Re smaller than 10^3 viscosity dominates the process. For values larger than about 10^3 viscosity is no longer capable of slowing down the motions and the character of the flows becomes very turbulent. For the bottom layer of the atmosphere the kinematic viscosity ν is around $10^{-5} \, m^2 s^{-1}$; a characteristic value of the velocity U is $10 \, m \, s^{-1}$ and of the length L is 10^3 m, if one is interested in the vertical dimensions. This would lead to a Reynolds number Re of 10^9. In the horizontal dimension one would take $L = 10^6$ m and find $Re = 10^{12}$. Thus one finds in both cases a very turbulent situation corresponding to experience.

If one studies the layer very close to the earth's surface (the lowest cm), the characteristic length L would be in the order of a mm and the velocity U of air around $0.3\,\mathrm{m\,s^{-1}}$.* This leads to a Reynolds number Re in the order of 30. Therefore viscosity dominates, leading to small wind velocities which vanish at the surface. This very thin laminar layer acts as a strong membrane around the surface where, for example, heat is transported by molecular diffusion and not by convection at all.

3.3 CLIMATE VARIATIONS AND MODELLING

The climate has been changing in the past, also without human interference, as will be obvious from the regular occurrence of glacial periods. This statement can be made more precise by looking at the temperature as one of the clearest characteristics of a climate. Figure 3.13 shows the temperatures during the last million years.

Before discussing the graph, it is interesting to mention that the Greenland temperatures in the top graph were measured by using the small mass difference between ^{16}O and ^{18}O atoms resulting in a little mass difference of the corresponding water molecules H_2O. The point is that the lighter molecules evaporate more easily and the heavier molecules condense more easily. Thus, when the oceans evaporate, the heavier molecules will tend to fall down again as rain but the lighter molecules will remain somewhat longer in the air, reaching the arctic regions where they fall as ice and snow. Therefore ice and snow are rather poor in heavy molecules and the oceans and its sediments rather rich. The effect is more noticeable at lower temperatures where evaporation is more 'difficult' than at higher temperatures. Therefore, accurate measurement of isotope ratios in old ice or in old sediments will show the temperature of the period when they were formed; its age of course should be estimated independently.

Let us first look at the top graph and ignore the details. One easily identifies the last two glacial periods. The incidence of glacial periods over the last few hundred thousand years is usually attributed to a change in the local insolation, not so much because the sun has changed over this relatively short period of time but rather because the orientation of the earth with respect to the sun has been changing.

The argument is due to the astronomer Milankovitch and is illustrated in Fig. 3.14. There are three physical phenomena determining the insolation.[†] The first is the eccentricity of the orbit, which varies with a period of 100 000 years, with the sun remaining in the focal point; the second factor is the tilt angle between

---

*This velocity $U = 0.3\,\mathrm{m\,s^{-1}}$ is the friction velocity defined in eq. (4.182) as $\sqrt{\tau/\rho}$. The tangential stress τ can be measured and with the well-known value of ρ one finds the value given.
[†] See Ref. 8. Readers may consult Figs. 4.20 and 4.21 of the present book to see the solar orientation more clearly.

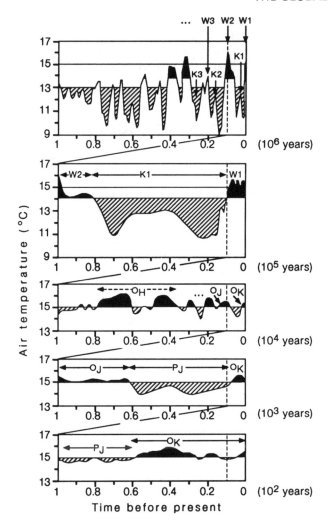

Figure 3.13 Trends in average surface temperatures during the last million years. The top curve gives the whole period; in the lower curves the time scales are a factor of 10 extended successively. The recent data are based on ice core drillings in Greenland (up to 75 000 years ago), while older periods use, among others, deep sea cores and pollen data. Note that the mean temperature is increasing from top to bottom. (Reproduced by permission of Deutsche Verlags-Anstalt, Stuttgart, from Christian-Dietrich Schönwiese and Bernd Diekmann, *Der Treibhaus Effekt*, Deutsche Verlags-Anstalt, Stuttgart, 1987, Fig. 9, p. 44. A comparable graph may be found in IPCC, 1990, p. 202)

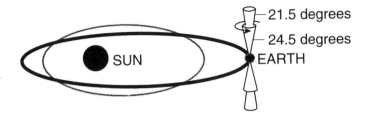

Figure 3.14   The three parameters that determine the insolation: the eccentricity of the orbit (illustrated by two orbits with different eccentricity), the tilt of the earth axis (changing between 21.5 and 24.5°) and the precession of the axis around the normal on the orbital plane

the earth's axis and the normal on the orbital plane, which varies between 21.5 and 24.5° with a period of 41 000 years. The last one is the precession of the earth's axis around the normal with a period of 23 000 years. The physical origin of this last effect is due to the deviation of the earth from a sphere which may be represented by an extra 'belt' at the equator of a spherical earth; the moon is exerting a torque on this 'gyroscope' which causes the precession. The two other effects are due to more complicated gravitational interactions with the rest of the solar system.

Let us consider the effects of the three periodic fluctuations. First, a larger tilt will cause hotter summers and colder winters at both hemispheres, as tilt determines the angle of a horizontal surface with the solar radiation. Second, look at the position farthest from the sun in a highly eccentric orbit. In the hemisphere where it is summer this season will be weakened and a half-year later the distance to the sun is smaller than for a spherical orbit, so the winter is less severe. For the other hemisphere both seasons will be more extreme.

Finally, the precession determines in what point of the orbit midsummer and midwinter are reached. If in a highly eccentric orbit midsummer in the Northern hemisphere occurs at a position far from the sun both seasons in the north will be weakened whereas in the south they will be more extreme. If this coincides with the largest tilt, the extremes in the south will be large indeed. The three effects taken together cause variations in the insolation at a certain location of some 10% around an average value. It should be kept in mind that the total insolation of the earth does not change except for a minor effect due to eccentricity variations. Thus the recurrence of ice ages should be attributed to an interaction of regional changes in insolation, temperature differences on earth and albedo changes due to glaciation.

Let us return to Fig. 3.13. Milankovitch-type calculations result in an interference of three long term effects. Thus they will at best reproduce only long term effects like the recurrence of ice ages and interglacial periods. It is clear that more effects are needed to reproduce the curves in more detail. The cold period of between 200 and 600 years ago in Fig. 3.13, for example, is referred to as the 'little ice age'. One usually interprets this cold age as a consequence of the non-linear character of the equations of motion, either of the sun or of the atmosphere, and as an example of the *chaos theory* which describes these phenomena. The non-linear character of the equations becomes apparent when we view the x component of the acceleration of a unit of mass following eq. (3.32):

$$\frac{d}{dt} u_x(x, y, z, t) = \frac{\partial u_x}{\partial x}\frac{\partial x}{\partial t} + \frac{\partial u_x}{\partial y}\frac{\partial y}{\partial t} + \frac{\partial u_x}{\partial z}\frac{\partial z}{\partial t} + \frac{\partial u_x}{\partial t}$$

$$= \left(u_x \frac{\partial}{\partial x} + u_y \frac{\partial}{\partial y} + u_z \frac{\partial}{\partial z} \right) u_x + \frac{\partial u_x}{\partial t} = (\mathbf{u} \cdot \nabla) u_x + \frac{\partial u_x}{\partial t} \quad (3.51)$$

Therefore, any change in u_x is squared in the first term on the right-hand side. When equations are linear this does not happen and a small change in initial conditions then only leads to a small change of state some time after, in other words cause and effect are proportional. For non-linear equations this no longer needs to be the case.

The inventor of the chaos theory, E. N. Lorenz, found in meteorological computer simulations that two initial positions which were only different in the fourth decimal led to two rather different physical situations after some time. In fact, Lorenz wrote, whether or not a butterfly in Brazil beat its wings could make all the difference to the weather in Texas a few days later. As the differential equations that govern weather and climate require some initial conditions for their solution, which will by necessity be incomplete, precise predictions of the future or reconstructions of the past will be impossible.

Another point, however, is to calculate the effects of some human-induced phenomena such as the increase of greenhouse gases in the atmosphere. One certainly could take a model with a certain concentration of greenhouse gases and the same model with another concentration and calculate the temperature difference in order to estimate the effects of human interference. The procedure outlined in Section 3.1, i.e. identifying several physical effects, assuming they may add up, is only the simplest approach. In better models interaction between the various effects do occur.

Before discussing how climate models work in practice, let us have a look at the temperature data since 1861, which were obtained by direct measurements and are shown in Fig. 3.15. The data are averaged over Northern and Southern hemispheres and over sea and land. It may be remarked that all individual data sets exhibit the same trends. On a time scale of a hundred years they correlate with the increase in CO_2 concentration, shown in Fig. 3.3 for the period since 1750. In fact, it is clear that the extensive burning of fossil fuels indeed started in the last century. Some authors therefore believe that the trend of Fig. 3.15 clearly shows the greenhouse effect, albeit with a time lag determined by the heat capacity of the oceans, as discussed in Section 3.1.

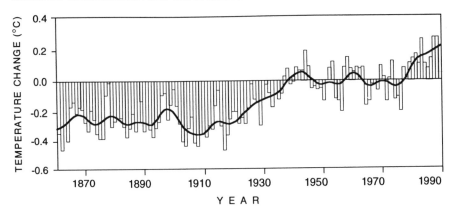

Figure 3.15 Global mean combined land–air and sea–surface temperatures 1861–1989, relative to the average for 1951–1980. It only roughly agrees with the bottom curve of Fig. 3.13 which applies to Greenland only. (Reproduced by permission of IPCC from J. T. Houghton, G. J. Jenkins and J. J. Ephraums (eds.), *The IPPC Scientific Assessment*, Cambridge University Press, 1990, 365 pp., Fig. 11 p. xxix)

The ingredients of three-dimensional climate modelling are given in Fig. 3.16. One notices that the structure of the earth's surface is represented by continents, mountains, ice, snow and oceans, whereas in the atmosphere evaporation and precipitation are taken into account with the change in cloud cover. For the calculations one needs the equations of motion which are essentially given in Section 3.2. Added is the radiation transfer as discussed in Section 3.1 as well as conservation of mass and the equation of state (3.22). The equations for water follow the discussion of wet and dry adiabats and the formation of clouds. Finally, the balance equations for energy and the first law of thermodynamics (3.25) express conservation of energy. The equations are written down for atmosphere and oceans separately and the air–ocean interaction is taken into account. An important aspect, for example, is the formation of sea ice which effectively insulates the air and ocean from each other. It is crucial therefore to calculate ocean temperatures near the freezing point of 271.2 K. Another point is the salinity which influences the equations of motion by the corresponding density ρ and influences the freezing point of water as well.

It goes without saying that climate models operate on a higher level of complexity than the present text. The so-called *general circulation models* (GCMs) aim at representing all aspects shown in Fig. 3.16 as accurately as possible. A summary of the physical processes depicted in Fig. 3.16 is shown in Table 3.4. It should be noted that the properties of the surface should be inserted in the models as a function of geographic position and the time of year.

One of the remaining problems in predicting climate change is lack of computer time. The computational procedure is therefore outlined below.

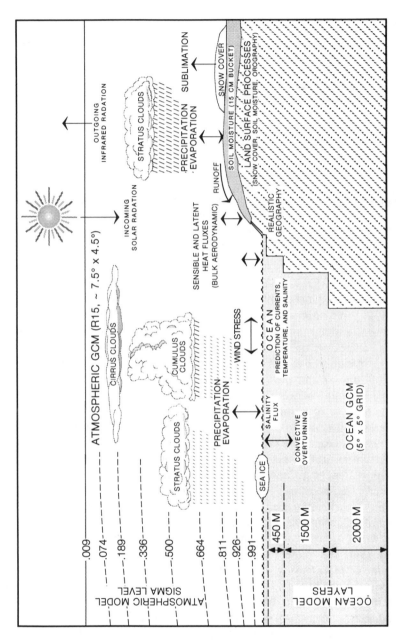

Figure 3.16 Representation of a general circulation model (GCM) with energy exchange, precipitation and evaporation on land and oceans, ice and snow, etc. (Reproduced by permission of American Institute of Physics from Warren M. Washington and Thomas W. Bettge, Computer simulation of the greenhouse effect, *Computers in Physics*, May/June 1990, Fig. 1, p. 241)

Table 3.4 Physics of general circulation models (GCM)[a]. (Reproduced by permission of C. J. E. Schuurmans, KNMI, The Netherlands, personal communication)

Atmosphere equations and variables	Processes
Equations of motion (u_x, u_y)	Friction, diffusion
Hydrostatic equation (u_z)	
First law (T)	Convention, radiation, condensation
Equations of state (p)	
Conservation of mass (ρ)	
Equations for water vapour (q)	Evaporation
Equations for water in clouds (l)	Precipitation

Earth surface properties	Need for modelling
Relief	Obstacles for flow
Roughness	Friction
Albedo	Radiative fluxes
Emissivity	Infrared radiation
Heat capacity	Soil heat currents
Heat conduction	Melting and freezing
Soil humidity	Evaporation
Ice and snow cover	Run-off
Salinity	Mixing in sea

[a]The processes on the top right cannot yet be included by using fundamental physical laws because of their scale which has not been resolved. They must be parametrized. The entities on the bottom left have to be either prescribed or computed.

Differential equations like (3.32) have to be solved numerically. Therefore one has to define a grid over the earth, choose initial conditions in each point and calculate the derivatives such as appear in eq. (3.51) by taking the values in two neighbouring points divided by their distance. One next computes the time derivative and finds the variables again for a finite, albeit small, time later.

This—much too simplified—summary of the numerical procedure should make it clear that, apart from the physics involved, the choice of grid points and of time steps are the decisive factors determining the computational quality of a calculation. In Section 3.2 (below eq. (3.48)) it has already been made clear that one requires at least two and preferably more vertical layers in order to connect a horizontal temperature gradient with a wind shear. Present-day models have some 10–20 layers with a thickness of 100 m close to the earth, increasing to a few km with a pressure of 10 hPa far above.

In the horizontal the best one can do at present is to use grid points with a distance of 200–300 km. This implies that the topology of the earth is taken into account only very roughly. The Rocky Mountains only appear as a single hump and important ocean currents like the Gulf are poorly represented. Still one is

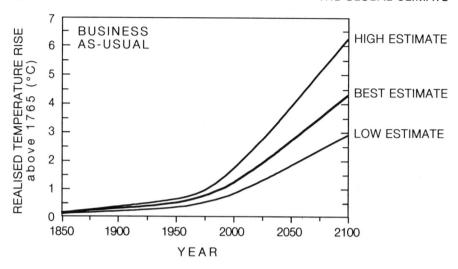

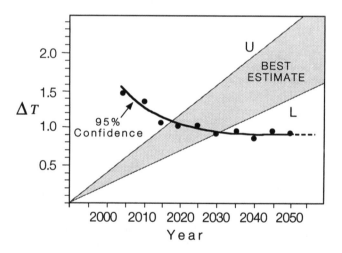

Figure 3.17 Global warming. On top one observes the increase in global mean temperature from 1850 to 1990 (observed) and 1990 to 2100 (predicted on the basis of a 'business-as-usual' scenario). On the bottom a similar prediction is shown for North Scandinavian summer temperatures with a 95% confidence curve, based on tree-ring research. (The top graph has been reproduced by permission of IPCC from J. T. Houghton, G. J. Jenkins and J. J. Ephraums (eds.), *The IPPC Scientific Assessment*, Cambridge University Press, 365 pp., Fig. 8, p. xxii; the bottom graph has been reprinted with permission from K. R. Briffa *et al.*, *Nature*, **346** (2 August 1990), Fig. 4, p. 438. Copyright 1990 Macmillan Magazines Limited)

able to reproduce seasonal variations within the climate, which gives some belief in the predictions of climate change.

In order to get some impression of the results of the calculations and their uncertainties we show on top of Fig. 3.17 the temperature rise in the future for the so-called 'business-as-usual' scenario; this assumes that economic growth and development will continue and that no measures are taken to reduce the emission of greenhouse gases.

It is clear that these calculations do not reproduce the small time scale variations of the global mean temperature shown in Fig. 3.15. It is questionable whether models will ever be able to do so. Many phenomenological parameters are required to represent motions on sub-grid scales and details of the atmosphere–ocean interaction. The careful change of parameters in controlled laboratory experiments is not possible in nature. One has to take the climate evolution as it is and interpret it in computer simulations. The fact that most models lead to similar results may rather mean that the models resemble each other than that they represent nature. Still, the greenhouse effect is real and curves such as Fig. 3.17 calculate the differences in temperature which may be more accurate than absolute predictions.

Natural climate variations will make it difficult to deduce with certainty global greenhouse warming from observed temperatures. Therefore one has to find the probability for a 'natural' climate variation. This has been done by analysing tree rings in North Scandinavia. One notices on the bottom of Fig. 3.17 the predicted summer temperature increase for North Scandinavia and a 95% confidence curve. If a measured average temperature rise is higher than the curve then, with a 95% confidence curve, global warming is occurring. We note in passing that the predicted temperature rise in the Arctic summer (the bottom of Fig. 3.17) is smaller than the global mean (the top of Fig. 3.17).

Exercises

3.1 Derive eqs. (3.1) and (3.2) and show that each term corresponds with a physical process. Note the term-by-term relation between both equations. Use Fig. 3.1 to deduce the albedo parameters given in Table 3.2.

3.2 Write a computer program for a PC (e.g. in BASIC) to find the temperatures T_a and T_s from eqs. (3.1) and (3.2). One could add both equations, express T_a in T_s, substitute in one of the equations and find the zero of the resulting function, e.g. by the bisection method. Check the examples given below eq. (3.2).

3.3 With the program of the previous exercise check the trends of the effects (a), (b) and (e) which reinforce a global warming, as given below eq. (3.6).

3.4 In the text below eq. (3.15) it was said that a doubling of CO_2 concentration would lead to a temperature rise of 4.2 K. From Fig. 3.3 top it may be deduced that between 1850 and 1990 the CO_2 concentration rose from 285 to 360 ppm. What would be the temperature rise you would expect when radiative forcing would be proportional to the CO_2 concentration and when there would be a logarithmic relationship, as models suggest?

3.5 Compare the results of Exercise 3.4 with the observed temperature rise shown in Fig. 3.15. This is usually put as 0.6 K. Blame the difference on the ocean heat capacity and not on inherent fluctuations in climate and deduce a value of the time lag τ_e defined in eq. (3.17).

3.6 For a mixture of gases like air the partial pressures of the components are p_i, the density ρ_i and the specific gas constant $R_i = 10^3 R'/M_i$. Define $\rho = \Sigma \rho_i$ and $x_i = \rho_i/\rho$ as the specific mass of component i. Assume that the total pressure is the sum of the partial pressures. Prove that in eq. (3.22) one should take $R = \Sigma x_i R_i$.

3.7 During morning hours solar heating raises the temperature of a layer of air of 300 m thickness with $2\,°C\,h^{-1}$. Use $\rho = 1.2\,kg\,m^{-3}$ and $c_p = 1004$ $J\,kg^{-1}\,K^{-1}$. Find the rate of input of solar heat and compare with the data given in Fig. 3.1. Comment. Argue why one may assume a constant pressure.

3.8 Find the Poisson equation relating T and p for dry adiabatic expansion from eq. (3.27). Why does this relation not hold for adiabatic expansion of moist air as well? And why does it hold for adiabatic compression of moist air?

3.9 Find the hydrostatic equation (3.18) from eq. (3.32) and subsequent equations by making the appropriate approximations.

3.10 During the flood of 1953 water was streaming into the dutch sea arms (width $W = 4.8\,km$) with a velocity of $1.25\,m\,s^{-1}$. Calculate the difference in sea level at both coasts of the estuary from the Coriolis force, using a latitude of 52° north. Note: the real difference was three times as high because of wind shear.

References

1. Ian M. Campbell, *Energy and the Atmosphere — A Physical-Chemical Approach*, John Wiley, London, 1977. An interesting and easy to read introduction.
2. José P. Peixoto and Abraham H. Oort, *Physics of Climate*, American Institute of Physics, New York, 1992. A book at the postgraduate level with much attention to data.
3. J. F. Kasting and O. B. Toon, Climate evolution, in *Origin and Evolution of Planetary and Satellite Atmospheres* (eds. S. K. Atreya, J. B. Pollack and M. S. Matthews University of Arizona Press, Tucson, AZ, USA, 1989, pp. 423–449.
4. J. E. Lovelock, *GAIA, A New Look at Life on Earth*, Oxford University Press, Oxford, 1979.
5. J. T. Houghton, G. J. Jenkins and J. J. Ephraums (eds.), *The IPCC Scientific Assessment*, Cambridge University Press, Cambridge, 1990, 365 pp; table 2.7 and fig. 2.4, p. 57.
6. M. E. Schlesinger, Equilibrium and transient climatic warming induced by increased atmospheric CO_2, *Climate Dynamics*, **1** (1986) 35–51.
7. N. A. Phillips, The general circulation of the atmosphere: A numerical experiment. *Q.J.R. Meteor. Soc.*, **82** (1956) 123–164.

8. Wallace S. Broecker and George H. Denton, What drives glacial cycles? *Scientific American*, January 1990, pp. 49–56.

Bibliography

Holton, James R., *An Introduction to Dynamic Meteorology*, Academic Press, London, 1992. A standard text.

Houghton, J. T., G. J. Jenkins, J. J. Ephraums (eds.), *Intergovernmental Panel on Climate Change, IPCC, United Nations Environmental Protection Agency*, Cambridge University Press, Cambridge, June 1990. The standard scientific text with many results and discussions.

McIlveen, Robin, *Fundamentals of Weather and Climate*, Chapman and Hall, London, 1992. A very readable book with extensive meteorological discussions and mathematically on a rather elementary level.

Schönwiese, Christian-Dietrich and Bernd Diekmann, *Der Treibhauseffekt*, Deutsche Verlags-Anstalt, Stuttgart, 1987.

Seinfeld, John H., *Atmospheric Chemistry and Physics of Air Pollution*, John Wiley, New York, 1986. This general book about air pollution contains some chapters on weather and climate, which are somewhat more elaborated than in McIlveen's book.

Sørensen, Bent, *Renewable Energy*, Academic Press, London, 1979. This book focuses on renewable energy but discusses solar radiation and its origins as well.

4

Energy for Human Use

Most authors agree that society should use energy as economically as possible. There is a limit on readily available fossil fuels; burning them will contribute to rapid climate change, as was illustrated in Chapter 3, and will also prohibit the use of fossil carbon as a source for producing new materials. Also, an increase in the standard of living in many developing countries will require energy and resources for their development; finally, it is wise to leave resources for coming generations as well. Another point is that energy consumption in whatever form is putting a burden on the environment and therefore economizing on energy consumption is seen as good policy.

In this chapter the first section is devoted to heat transfer as this is a factor in virtually all devices for energy conversion. The bulk of the chapter is devoted to a discussion of the production of usable mechanical energy, of electricity and of so-called renewable energy sources. Although in some countries nuclear power is at present not seen as an acceptable way of producing energy, it is clear that nuclear power stations will be around for a long time. This is sufficient reason to discuss them and the risks their operations pose to society. Another reason is that many policy makers still regard nuclear power as a viable option to produce usable energy; therefore an understanding of the physics and in particular of the concept of an 'inherently safe' reactor deserves attention. Of course the management of nuclear waste and the health effects of radiation will be discussed as well. In passing, a short section is devoted to economical calculations of costs.

It will become clear that each method to produce usable energy has its own problems and its own costs. Risks and strategies, however, are common to any choices society has to make. Therefore they are discussed in the final chapter of this book.

4.1 HEAT TRANSFER

In using heat one encounters two related problems. The first occurs when one wants to keep heat inside a house, a hall, a plant, a heat container, etc. In this case the transfer of heat to the surroundings should be as small as possible. The second occurs when the heat transfer needs to be as great as possible. This occurs in all engines with heat reservoirs and a working fluid absorbing and rejecting heat, or in collectors of solar heat in a reservoir, etc. Heat transfer is therefore discussed below in a rather general way.

The Three Mechanisms of Heat Transfer

Heat transfer always happens from a position with a higher temperature to a location with a lower temperature. There are three mechanisms to be distinguished. The first is *conduction*. This happens through the bulk of a material and is due to collisions between atoms by which kinetic energy is being transferred. The heat current density q'' is defined as the amount of heat in joules which flows through 1 m^2 per second; its dimension is therefore $J\,m^{-2}\,s^{-1}$. When the total surface area under consideration is A the total heat transfer per second q follows as

$$q = Aq''$$

According to *Fourier's law* there is a simple relation between the heat flow q'' and the temperature gradient within a homogeneous material, expressed as

$$q'' = -k\nabla T \tag{4.1}$$

The *thermal conductivity* k is measured in units of $W\,m^{-1}\,K^{-1}$. It is dependent on the kind of material, its temperature, density and humidity. For inhomogeneous materials k may depend upon position. In Table 4.1 values of k are given for some common materials. In general, the k values are strongly temperature dependent. The last two columns of Table 4.1 will be discussed later.

The second means of heat transfer is *convection*. This happens especially when a material is in contact with a fluid (a liquid or gas). When the fluid is in motion the fluid particles in the boundary layer will again exchange heat with the surface through the transfer of kinetic energy, but now the hotter or colder particles disappear in the fluid; this is called forced convection. If the fluid initially is at rest it will start moving as heat exchange with the surface will result in density changes; this is called free convection and was discussed in Section 3.2 for the earth surface–air interaction. According to *Newton's law of cooling* the heat exchange is proportional to the temperature difference at the interface:

$$q'' = h(T_s - T_\infty) \tag{4.2}$$

The heat exchange is again measured in $W\,m^{-2}$. The material surface temperature is denoted by T_s and the fluid temperature by T_∞. Finally h is the *convection heat transfer coefficient*. Some typical values of this coefficient are given in Table 4.2. It will be noted that an enormous variation in values is indicated. In practice the convection will depend not only on the materials and fluids concerned but will

Table 4.1   Heat transport properties of some common materials at $T = 300\,K$ and normal conditions. (The values for ρ, k and a may be found, for example, in Ref. 1, Appendix A; the value of b were calculated. The last four values of b were taken from unpublished lecture notes of the Delft Technological University.)

Material	Density ρ $(kg\,m^{-3})$	Thermal conductivity k $(W\,m^{-1}\,K^{-1})$	Fourier coefficient a $(10^{-7}\,m^2\,s^{-1})$	Contact coefficient b $(J\,m^{-2}\,K^{-1}\,s^{-1/2})$
Insulators				
Air	1.161	0.026	225	
Glass fibre (loose fill)	16	0.043	32	24
Urethane foam	70	0.026	3.6	44
Cork	120	0.039	1.8	92
Mineral wool granules	190	0.046		
Paper	930	0.180	1.4	470
Glass	2500	1.4	7.5	1620
Gypsum plaster	1680	0.22	1.21	635
Building materials				
Cement mortar	1860	0.72	4.96	1020
Soft wood	510	0.12	1.71	290
Hard wood	720	0.16	1.77	380
Oak wood	545	0.19	1.46	499
Brick	1920	0.72	4.49	1075
Concrete	2300	1.4	6.92	1680
Metals				
Iron	7870	80.2	228	17000
Aluminium	2700	237	972	24000
Steel (C, Si)	7800	52	149	13500
Copper	8933	401	1166	37000
Miscellaneous				
Sand	1515	0.27	2.23	572
Soil	2050	0.52	1.38	1400
Soft rubber	1100	0.13	0.59	540
Cotton	80	0.06	5.77	80
Porcelain				1610
Human skin		0.37		1120
Linoleum				525
Carpetwool				105

also be a function of the position at the interface. In fact, it is a major challenge to calculate the values of the convection coefficient in different circumstances. If these calculations are lacking an empirical value is often taken.

A third mechanism of heat transfer is by *radiation*. A body will radiate at a maximum rate determined by the black body radiation at the given temperature T_s (cf. Section 2.1.1). A *black body* is defined as a body with a surface which

Table 4.2 Typical values for the heat transfer coefficient h in $Wm^{-2}K^{-1}$. (Reproduced by permission of John Wiley & Sons from F.P. Incropera and D.P. DeWitt, *Introduction to Heat Transfer*, Wiley, New York, 1990, p. 9.

Phase	Free convection	Forced convection
Gases	2–25	25–250
Liquids	50–1000	50–20 000

Phase change = boiling or condensation 2500–100 000.

absorbs all incoming radiation and therefore has an absorptivity $\alpha = 1$; therefore its reflectivity $\rho = 0$. *Kirchhoff's law* says that the ratio of absorption to emission of a body only depends on wavelength and temperature and not on any other properties of the body. If the absorption is maximal, i.e. $\alpha = 1$ the total emission is at maximum:

$$q'' = \sigma T^4 \tag{4.3}$$

If the body is not black the *emissivity* ε is defined as the fraction of black body radiation that is in fact emitted, leading to

$$q'' = \varepsilon \sigma T^4 \tag{4.4}$$

This simplified discussion holds for the case where α and ε do not depend on the wavelength λ, a so-called *grey body*. If they do depend on the wavelength one has to introduce quantities α_λ and ε_λ and to modify eqs. (4.3) and (4.4).

Let us return to heat transfer by radiation. Besides the emission $q'' = \varepsilon \sigma T^4$ of the body itself, the surroundings with a temperature T_∞ will radiate back. In fact, if the surroundings completely enclose the body, the surroundings will always act as a black body radiating σT_∞^4. Of this only the fraction $\varepsilon \leqslant 1$ will be absorbed by the body as it is not (necessarily) black.

If absorption and emission are independent of the wavelength, its heat exchange may be simply written as

$$q'' = \varepsilon \sigma (T_s^4 - T_\infty^4) \tag{4.5}$$

In the remainder of this section radiation transfer will essentially be ignored and the convection will be taken into account by a single convection coefficient h.

Analogy with Electric Resistance

Consider a homogeneous extended sheet, extended far into the y and z directions and with a thickness d in the x direction. On the left at $x = x_1$ it has a temperature T_1 and on the right at $x = x_2$ it has a temperature T_2, as shown in the left picture of Fig. 4.1. Assume $T_2 > T_1$ and a stationary situation.

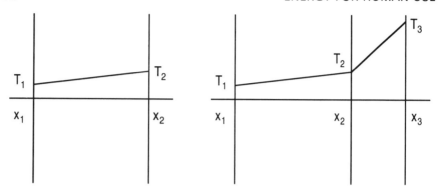

Figure 4.1 Simplest example of heat conductivity: on the left are two parallel boundaries of a sheet with thickness d. Indicated is the temperature T for $T_2 > T_1$. On the right the situation of two adjacent sheets is indicated

The problem is essentially one dimensional with variable x. The heat flow q will be independent of x, as temperatures will be constant in time (a stationary situation). From Eq. (4.1) it follows that, for a surface area A,

$$q = -kA\frac{dT}{dx} = \text{constant} \qquad (4.6)$$

Therefore $T(x)$ is a straight line between (x_1, T_1) and (x_2, T_2) and

$$T_2 - T_1 = q\frac{d}{kA} = qR \qquad (4.7)$$

The analogy with Ohm's formula for electrical current is apparent when $T_2 - T_1$ is analogous to $V_2 - V_1$, the heat current q with the electric current i and therefore

$$R = \frac{d}{kA} \qquad (4.8)$$

is called the *heat resistance* of the sheet in units of $W^{-1}\,K$. The thickness of a sheet with area $1\,m^2$ and with heat resistance $R = 1$ follows as $d = k$. So the column for k in Table 4.1 may be interpreted as the thickness of a sheet with $A = 1$ and resistance $R = 1$. It is immediately clear that most building materials are bad insulators. One has to combine them with mineral wool or glass fibre to improve insulation.

Let us consider a stack of sheets, each with the same surface area A and making excellent thermal contact; then the flow out of the first sheet will enter the second sheet and the temperature at both adjacent surfaces will be equal. In that case, indicated on the right in Fig. 4.1 the value q of eq. (4.6) is the same for all sheets.

They may have a different thickness d and thermal conductivity k. For two sheets the following equations hold:

$$T_3 - T_2 = q\frac{d_2}{k_2 A} = qR_2$$

$$T_2 - T_1 = q\frac{d_1}{k_1 A} = qR_1 \qquad (4.9)$$

$$T_3 - T_1 = q(R_1 + R_2) = qR$$

where the last equation shows that thermal resistances may be added in series just like electric resistances.

A heat resistance may also be associated with convection and radiation. From Newton's law of cooling (4.2) it follows that the heat current from a surface with temperature T_s to a fluid with temperature T_∞ is

$$q = hA(T_s - T_\infty) \qquad (4.10)$$

or

$$T_s - T_\infty = q\frac{1}{hA} = qR \qquad (4.11)$$

The heat resistance connected with convection may therefore be expressed as $R = 1/(hA)$.

For radiation the situation is a little more complicated, as one must write eq. (4.5) as a temperature difference $T_s - T_\infty$, where T_∞ is the temperature of the surroundings. This can be done by

$$T_s - T_\infty = q\frac{1}{A\varepsilon\sigma(T_s + T_\infty)(T_s^2 + T_\infty^2)} = qR \qquad (4.12)$$

It is clear that the heat resistance now contains the temperatures itself. Ignoring this complication one might work with some effective resistance R. In practice surface convection and radiation are combined into a single resistance $1/(hA)$.

The above may be summarized by writing down the total heat resistance for a wall consisting of two parallel sheets and two outer surfaces as

$$R = \frac{1}{h_1 A} + \frac{d_1}{k_1 A} + \frac{d_2}{k_2 A} + \frac{1}{h_2 A} \qquad (4.13)$$

where h_1 and h_2 summarize the losses (or gains) due to convection and radiation on both sides. It should be remarked that in practice the contact between two adjacent surfaces will not be ideal, due to the surface roughness. This will result in a slight temperature difference between both surfaces and correspondingly with a contact resistance. This resistance should be added to the total resistance of eq. (4.13). In practical applications where temperatures are given and fixed, one may calculate the heat flux by equations such as (4.11) and calculate intermediate temperatures by eqs. (4.9).

The Heat Equation

The temperature in a substance will in general depend both on time t and position $\mathbf{r}$, so $T = T(\mathbf{r}, t)$. The equation that governs the rate of change of the temperature T follows from the conservation of heat, where only conduction is studied. Consider a volume element dV. Then for a unit of time the following relation holds:

$$\text{increase of heat content} = \text{net inflow by heat conduction}$$
$$+ \text{ inner heat production}$$

which leads to

$$\frac{\partial}{\partial t}(\rho c_p T \, dV) = -\operatorname{div} \mathbf{q}'' \, dV + \dot{q} \, dV$$

or

$$\rho c_p \frac{\partial T}{\partial t} = -\operatorname{div} \mathbf{q}'' + \dot{q} \tag{4.14}$$

Here, c_p is the specific heat at constant pressure and $\dot{q}$ is the heat production per unit of time inside the volume element. This may be due to chemical or nuclear reactions or to the presence of a heat sink. The minus sign in front of the divergence term originates from the fact that divergence is an outflow instead of the required inflow.

Using Fourier's equation (4.1) leads to

$$\rho c_p \frac{\partial T}{\partial t} = \operatorname{div}(k \operatorname{grad} T) + \dot{q} \tag{4.15}$$

This equation is known as the *heat diffusion equation* or in short the *heat equation*. It may be simplified when the thermal conductivity k is independent of position, which gives

$$\rho c_p \frac{\partial T}{\partial t} = k \operatorname{div} \operatorname{grad} T + \dot{q} \tag{4.16}$$

or

$$\frac{\partial T}{\partial t} = a \Delta T + \frac{1}{\rho c_p} \dot{q} \tag{4.17}$$

in which the Laplace operator Δ appears. This operator is written out for several coordinate systems in Appendix B. The coefficient $a = k/(\rho c_p)$ in eq. (4.17) is a material property called the *Fourier coefficient*. It is given in Table 4.1 for a few common materials. When there are no heat sources or heat sinks eq. (4.17) can be further simplified to

$$\frac{\partial T}{\partial t} = a\Delta T \qquad (4.18)$$

In two or three dimensions this equation can be solved by writing the Laplacian in appropriate coordinates. One may get an impression of the physics involved from some one-dimensional examples, which are discussed below. In that case the heat equation reduces to

$$\frac{\partial T}{\partial t} = a\frac{\partial^2 T}{\partial x^2} \qquad (4.19)$$

Sinusoidal Boundary Conditions in One Dimension

Consider a semi-infinite homogeneous medium with x as the only variable perpendicular on the boundary at $x = 0$. The medium extends from $x = 0$ to $x = \infty$. Assume that at $x = 0$ a periodic heat wave enters due to daily or annual variations in temperature. These boundary conditions may be summarized by

$$\begin{aligned} x = 0: \quad & T = T_{av} + T_0 \cos \omega t \\ x = \infty: \quad & T = T_{av} \end{aligned} \qquad (4.20)$$

Of course the temperature variations will not be felt an infinite distance away.

The solution of the heat equation (4.19) with boundary conditions (4.20) will be a damped sinusoidal wave travelling in the positive x direction:

$$T = T_{av} + T_0 e^{-Ax} \cos \omega \left(t - \frac{x}{v} \right) \qquad (4.21)$$

Substitution into eq. (4.20) gives both sinus and cosine terms which should vanish identically, giving the two parameters A and v as

$$\begin{aligned} v &= \sqrt{2a\omega} \\ A &= \sqrt{\omega/(2a)} \end{aligned} \qquad (4.22)$$

The damping depth $x_e = 1/A$ is the depth at which the amplitude of the wave is reduced to $1/e$; the delay time τ is the time the heat wave needs to travel 1 m: $\tau = 1/v$. By using the data from Table 4.1 it appears that for a concrete wall the damping depth of the daily temperature variation is some 14 cm and the delay time more than a day. One may also calculate how deep one should bury water pipes in order to prevent freezing.

A Sudden Change in Temperature

A similar half infinite medium as described above now experiences a sudden temperature change at the surface from T_0 to T_1 (e.g. cooling of a hot chimney by rain or heating of a stone floor by a fire). The boundary conditions are

$$T(x = 0, t \geqslant 0) = T_1$$
$$T(x > 0, t = 0) = T_0 \tag{4.23}$$

The solution of the heat equation (4.19) with these conditions leads to the error function defined in eq. (A.5) of Appendix A. This leads to

$$T = A + Bx + C \operatorname{erf}\left(\frac{x}{2\sqrt{at}}\right), \qquad t \geqslant 0 \tag{4.24}$$

or

$$T = T_1 + (T_0 - T_1)\operatorname{erf}\left(\frac{x}{2\sqrt{at}}\right), \qquad t \geqslant 0 \tag{4.25}$$

It should be mentioned that from the temperature field $T(\mathbf{r}, t)$ one may always deduce the heat current density $\mathbf{q}''$ by differentiation as shown in eq. (4.1). In the present case this would lead to

$$\mathbf{q}'' = \mathbf{e}_x \frac{k(T_1 - T_0)}{\sqrt{\pi at}} e^{-x^2/(4at)} \tag{4.26}$$

or at $x = 0$, omitting the unit vector $\mathbf{e}_x$:

$$q'' = \frac{k}{\sqrt{a}} \frac{T_1 - T_0}{\sqrt{\pi t}} = \frac{b}{\sqrt{\pi}} \frac{T_1 - T_0}{\sqrt{t}} \tag{4.27}$$

with $b = \sqrt{k\rho c_p}$, a material constant displayed in the last column of Table 4.1.

Contact Temperature

Suppose that one brings two half infinite materials with temperatures T_1 and T_2 ($T_1 > T_2$) into ideal thermal contact at time $t = 0$. The contact plane will acquire a *contact temperature* T_c in between. This temperature follows from eq. (4.27) by requiring that the flow out of the hotter surface equals the flow into the cooler one. One finds

$$\frac{b_1}{\sqrt{\pi t}}(T_1 - T_c) = \frac{b_2}{\sqrt{\pi t}}(T_c - T_2) \tag{4.28}$$

giving

$$T_c = \frac{b_1 T_1 + b_2 T_2}{b_1 + b_2} \tag{4.29}$$

The application of eq. (4.29) for this concept is particularly interesting when the contact of human skin with a hot or cold material is studied. The human foot skin has a temperature of about 30 °C. On an oak wooden floor with a temperature of 15 °C a naked foot feels comfortable. Using the data of Table 4.1 yields a contact temperature of 25.4 °C. One now can deduce the temperatures of floors of different materials which give the same contact temperature of 25.4 °C and subsequently feel equally pleasant. It follows that materials with low b values should be taken. For cork, for example, the surface temperature might be 31 °C below zero, whereas for concrete it would be 22 °C above zero.

A similar argument may be given for the materials which the human fingers can handle. The finger temperature appears to be 32 °C and the maximum contact temperature 45 °C. For iron one could handle 46 °C and for soft wood 95 °C. It is obvious that fingers should not be used as a thermometer.

Heat Generation in a Cylindrical Tube

In a system with cylindrical symmetry and infinitely long in the direction of the axis the temperature T will be a function of the radial coordinate r only. According to eq. (B.2) the heat equation (4.16) then reduces to a one-dimensional equation in the coordinates r and time t. Even simpler is the stationary problem where the time derivative vanishes as well. Then

$$k \frac{1}{r} \frac{\partial}{\partial r} \left(r \frac{\partial T}{\partial r} \right) + \dot{q} = 0 \tag{4.30}$$

As an example we consider the heat generation due to homogeneous heat production $\dot{q}$ within a long cylindrical tube with radius r_0, which may be a current-carrying wire. The tube is cooled at the outside by a fluid with temperature T_∞ and a convection rate h. The question is to determine the surface temperature T_s for steady state conditions as a function of the convection rate h.

Integration of eq. (4.30) leads to

$$T(r) = -\frac{\dot{q}}{4k} r^2 + C_1 \ln r + C_2 \tag{4.31}$$

The integration constants are determined by the boundary conditions

$$\frac{dT}{dr} = 0 \qquad \text{at } r = 0$$

$$T(r_0) = T_s \tag{4.32}$$

The first condition expresses the symmetry at $r = 0$, the axis of the cylinder. The second one follows from the definition of the problem. This leads to the solution

$$T(r) = \frac{\dot{q} r_0^2}{4k} \left(1 - \frac{r^2}{r_0^2} \right) + T_s \tag{4.33}$$

The energy produced within a length L should be equal to the energy lost by convection at the surface which follows from eq. (4.2). This gives

$$\dot{q}(\pi r_0^2 L) = 2\pi r_0 Lh(T_s - T_\infty) \tag{4.34}$$

which simplifies to

$$\frac{\dot{q}r_0}{2h} = T_s - T_\infty \tag{4.35}$$

The case of a *point source* with spherical symmetry will be discussed in Section 4.5.4 in connection with storage of high level waste.

Heat Exchange in Fins

Improving heat exchange is one of the ways to economize on energy consumption. As a simple one-dimensional example consider the case of a fin of constant cross-section A_c and length L that transports heat from an engine at $x = 0$ to a fluid with convection coefficient h. The conduction within the fin is determined by a thermal conductivity k. The set-up is sketched in Fig. 4.2 and may be understood as the fins connected on a motorcycle engine block. For simplicity it is assumed that the temperature across a section of the fin is constant, so the temperature T is a function of x only.

The easiest way to derive the relevant equations is to look at a cross-section between x and $x + dx$, as indicated on the right in Fig. 4.2. The fact that for a stationary situation the heat entering equals the heat leaving leads to

$$q''(x)A_c = q''(x + dx)A_c + (dA_s)h(T_s - T_\infty) \tag{4.36}$$

where the heat losses on the right are due to the heat current conducted out

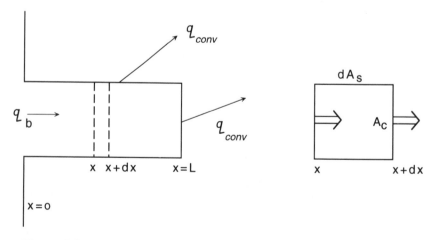

Figure 4.2 Conduction and convection in a fin of constant cross-section

added to the convection loss at the surface element dA_s. At the end with $x = L$ there is only convection loss

$$q''(x = L)A_c = hA_c(T(L) - T_\infty) \qquad (4.37)$$

Equation (4.36) leads to

$$-k\frac{dT(x)}{dx}A_c = -k\frac{dT(x + dx)}{dx}A_c + (dA_s)h(T_s - T_\infty) \qquad (4.38)$$

As the cross-section A_c is constant and dA_s/dx is a constant P, the equation reduces to

$$\frac{d^2T}{dx^2} - \frac{hP}{kA_c}(T_s - T_\infty) = 0 \qquad (4.39)$$

with the boundary condition at $x = L$ from eq. (4.37):

$$-k\frac{dT}{dx}(x = L) = h(T(L) - T_\infty) \qquad (4.40)$$

Equation (4.39) is solved by introducing an excess temperature

$$\Theta(x) = T(x) - T_\infty \qquad (4.41)$$

and using $T_s = T(x)$. This leads to

$$\frac{d^2\Theta}{dx^2} - m^2\Theta = 0; \qquad m^2 = \frac{hP}{kA_c} \qquad (4.42)$$

and gives the solution

$$\Theta(x) = C_1 e^{mx} + C_2 e^{-mx} \qquad (4.43)$$

The integration constants C_1 and C_2 are found from the boundary conditions at the base of the fin $x = 0$ where the temperature is fixed:

$$\Theta(0) = T_b - T_\infty = \Theta_b \qquad (4.44)$$

and at the end of the fin with $x = L$. There are several possibilities, but here it is assumed that there is convection loss as indicated in eq. (4.40). This gives

$$-k\frac{d}{dx}(\Theta)(x = L) = h\Theta(L) \qquad (4.45)$$

The final solution then becomes

$$\frac{T(x) - T_\infty}{T_b - T_\infty} = \frac{\cosh m(L - x) + \dfrac{h}{mk}\sinh m(L - x)}{\cosh mL + \dfrac{h}{mk}\sinh mL} \qquad (4.46)$$

The total heat transfer through the fin to the fluid equals the heat passing at $x = 0$, which is easily found by substitution of eq. (4.46) into

$$q_b = - kA_c \frac{dT}{dx} \qquad \text{at } x = 0 \tag{4.47}$$

Heating

The human body must remain at a temperature of 37 °C to survive and one knows from experience that an air temperature of 20 °C within a shelter is adequate for that purpose. The shelter radiates from its walls and it reduces air circulation, resulting in a low value of the heat convection coefficient h. In most climates additional heat is required to keep the inside temperature at 20 °C during winter times, whereas during the summer it might be necessary to provide cooling.

In heating of homes, offices and industrial buildings the first point will be to produce the heat as efficiently and cleanly as possible. From first principles one would reject the use of electricity for this purpose and rather use gas or waste heat.

Second, one will try to make the heat transfer through the walls as small as possible by means of good insulation and one will not heat or cool rooms at times when they are not in use. From the previous discussion it follows that it is not the thermal conductivity k but rather the Fourier coefficient a that determines heat transfer by conduction (cf. eq. (4.17)).

Finally, one may design buildings in such a way that the energy requirements during summer and winter are as low as possible. This can be done by making an energy balance. It may be interesting to note that in an average Dutch home 12% of the heat is provided by the sun shining through the windows and still as much as 4% by human body heat.

Translucent Thermal Insulation

An interesting development in insulation is the translucent thermal insulation, the principle of which is displayed in Fig. 4.3. The solar radiation comes in from the left and is transmitted through a layer of translucent insulation materials (TIM).

The radiation is absorbed by the blackened wall, which therefore is heated. The TIM layer has a lower conductivity than the wall so that the bulk of the solar heat flows to the inside. There it arrives some hours later, during the cool evening hours, due to the damping effects discussed in eqs. (4.21) and (4.22).

4.2 ENERGY FROM (MAINLY) FOSSIL FUELS

Thermodynamics is understood as that part of physics which from a macroscopic point of view describes the conversion of heat into mechanical energy and vice

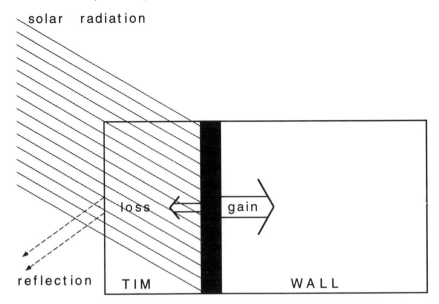

Figure 4.3 Translucent thermal insulation. The wall on the right is covered with translucent insulation materials (TIM)

versa. In this section the major thermodynamic variables are discussed. Examples of conversion devices are given where we distinguish between heat engines in which the heat is provided by an external source and internal combustion engines in which the heat is provided internally. Heat pipes are discussed as an efficient means of transporting heat.

4.2.1 THERMODYNAMIC VARIABLES

In the simplest case, a homogeneous quantity of matter in a single phase, the macroscopic state of the system is defined by two independent variables; usually the pressure p and either volume V or temperature T are taken. For a mixture of chemical substances in perhaps several of the phases solid, fluid, gas the number of moles n_i^φ of chemical i in phase φ must be specified as well.

The *first law of thermodynamics* for an infinitesimal change of state reads as

$$\delta Q = \mathrm{d}U + \mathrm{d}W \tag{4.48}$$

That is, the amount of heat δQ added to the system is used for an increase δU in internal energy and for performing work $\mathrm{d}W$. Often, however, $\mathrm{d}W$ is defined as the work done *on* the system; in that case it has a different sign. For a quasi-static reversible process where expansion is the only form of work we write in our convention $dW = p\,dV$, leading to

$$\delta Q = dU + p\,dV \tag{4.49}$$

In integral form these equations may be written as

$$Q = U_2 - U_1 + W_{1 \to 2} \tag{4.50}$$

and

$$Q = U_2 - U_1 + \int_1^2 p\,dV \tag{4.51}$$

If the system not only expands with infinitesimal work $p\,dV$ but also performs other work (e.g. electromagnetic) of form dW_e, this should be added on the right-hand side of eqs. (4.49) and (4.51), giving

$$\delta Q = dU + p\,dV + dW_e \tag{4.52}$$

and

$$Q = U_2 - U_1 + \int_1^2 p\,dV + \int_1^2 dW_e \tag{4.53}$$

In the following we assume that only expansion work is done by the system, unless stated otherwise.

The second law may be formulated by the use of the *entropy* function S. Its increase dS is defined as

$$dS = \frac{\delta Q}{T} \tag{4.54}$$

where the reversibly added heat δQ is divided by the absolute temperature T. More generally, including irreversible processes, the *Clausius inequality* reads

$$dS \geqslant \frac{\delta Q}{T} \tag{4.55}$$

where the equality sign holds for reversible processes. The *second law* may then be expressed by writing that for a process in a closed system the entropy does not decrease:

$$dS \geqslant 0 \tag{4.56}$$

The equality holds for reversible processes and the inequality sign for irreversible ones.

In practice several, mutually dependent, functions are needed to describe physical processes. Consider the *enthalpy H*, defined as

$$H = U + pV \tag{4.57}$$

For an infinitesimal change of state it follows that

$$dH = dU + p\,dV + V\,dp$$
$$dH = \delta Q + V\,dp \tag{4.58}$$

For a process at a constant volume V the amount of heat δQ added would cause an increase in internal energy dU, following eq. (4.49). However, it is much more common for processes to be *isobaric*, with $dp = 0$. Then the added heat is equal to the increase in the enthalpy of the system. This also holds when chemical reactions take place, or in a change of phase (vaporization, condensation, melting, etc.). Consequently, in tables of heating values or calorific values of fuels one will find the enthalpy change ΔH for 1 kg or for one mole reacting with oxygen. The tabulated values will be the energy released, which is $-\Delta H$.

Studying these tables carefully, often two values of $-\Delta H$ are found. This corresponds to the phase in which released water is left. If the water is condensed one will find a higher value of $-\Delta H$ than when it is released as vapour, as in the latter case the latent condensation heat of the vapour is lost. It should be stressed that the enthalpy H is a function sharply defined by eq. (4.57), whereas the 'heat Q' cannot be defined. Added heat δQ is therefore not a total differential whereas added enthalpy dH is.

The enthalpy H is a useful concept in engineering when one regards flows. We shall encounter several times so-called *throttles*, narrow places in a flow, where a certain amount of mass is allowed to expand from p_1, V_1 to p_2, V_2. The situation is illustrated in Fig. 4.4. Assume that this happens without exchange of heat with the surroundings, so $Q = 0$. We note that the mass flows through the throttle with constant pressures p_1 and p_2 on both sides. We write, using eq. (4.50),

$$Q - W = U_2 - U_1 \tag{4.59}$$

where $-W$ is the work done on the system. As $Q = 0$, it follows that

$$U_2 - U_1 + W = 0 \tag{4.60}$$

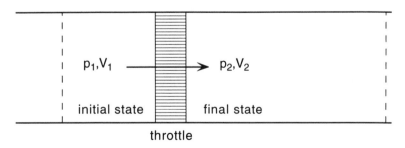

throttle

Figure 4.4 A throttle where a certain mass expands through narrow holes from the left to the right without heat exchange with the surroundings

The work W is the sum of two terms:

$$W = \int_{V_1}^{0} p_1 \, dV + \int_{0}^{V_2} p_2 \, dV = -p_1 V_1 + p_2 V_2 \tag{4.61}$$

Together with eq. (4.60) this gives

$$H_1 = U_1 + p_1 V_1 = U_2 + p_2 V_2 = H_2 \tag{4.62}$$

which implies that the enthalpy in both situations is the same. From eq. (4.61) it is clear that the term pV is the result of work done by the system.

Another useful function is the *free energy* F, defined as

$$F = U - TS \tag{4.63}$$

For a reversible infinitesimal change of state it follows that

$$\begin{aligned} dF &= dU - T \, dS - S \, dT \\ &= dU - \delta Q - S \, dT \\ &= -dW - S \, dT \end{aligned} \tag{4.64}$$

For a reversible isothermal process with $dT = 0$ it follows that the work done *on* the system, being $-dW$, equals the increase in free energy. Conversely, the decrease in free energy for an isothermal process equals the work that can be delivered, hence the name 'free'.

A similar meaning can be attached to the *Gibbs free energy* G, defined by

$$G = H - TS \tag{4.65}$$

For a reversible, infinitesimal process one finds, using eqs. (4.58) and (4.54),

$$\begin{aligned} dG &= dH - T \, dS - S \, dT = dU + p \, dV + V \, dp - T \, dS - S \, dT \\ &= \delta Q - dW_e + V \, dp - \delta Q - S \, dT \\ &= V \, dp - S \, dT - dW_e \end{aligned} \tag{4.66}$$

It first follows that for isobaric, isothermal processes

$$dG = -dW_e \tag{4.67}$$

so non-expansion work dW_e will decrease the Gibbs free energy by the same amount. This equation applies to the *fuel cell*, to be discussed later.

We are now interested in a system with n_i moles of chemical i with $i = 1, 2, \ldots$ Then one could write $G = G(p, T, n_i)$. The differential dG may be written as

$$dG = \left.\frac{\partial G}{\partial p}\right| dp + \left.\frac{\partial G}{\partial T}\right| dT + \sum_i \left.\frac{\partial G}{\partial n_i}\right| dn_i \tag{4.68}$$

Each term on the right assumes all variables but one to be constant. Thus, the first two terms of eq. (4.68) should give back the first two terms of eq. (4.66) while the last term should describe the non-expansion work and will be equal to $-dW_e$. Equation (4.68) may be written as

$$dG = V\,dp - S\,dT + \sum_i \mu_i\,dn_i \qquad (4.69)$$

The quantities μ_i with the dimension of energy are called *chemical potentials*. If need be, in eqs. (4.68) and (4.69) sums over phases may be added, which leads to complications that will not be discussed here.

Consider a chemical reaction taking place at a constant temperature T and pressure p. The increase in Gibbs free energy G following the reaction only depends on dn_i and can easily be deduced from tabulated values of the chemical potential μ_i, using eq. (4.69) or directly from tables showing ΔG per mole.

Finally, it may be mentioned that for finite, reversible changes one simply has to take the integrals of the relevant quantities; for many purposes one just writes ΔW instead of dW, etc.

4.2.2 CONVERSION OF HEAT INTO WORK AND VICE VERSA; AVAILABLE WORK

The first law of thermodynamics, eq. (4.50), expresses conservation of energy. The second law limits the amount of *useful energy*, i.e. energy that can be converted into mechanical work. The argument is as follows.

Consider a heat engine in which an amount of heat Q is extracted from a reservoir at a high temperature T_H and which performs an amount of work W while an amount of heat $Q - W$ (in accordance with the first law) is delivered to a colder reservoir T_C. Then the total entropy increase of the total system is found by adding the entropy increases, eq. (4.54), for both reservoirs:

$$\frac{Q - W}{T_C} - \frac{Q}{T_H} \geq 0 \qquad (4.70)$$

The ≥ 0 on the right-hand side expresses the second law of thermodynamics, eq. (4.56), which states that the total entropy of an isolated system does not decrease. Equation (4.70) leads to the inequality

$$W \leq Q\left(1 - \frac{T_C}{T_H}\right) \qquad (4.71)$$

The best one can achieve by technical means is the equality. Therefore, the maximum work W_{max} becomes

$$W_{max} = Q\left(1 - \frac{T_C}{T_H}\right) \qquad (4.72)$$

It is clear that $T_C < T_H$ is required to be able to perform work W_{max}. As in practice the temperature T_C of the cold reservoir is well above the absolute zero, the useful energy, i.e. the maximum amount of mechanical work, is well below the amount of extracted heat energy Q. The theoretical maximum of the work to be performed corresponds with the Carnot cycle, to be discussed later.

The *thermal efficiency* η is defined as

$$\eta = \frac{\text{work output}}{\text{heat input}} = \frac{W}{Q} \tag{4.73}$$

From eq. (4.72) it follows that the maximum efficiency η_{max} of heat engines may be written as

$$\eta_{max} = 1 - \frac{T_C}{T_H} \tag{4.74}$$

In practice one might distinguish two extreme types of *heat engines*, one in which the temperature T_H is much higher than the ambient temperature T_{amb} at which the lower temperature reservoir is held ($T_C = T_{amb}$) and one in which the higher temperature equals the ambient temperature $T_H = T_{amb}$. These cases are displayed as Fig. 4.5a and 4.5c respectively. The devices shown in Fig. 4.5b and d are the equivalents of those shown in Fig. 4.5a and c, except that they operate the other way round.

In the *refrigerator* of Fig. 4.5d work W_{in} is applied to extract heat Q_C from the cold reservoir and supply heat $Q_H = Q_C + W_{in}$ to the hot reservoir. As the total entropy does not decrease one may write down

$$\frac{Q_H}{T_H} - \frac{Q_C}{T_C} \geqslant 0 \tag{4.75}$$

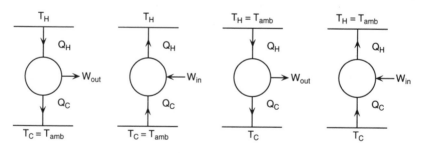

a. Heat engine   b. Heat pump   c. 'Cold' heat engine   d. Refrigerator

Figure 4.5 Four heat engine roles. One observes: (a) a heat engine, (b) a heat pump, (c) a 'cold' engine—a heat engine working with a reservoir below ambient temperature, (d) a refrigerator. Here T_H is the temperature of a *hot* reservoir, T_C the temperature of a *cold* reservoir, T_{amb} the *ambient* temperature, Q_H the heat exchanged with the *hot* reservoir, Q_C the heat exchanged with the *cold* reservoir, W_{out} the work *output* and W_{in} the work *input*. (Reproduced by permission of Peter Piregrinus from P. D. Dunn, *Renewable Energies*, Peter Peregrinus, 1986, p. 76)

which leads to

$$W_{in} \geq Q_C\left(\frac{T_H}{T_C} - 1\right) \tag{4.76}$$

The *coefficient of performance* (COP) is defined as the ratio of the heat Q_C extracted to the required energy W_{in}:

$$COP = \frac{Q_C}{W_{in}} = \frac{Q_C}{Q_H - Q_C} \tag{4.77}$$

We come back to refrigeration in Section 4.2.9.

Similar equations hold for the *heat pump* of Fig. 4.5b. There the cold reservoir might be the cold air outside a building and the hot reservoir the inside. Then the work W_{in} takes heat Q_C from the outside to produce heat $Q_H = Q_C + W_{in}$ inside. The interesting quantity is the coefficient of performance (COP) of the heat pump, which is the amount of heat Q_H delivered at the higher temperature divided by the work W_{in} required to bring it there. From eq. (4.75) which holds here as well it is found that

$$COP = \frac{Q_H}{W_{in}} = \frac{Q_H}{Q_H - Q_C} = \left(1 - \frac{Q_C}{Q_H}\right)^{-1} \leq \left(1 - \frac{T_C}{T_H}\right)^{-1} = \eta_{max}^{-1} \tag{4.78}$$

As T_C and T_H are measured on a Kelvin scale, T_C/T_H may be close to 1 and the maximum value of COP may be as big as 5 or 6. It must be stressed that the formulas given above are limits for reversible, idealized systems. In practice efficiencies will be much lower.

Second Law Efficiency; Available Work

The thermal efficiency η defined in eq. (4.73) is a measure connected with the conversion of heat into mechanical energy. More generally, efficiency for an energy transfer process would be defined by

$$\eta = \frac{\text{energy output (heat or work) by a device}}{\text{energy input required}} \tag{4.79}$$

In this way, besides the thermal efficiency the COP for the heat pump is also included in the definition, and the efficiency (4.79) may be either smaller or bigger than 1.

It must be stressed that a high efficiency η applies to the performance of a device and does not imply an intelligent use of energy. Heating of a human being inside a house, for example, may be achieved in several ways:

(a) by heating the house with electric heaters with an efficiency close to 1;
(b) by heating the house with heat pumps, electrically driven with COP bigger than 1;

(c) by gas heaters in each room with an efficiency of 0.8 but only burning at full capacity when somebody uses the room;

(d) by heating the house with a gas furnace driven central heating system;

(e) by wrapping the person concerned in an electric blanket.

If the aim is to make a person comfortable with as little energy consumption as possible – and option (e) is discarded, all aspects should be taken into account: the way in which the electricity or gas is produced, the losses during transport to the building, etc. In comparing options (c) and (d) it is clear that option (c) will be the most economic as it is geared to the *task* that has to be fulfilled.

The point of the foregoing is that in studying energy efficiencies one should look at the task to be performed and not solely at the device. To do so, a so-called *second law efficiency* ε for a single output (either work or heat) is defined as follows:

$$\varepsilon = \frac{\text{useful heat/work output by a device/system}}{\text{maximum possible by any device/system for the same energy input}} \quad (4.80)$$

This number will always be smaller than 1 and focuses on the task of the system. The maximum in the denominator refers to the maximum permitted by the laws of thermodynamics.

Example: Home Heating

Applying the concept of second law efficiency to home heating, a certain amount of heat Q_2 is delivered to the home. With a gas furnace a certain amount of gas is needed, the combustion of which corresponds to a change in enthalpy $|\Delta H|$. The efficiency η of eq. (4.79) is

$$\eta = \frac{Q_2}{|\Delta H|} \quad (4.81)$$

For the second law efficiency ε of eq. (4.80) the numerator is the same Q_2. To find the denominator, we assume that gas is also used; the best way then would be to use the gas to drive a heat pump. The gas therefore has to perform work between the cold atmosphere and the warm home. It is useful to introduce the concept of *available work B*. This is defined as

> The maximum work that can be provided by a system or a fuel
> as it proceeds by any path to a specified final state in   (4.82)
> thermodynamic equilibrium with the atmosphere.

The atmosphere is taken as it is the obvious reservoir in applications although other reservoirs may be introduced in the definition. Work done against the atmosphere is not counted in the definition of B since it cannot be recovered.

Therefore, in using a heat pump for home heating, the maximum work output of the fuel will be B. The maximum COP for running the heat pump is given by eq. (4.78), where $T_C = T_0$ equals the temperature of the atmosphere and $T_H = T_2$

equals the temperature of the radiators, heating the home. This gives

$$\varepsilon = \frac{Q_2}{B \times \text{COP}} = \frac{Q_2}{B/(1 - T_0/T_2)} = \frac{Q_2}{B}\left(1 - \frac{T_0}{T_2}\right) \tag{4.83}$$

as the second law efficiency of the gas furnace heating. It appears that B and $|\Delta H|$ usually differ by less than 5%. Ignoring this difference it is clear that the second law efficiency of a gas furnace is smaller than the efficiency (4.79) by the factor $1 - T_0/T_2$. Assuming the outside air around the freezing point, $T_0 = 273$ K and radiators at $T_2 = 343$ K, this factor amounts to 20%. Thus, one certainly should use a heat pump for the task of home heating.

The important point to note in the definition of B is that it applies to *work*. The reason is that work can be considered as the highest form of energy. Work can be converted into heat with 100% efficiency; this is not possible the other way round. For a weight with mass m raised to a height h the available work B equals mgh. For heat Q extracted from a hot reservoir at temperature T_H the available work becomes, according to eq. (4.72),

$$B = Q\left(1 - \frac{T_0}{T_H}\right) \tag{4.84}$$

where again T_0 is the atmospheric temperature. It is clear that high temperature heat is very valuable as its available work may be close to its heat energy.

An Expression for Available Work

In the following, an expression is derived for the available work B in order to express the second law efficiency (4.80) in terms of entropy. Consider a system characterized by its energy U, entropy S and volume V. The atmosphere has temperature T_0 and pressure p_0. The system may exchange heat and work with the atmosphere and will reach equilibrium with the atmosphere at energy U_f, entropy S_f and volume V_f. The atmosphere is so big that it does not change. Define a function B (which is to become the available work) by

$$B = (U - U_f) + p_0(V - V_f) - T_0(S - S_f) \tag{4.85}$$

In order to show that this B obeys definition (4.82) one considers a change of the system by ΔU, ΔV and ΔS. There is a heat flow Q' to the atmosphere; useful work W is done on other systems and non-useful work $W' = p_0 \Delta V$ on the atmosphere. The first law (4.50) may be written as

$$-Q' = \Delta U + W + W' \tag{4.86}$$

The change of entropy (4.54) is calculated for the atmosphere, of which the temperature T_0 does not change:

$$(\Delta S)_{\text{system + atm}} - (\Delta S)_{\text{system}} \equiv (\Delta S)_{\text{sa}} - \Delta S = \frac{Q'}{T_0} \tag{4.87}$$

Here, the system + atmosphere is denoted by the index sa but the quantity for the system itself does not have an index attached. From eqs. (4.86) and (4.87) one finds

$$
\begin{aligned}
W &= - Q' - \Delta U - W' \\
&= - T_0(\Delta S)_{\text{sa}} + T_0 \Delta S - \Delta U - p_0 \Delta V \\
&= - T_0(\Delta S)_{\text{sa}} - (\Delta U + p_0 \Delta V - T_0 \Delta S)
\end{aligned}
\tag{4.88}
$$

In introducing B in eq. (4.85) it was remarked that U_f, V_f and S_f apply to the *final* state of the system. Therefore, when $\Delta U, \Delta V$ and ΔS refer to a change to that final state

$$
W = - T_0(\Delta S)_{\text{sa}} + B
\tag{4.89}
$$

or

$$
B = W + T_0(\Delta S)_{\text{sa}}
\tag{4.90}
$$

Assume that all other systems besides the system and the atmosphere are passive receivers of work. Then according to the second law of thermodynamics (4.56) $\Delta S_{\text{sa}} \geqslant 0$. For reversible processes the equality sign will hold and it follows that B is the maximum possible work, i.e. the available work of eq. (4.82). In this way the equivalence of eqs. (4.82) and (4.85) is shown.

It is now possible to rewrite definition (4.80) of the second law efficiency. In producing work W the numerator will be W and the denominator will be more than W by the wasted entropy increase (sometimes called the *lost work*), giving

$$
\varepsilon = \frac{W}{W + T_0(\Delta S)_{\text{sa}}}
\tag{4.91}
$$

Finally it may be remarked that for an isobaric isothermal process at atmospheric pressure p_0 and temperature T_0 comparison of eq. (4.85) with the first line of eq. (4.66) shows that then the change in Gibbs function gives the available work. As the change in Gibbs function does not differ very much from the change in enthalpy H, this justifies putting $B = |\Delta H|$ in eq. (4.83).

Although the concept of second law efficiency is useful as an intellectual tool, its practical use is mainly restricted to cases which compare different applications of the same kind of fuels. For home heating applications of gas, oil and coal may be compared. It would be somewhat artificial to compare these with solar heat converters. The point can always be made, however, that one should make the best use of the available work, once it has been decided which source of work to use.

Exergy

Sometimes the concept of *exergy* is used, which is defined as the maximum work to be gained from a certain amount of heat. This quantity therefore is the available heat Q, multiplied by the Carnot factor of eq. (4.74), resulting in eq. (4.84). The ambient temperature T_0 is taken as 298 K and T_H the temperature of the heat. In a similar vein for electricity the exergy factor is taken as one, since essentially all energy is convertible into work.

The concept of exergy is often used to argue that high temperature heat should be converted into electricity, the remaining heat used as process heat, and when the temperature becomes lower the remaining heat should be used for home heating. The problem here is that such a use of energy requires an expensive infrastructure and does not give many incentives to energy economizing on the individual scale, for the heat is being produced and delivered anyway.

4.2.3 HEAT ENGINES: CONVERSION OF HEAT INTO WORK

An engine usually contains an amount of matter that undergoes compression and expansion, possibly with change of phase, but in such a way that after performing a complete cycle the state of the matter has returned to the starting situation. The net effect then is conversion of heat into mechanical work with the theoretical limits discussed in the previous section.

A first cycle, important from a theoretical point of view, is the so-called *Carnot cycle*. For an idealized situation it is depicted in Fig. 4.6, both in a pV diagram (left) and a TS diagram (right). From $1 \rightarrow 2$ a gas is adiabatically compressed $(dS = 0)$; from $2 \rightarrow 3$ it expands isothermally and absorbs an amount of heat $Q_H = T_H(S_3 - S_2)$; from $3 \rightarrow 4$ it expands adiabatically; from $4 \rightarrow 1$ it is compressed isothermally releasing an amount of heat $Q_C = T_C(S_4 - S_1) = T_C(S_3 - S_2)$.

From the first law (4.48) it follows that the work W can be expressed as

$$Q_H - Q_C = W = (T_H - T_C)(S_3 - S_2)$$

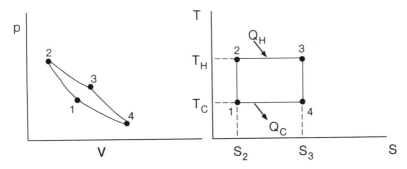

Figure 4.6 The idealized Carnot cycle. On the left it is illustrated in a *pV* diagram and on the right in a *TS* diagram

The thermal efficiency η from eq. (4.73) becomes

$$\eta = \frac{W}{Q_H} = \frac{Q_H - Q_C}{Q_H} = \frac{T_H - T_C}{T_H} = 1 - \frac{T_C}{T_H} \tag{4.92}$$

which is precisely the theoretical maximum (4.74).

The work W can also be found from the pV diagram. The work performed by the system from state 2 to state 3 can be written as

$$W_{2 \to 3} = \int_2^3 p\,dV$$

For the total cycle one obtains

$$W = \int_1^2 p\,dV + \int_2^3 p\,dV + \int_3^4 p\,dV + \int_4^1 p\,dV = \oint p\,dV \tag{4.93}$$

Of course the contributions $4 \to 1$ and $1 \to 2$ are negative, leading to the closed integral $\oint p\,dV$, which is equal to the surface area inside the curve in the pV diagram.

In practice one does not design Carnot cycles. It appears that the work done during a cycle is relatively small and friction and other losses count heavily.

It must be stressed that in a heat engine the source of heat does not matter. The heat may come from solar heating or from the waste heat in a nuclear power station, or it may use the flame of the external combustion of oil, gas, wood or coal. This latter combustion process may be done with some precautions against pollution. Finally, the heat may be transported over some distance by means of heat pipes (cf. Section 4.2.5).

Much research has been done on the *Stirling* engine, called after a Scottish theologian and inventor in the 1830s. The idealized pV and TS diagrams are shown in Fig. 4.7. A gas (air, helium or hydrogen) is isothermally compressed from $1 \to 2$ releasing heat Q_C; from $2 \to 3$ the gas is pressurized without change

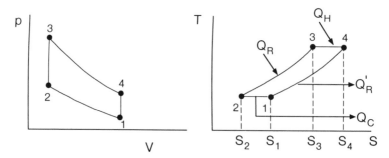

Figure 4.7 The idealized cycle for a Stirling engine

of volume, giving a rise of temperature with absorption of heat Q_R; from $3 \rightarrow 4$ the gas expands isothermally and absorbs heat Q_H; from $4 \rightarrow 1$ the pressure falls and an amount of heat $Q_{R'}$ is released.

In practice the values Q_R and $Q_{R'}$ cancel; this may be understood by noting that they equal $\int c_V \, dT$ where the specific heat c_V at constant volume will not be too different for the transitions $2 \rightarrow 3$ and $4 \rightarrow 1$. In that case $\int T \, dS$ for both transitions will be the same as well, leading to approximately $S_3 - S_2 = S_4 - S_1$ or $S_4 - S_3 = S_1 - S_2$. Following the argument leading to eq. (4.92), one finds the Carnot efficiency for the Stirling cycle as well. The crucial practical point is that the released $Q_{R'}$ can be 'parked' and picked up when required for Q_R, so that $Q_{R'}$ is not lost. This is done by a 'regenerator' that keeps 98% of the heat stored.

Much research has been done and is still going on in improving the Stirling engine. The Dutch electronics company Philips spent much effort in applying the engine to cars. It may be, however, that other applications of Stirling engines will become at least as important, as the engine only needs heat to run. The source of the heat is unimportant.

Instead of gas one may use water and water vapour to perform the cycle, resulting in a steam engine with the Rankine cycle. In Fig. 4.8 a schematic representation of the Rankine cycle is given together with its idealized TS diagram. At point 1 liquid water is present which is compressed adiabatically by the pump costing work W_p and is transported to the boiler (point 2). There heat Q_H is added isobarically, first increasing the temperature to the boiling point at the pressure p_2; during boiling the temperature remains constant and when all liquid is evaporated the temperature rises again (point 3). The vapour is allowed to expand adiabatically in a turbine (or an old-fashioned cylinder with a piston) performing work W (point 4), while after the turbine a low pressure is maintained by condensing the water and rejecting heat Q_L between points 4 and 1. Then the same water may be used all over again or fresh water may be inserted.

These thermodynamic processes are best described by using the enthalpy H defined in eq. (4.57) and elaborated in eq. (4.58). Going along the cycle we note the following. In step $1 \rightarrow 2$, $\delta Q = 0$, but there is an increase in enthalpy of the system because of the work $W_p = -W_{12}$ done by the pump. The minus sign takes into account the fact that W_{12} is defined as work done by the system. Thus

$$-W_{12} = H_2 - H_1 = \int_1^2 V \, dp = V_1(p_2 - p_1) \tag{4.94}$$

where it is assumed that the pumping increases the pressure but not the volume because of the incompressibility of water.

In step $3 \rightarrow 4$ the system performs positive work itself losing enthalpy, which leads to

$$W_{34} = H_3 - H_4 \tag{4.95}$$

The net work performed by the working fluid, the water, during the cycle there-

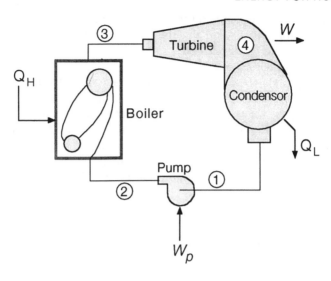

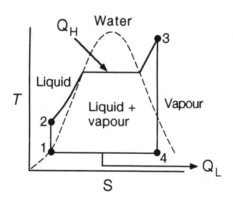

Figure 4.8 Idealized Rankine cycle. The points indicated by 1, 2, 3 and 4 (top) correspond to the points in the TS diagram, below. The dotted line describes the phase diagram underneath in which an equilibrium mixture of liquid–vapour exists

fore becomes

$$W = W_{12} + W_{34} = H_1 - H_2 + H_3 - H_4 \qquad (4.96)$$

Step $2 \rightarrow 3$ is performed isobarically at pressure p_2 and step $4 \rightarrow 1$ is again done isobarically at the lower pressure p_1. For these isobaric processes the added heat equals the increase in enthalpy (following eq. (4.58)); therefore

$$Q_H = H_3 - H_2 > 0 \qquad (4.97)$$

$$Q_L = H_4 - H_1 > 0 \qquad (4.98)$$

consistent with $W = Q_H - Q_L$. The (maximum) thermal efficiency of the process then becomes

$$\eta = \frac{W}{Q_H} = \frac{H_3 - H_4 + H_1 - H_2}{H_3 - H_2} = 1 - \frac{H_4 - H_1}{H_3 - H_2} = 1 - \frac{Q_L}{Q_H} \qquad (4.99)$$

It is, of course, advantageous to make Q_H as high as possible. This is done by boiling under high pressure p_2 and high temperature T_2. The heats Q_L and Q_H are difficult to measure directly. Therefore one uses tabulated values of the enthalpy H as a function of temperature and pressure to calculate the maximal η for a device.

4.2.4 INTERNAL COMBUSTION ENGINES: CONVERSION OF CHEMICAL ENERGY INTO WORK

In an internal combustion engine air is mixed with a vaporized fuel that is ignited by a spark (e.g. the Otto cycle) or is ignited by the temperature rise from compression (the diesel cycle). The engines are no heat engines in the sense discussed before: neither are there two external heat reservoirs and nor is there a single gas changing its thermodynamic state during the cycle. In fact, exhaust gases are expelled, fresh air supplied and fuel injected.

In practice, however, the air/fuel ratio is around 14 in mass for an Otto cycle and around 20 for a diesel engine. Also the nitrogen in the air remains essentially unchanged, which makes it acceptable to approximate the cycle of an internal combustion engine in a pV or TS diagram. The lower temperature may be taken as the ambient air temperature and the higher temperature as the temperature of the fuel/air/exhaust mixture after ignition.

In Fig. 4.9 the Otto cycle is depicted. From $1 \rightarrow 2$ the ideal gas (n_1 moles of air) experiences adiabatic compression. This means that no heat is exchanged ($\delta Q = 0$) and the entropy remains constant. This adiabatic process shows up as a vertical line in the TS diagram of Fig. 4.9. For an adiabatic process and an ideal gas it is well known that

$$T_1 V_1^{\kappa - 1} = T_2 V_2^{\kappa - 1} \qquad (4.100)$$

in which $\kappa = c_p/c_V$, the ratio of the specific heats with constant pressure or volume respectively.

In step $2 \rightarrow 3$ the pressure rises with a constant volume; an amount of heat Q_H is absorbed, simulating the explosion in the gasoline engine. Assuming a constant

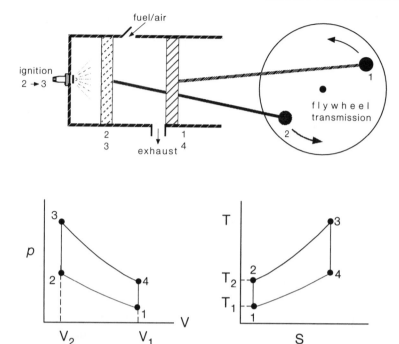

Figure 4.9 The idealized Otto cycle. Besides the pV and TS diagram a cylinder with a piston is also shown

c_V we find

$$Q_H = c_V(T_3 - T_2) \tag{4.101}$$

In step $3 \to 4$ the n_1 moles of air expand adiabatically:

$$T_3 V_3^{\kappa-1} = T_4 V_4^{\kappa-1} = T_4 V_1^{\kappa-1} \tag{4.102}$$

In step $4 \to 1$ an amount of heat Q_C is released to a cold reservoir, simulating the exhaust part of the cycle. As above,

$$Q_C = c_V(T_4 - T_1) \tag{4.103}$$

Notice that the n_1 moles of air are released after transition $4 \to 1$, giving a volume zero, while the same amount of fresh air is next sucked up. It is assumed that these effects cancel out. With the thermal efficiency η of a cycle defined in eq. (4.73) as the ratio of (useful) work and supplied heat, it is found for the Otto cycle that

$$\eta = \frac{W}{Q_H} = \frac{Q_H - Q_C}{Q_H} = 1 - \frac{Q_C}{Q_H} = 1 - \frac{T_4 - T_1}{T_3 - T_2} \tag{4.104}$$

With eqs. (4.100) and (4.102), using $V_3 = V_2$, it follows that

$$\eta = 1 - \frac{1}{r^{\kappa-1}} \tag{4.105}$$

in which $r = V_1/V_2$ is the *compression ratio* of the engine.

It follows that a bigger compression ratio r will lead to a higher efficiency of the engine. An upper limit for r is given by the fact that for a high compression the engine will pre-ignite or cause pinging or knocking of the engine. This effect can be prevented by 'doping' the fuel with, for example, lead. In practice the upper limit of the compression ratio r for internal combustion engines is around $r = 10$.

In a diesel engine these problems are overcome by making step $2 \rightarrow 3$ horizontal in the pV diagram, which is illustrated in Fig. 4.10. In step $1 \rightarrow 2$ air is compressed adiabatically, so that eq. (4.100) holds again. The compression is so high that when the fuel is injected at point 2 the mixture ignites. This happens isobarically, so in step $2 \rightarrow 3$ the piston moves out and the heat absorbed equals

$$Q_H = c_p(T_3 - T_2) \tag{4.106}$$

Steps $3 \rightarrow 4$ and $4 \rightarrow 1$ are the same as in the Otto cycle. In order to find an expression for the efficiency η we apply the ideal gas law $p_2 V_2/T_2 = p_3 V_3/T_3$ with $p_2 = p_3$ and an extra *cut-off ratio*

$$r_{cf} = \frac{V_3}{V_2} \tag{4.107}$$

At volume V_3 the fuel addition is cut off and the expansion stops, whence the name. From eqs. (4.103), (4.100), (4.106), (4.107) and the ideal gas law it follows that

$$\eta = 1 - \frac{Q_C}{Q_H} = 1 - \frac{1}{\kappa}\frac{T_4 - T_1}{T_3 - T_2} = 1 - \frac{1}{\kappa}\frac{1}{r^{\kappa-1}}\frac{r_{cf}^{\kappa} - 1}{r_{cf} - 1} \tag{4.108}$$

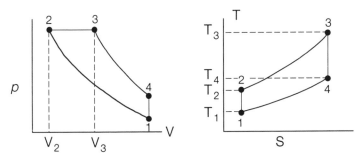

Figure 4.10 Idealized diesel cycle. The essential difference with the Otto cycle is that step $2 \rightarrow 3$ goes isobaric

Besides the factor containing the cut-off ratio r_{cf}, this expression is very similar to the efficiency η for the Otto cycle in eq. (4.105). The compression ratio r for the diesel cycle will be between 18 and 25, which gives this cycle the possibility to be more efficient than the spark ignition engine. The drawback lies in the last factor of eq. (4.108) with the cut-off ratio both in the numerator and denominator. As κ is fixed r_{cf} could be minimized, making the ratio close to one. This would, however, reduce the work per cycle. In practice, therefore, both parameters r and are r_{cf} used to optimize the performance of the engine.

We mention that combustion of methanol CH_3OH in a diesel-type cycle is being developed. As the fuel is the simplest carbon compound usable combustion can be more complete and more efficient than in gasoline powered vehicles where more complicated compounds like C_8H_{18} are used. A drawback, however, might be the exhaust of poisonous compounds like formaldehyde.

4.2.5 HEAT PIPES

A heat pipe has the purpose to transport heat without appreciable losses. It can be used to transport heat from a source to one or more heat engines. The design is based on the idea of a thermosyphon, as shown in Fig. 4.11. Here a liquid at the bottom evaporates, picking up latent heat. It condenses on top where the syphon is cooled; the latent heat is taken away and the liquid goes down by gravitation.

A heat pipe should work without the need for gravitational assistance. Therefore the device on the right-hand side of Fig. 4.11 is put horizontally.

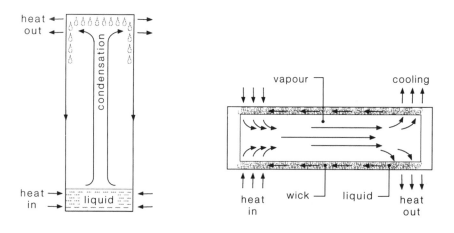

Figure 4.11 Transport of heat by (a) a thermosyphon and (b) a heat pipe

Capillary forces should return condensing vapour on the right to the heat source on the left. In practice the transport is done by a gauze to which the working fluid adsorbs. The fluid will go into the direction of the negative gradient of its concentration and thus into the direction where it is evaporated.

4.2.6 ELECTRICITY

The great majority of the electricity in the world is produced in generators. The principle is based on Faraday's law which says that a rotating winding in a constant magnetic field produces an alternating electric current. In this way, mechanical energy of rotation is converted into electric power.

For small instruments like a bicycle dynamo permanent magnets are used to produce the static magnetic field. In power stations a separate winding is usually added to the big rotating coil, which produces its own alternating electric field; this is then rectified to produce dc power for the magnets.

The rotational energy to drive the coils is sometimes produced by hydropower, to be discussed at the end of this section, but in the majority of cases it is produced by engines that follow the principles discussed in Sections 4.2.3 and 4.2.4. Instead of letting the working gases or fluids expand in cylinders which drive reciprocating pistons, one often lets them perform work against a turbine which immediately produces rotational energy.

The loss of energy in converting mechanical to electrical energy is not negligible (up to 10% for big generators), but the really big losses occur because of the laws of thermodynamics, which restrict the efficiency to values below the upper limit of the Carnot efficiency

$$\eta_{max} = 1 - \frac{T_C}{T_H} \qquad (4.109)$$

Here the lower temperature T_C is determined by the cooling facilities, ambient air in cooling towers or water from lakes or rivers, both not lower than 290 K. The higher temperature T_H should be well below the melting point of the materials and metals used and is often much lower, as will be discussed later.

In order to understand the operation of a power plant, or rather the system of producing electricity, one should consider the variation of the demand of electricity during the day–night cycle, keeping in mind the differences between summer and winter. They are shown in Fig. 4.12 for a number of industrialized countries.

The mission of the electricity utilities is to produce the electricity against as low as possible overall costs. This is achieved by dividing the demand in a base load (the lower levels of Fig. 4.12) and the peak demand. It so happens that power plants may be divided into types that have a low fuel cost but a high capital cost (cf. Section 4.3 about this concept) and types where it is just the other way round.

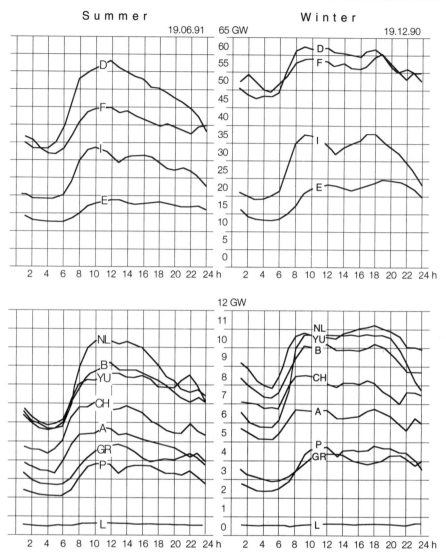

Figure 4.12 Grid load of electricity during the day–night cycle for a number of industrial countries, indicated by their car identification. On the left the midsummer graphs are given and on the right the midwinter graphs. The total electricity use will be higher because of self-generation by big industries. (Reproduced by permission of UCPTE from half-yearly reports I-1991, p. 17, and II-1991, p. 17, of the Union for the Coordination of Production and Transmission of Electricity)

For the base load plants will be used which are expensive to build but cheap in fuel expense. They will operate continuously (except for maintenance) so that the high capital costs are divided over a high number of kW h, resulting in an acceptable capital cost per kW h. For the peak loads plants will be used that are expensive in fuel but cheap in capital. Of course the total capacity in power stations should exceed the maximum loads indicated in Fig. 4.12.

Large scale power plants often have several units. The biggest are those where fossil fuel is burned producing steam or where the steam is generated in a nuclear power station. The cycles are similar to the Rankine cycle shown in Fig. 4.8. The higher temperature T_H is determined by the corrosion of the materials which increase with temperature. It is then a matter of optimizing costs which determine T_H as around 800 K, resulting in a maximum efficiency of 63%. At present the best fossil fuel fired power stations have an efficiency of 40%, whereas for nuclear power stations they reach some 33%. The reason is that preventing corrosion in the nuclear fuel elements requires lower temperatures T_H. The large steam power plants have medium to large capital costs but the fuel cost per kW h produced is relatively low.

The other type of units are the gas turbines. They use natural gas as a fuel and may be operated in a so-called Brayton cycle, both with internal and external combustion. The fuel is relatively expensive but gas turbines are cheap to build and can quickly be turned on and off, which makes them ideal for peak load use. The temperature T_H could be as high as 1700 K but usually does not exceed 1400 K. The temperature T_C would correspond to the outlet temperature of the gases, usually around 900 K. These two numbers result in a maximal efficiency of 35%. The most advanced gas turbines have a $T_H = 1530$ K and a $T_C = 850$ K, resulting in a maximum efficiency of 44%.

The numbers given for the input and output temperatures immediately suggest the introduction of so-called combined cycle power systems. Here the exhaust of a gas turbine is used to heat the steam of a steam turbine. Figure 4.13 shows the scheme.

Assume that the heat input in the gas turbine is q_A, its efficiency η_A and its output $(1 - \eta_A)q_A$. Let some extra heat q_B be added to the steam turbine B with efficiency η_B. The total efficiency of the combined cycle becomes

$$\eta_{tot} = \eta_B + \frac{\eta_A q_A (1 - \eta_B)}{q_A + q_B} \tag{4.110}$$

For the case that $q_B = 0$ the efficiency simplifies into

$$\eta_{tot} = \eta_A + \eta_B - \eta_A \eta_B \tag{4.111}$$

If individual efficiencies are 30% and 20% the overall efficiency adds up to 44%,

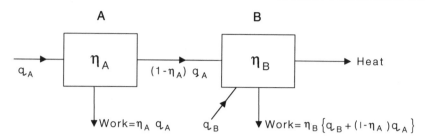

Figure 4.13 The combined power cycle: the output of a gas turbine is used to heat the steam of a steam turbine

an appreciable gain. These combinations are used as intermediate load units between the base load and peak loads.

Thermoionic Energy Conversion

Because of the inherent loss in efficiency when one goes from heat to mechanical energy and then to electrical energy, as displayed in eqs. (4.73) and (4,74), much research has been done on direct conversion of heat into electrical energy. One possibility is so-called thermoionic conversion. An electrode is heated to a high temperature so that it emits electrons which are collected on the other electrode and return to the first electrode via an external load.

The first problem is that the binding energy of most electrons in a solid is of the order of 1 eV, whereas at room temperature the kinetic energy of electrons is of the order of $kT = 0.025$ eV, so high temperatures are needed. It is not required to go to 40 times 300 K in order to have electrons emitted as there will be a distribution of energies around the average, which is the Fermi level of the material of the electrode. Nevertheless, high temperatures will be needed, and the second electrode has to be cooled, and it must be certain that its Fermi level is below that of the first electrode in order to get a gain in energy.

A second problem is that soon after the first electrons have been emitted a space charge will build up between both electrodes. In practice caesium is evaporated which ionizes because of the high temperatures, after which the positively charged Cs ions pass the diode and neutralize the space charge.

Experiments show that a power density of $50 \, kW \, m^{-2}$ could be tapped when using ceramic materials to withstand the high temperatures. The air cooling the second electrode could be used to power turbines or heat engines. The idea is shown in Fig. 4.14. It has been applied in space travelling far from the sun, where solar panels do not work. In that case the heat is provided by radioactive decay of ^{238}Pu. Besides these exotic applications, the method has not yet found large scale use as efficiencies are low and the technology expensive.

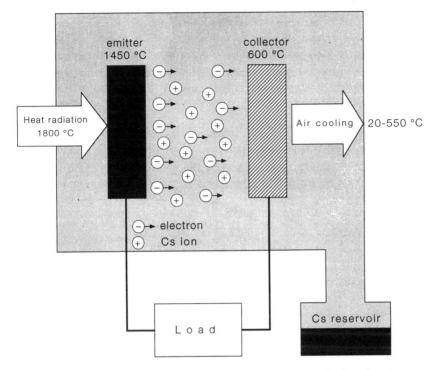

Figure 4.14 The principle of thermoionic conversion for producing electric power

4.2.7 ENERGY STORAGE AND ENERGY TRANSPORT

In discussing Fig. 4.12 it was noted that the demand for electrical energy fluctuates, resulting in the need to build enough power plants to satisfy the maximum needs. However, during quiet hours these machines just cost capital without earning anything. Therefore utilities have been looking for ways to store energy. Another incentive to look for ways to store energy is that many renewable energies like wind or solar, to be discussed in later sections, are being produced intermittently. In these cases a storage facility would be practical. Finally, when driving a car in the rush hour much kinetic energy is wasted by stopping all the time. Even outside the rush hour this is true for buses which have to stop frequently to load or unload passengers. Again, energy storage might save energy here.

Below a very brief discussion is given of some of the ways energy is stored that are in use or under consideration. In sections on renewables a few of these will be looked at again.

Gravitational Storage

Consider two reservoirs of water on different altitudes. Then water may be pumped up during idle hours and the hydropower (cf. Section 4.4.5) used during peak hours. This is applied in practice by utilities where the geography of a region is employed or where empty mines are available to be used as a low-lying reservoir.

Storage in Flywheels

The kinetic energy of a flywheel is given by the relation for a rigid body, which for the case where all mass is concentrated in the outside rim with radius R becomes

$$K = \frac{1}{2}I\omega^2 = \frac{1}{2}mR^2\omega^2 \tag{4.112}$$

The stress σ in the material of the rim is the force per unit area working in the rim. The acting force there is the centrifugal force

$$m\omega^2 R \tag{4.113}$$

which per unit of volume becomes

$$\rho\omega^2 R \tag{4.114}$$

As the stress has the dimension of the centrifugal force per unit of volume times a length, the relation between the material stress σ and the centrifugal force $\rho\omega^2 R$ can be assumed to be

$$\sigma = \rho\omega^2 R \times R = \rho\omega^2 R^2 \tag{4.115}$$

Another way of expressing this relation is that the change of stress over a spoke of the wheel, which is σ/R, should equal the centrifugal force at the end of the spoke per unit of volume.

The kinetic energy per unit of mass is called the specific energy and follows from eqs. (4.112) and (4.115) as

$$\frac{K}{m} = \frac{1}{2}R^2\omega^2 = \frac{\sigma}{2\rho} = K_w\frac{\sigma}{\rho} \tag{4.116}$$

Here the factor $K_w = 0.5$. This equation also holds for flywheels where the mass is more evenly distributed and may become as large as 1.0 for certain designs. One notices that the maximum specific energy to be stored is proportional to the maximal material strength before distortions occur and inversely proportional to the density. This is the reason why much research is being done with fibre composites.

Batteries

Electrical energy is traditionally stored in batteries, where essentially chemical reactions store the energy. They are being considered for use in electric cars, but their capacity is limited.

Chemical Storage in Fuel Cells

Another method of chemical storage is to produce hydrogen by a chemical reaction and store the hydrogen until it is used to react with oxygen, producing energy. This happens in fuel cells, to be discussed in Section 4.4.5. For use in cars, the problem then is to store the hydrogen. Besides storing it as a gas, possibilities are to bind the hydrogen to a crystal lattice from which it can be easily regained.

In Table 4.3 a summary is given of the storage possibilities of various systems. It is clear that gasoline is one of the most readily available forms.

Energy Transport

The cost of energy transport per kJ will be dependent on the cost of transporting the energy carrier. For the materials shown in Table 4.3 it is clear that the transport price of uranium per kJ will be much lower than that of gasoline or, even worse, than that of batteries.

Much of the transport of oil and gas is by pipelines. Besides the economic costs one should consider the environmental costs in terms of leakage, particularly near the pumping stations and at the input and output.

Let us calculate the power P required to move a volume V of fluid per second

Table 4.3 Specific energy storage of various materials (kJ/kg). For hydrogen, methane and gasoline the lower heating values are given, assuming that the latent heat in the vapour is not being used. (Reproduced by permission of McGraw-Hill from A. W. Culp Jr, *Principles of Energy Conversion*, McGraw-Hill, New York, 1991, Table 9.1, p. 482.)

Deuterium (D–D fusion reaction)	3.3×10^{11}	Silver oxide–zinc battery	437
Uranium-235 (fission reaction)	7.0×10^{10}	Lead–acid battery	119
Heavy water (fusion reaction)	3.5×10^{10}	Flywheel (uniformly	
Reactor fuel (2.5% enriched UO_2)	1.5×10^{9}	stressed disc)	79
Natural uranium	5.0×10^{8}	Compressed gas (spherical	
95% Po-210 (radioactive decay)	2.5×10^{6}	container)	71
80% Pu-238 (radioactive decay)	1.8×10^{6}	Flywheel (cylindrical)	56
Hydrogen (LHV)	1.2×10^{5}	Organic elastomer	20
Methane (LHV)	5.0×10^{4}	Flywheel (rim-arm)	7
Gasoline (LHV)	4.4×10^{4}	Torsion spring	0.24
Lithium hydride (at 700 °C)	3.8×10^{3}	Coil spring	0.16
Falling water ($\Delta z = 100\,\mathrm{m}$)	9.8×10^{2}	Capacitor	0.016

through a cylindrical pipe with radius R. There are no accelerations and the pressure drop per metre that is pushing the fluid will be $G = -dp/dx$, with x the coordinate in the direction of the flow. Equation (3.32) will hold, but the left-hand side is zero, there are no Coriolis forces present and the complication of gravity is ignored for horizontal flows. The flow is assumed to be laminar and axially symmetric, the only variation being $u = u(r)$, with all other velocity components vanishing. Therefore,

$$0 = \mathbf{F}_{\text{press}} + \mathbf{F}_{\text{viscous}} \qquad (4.117)$$

The situation is sketched in Fig. 4.15, where $u(r)$ is the horizontal velocity of the fluid.

Consider a cylindrical ring with thickness dr and length δx. The viscosity force on the inside of the ring will be, following the discussion around Fig. 3.11,

$$-\mu \left.\frac{du}{dr}\right|_{r} 2\pi r\, \delta x \qquad (4.118)$$

Similarly, the force on the outside will be

$$+\mu \left.\frac{du}{dr}\right|_{r+dr} 2\pi(r + dr)\delta x \qquad (4.119)$$

with the net force

$$\mu\left[\left(r\frac{du}{dr}\right)_{r+dr} - \left(r\frac{du}{dr}\right)_{r}\right]2\pi\, \delta x = 2\pi\mu\, \delta x\, dr\, \frac{d}{dr}\left(r\frac{du}{dr}\right) \qquad (4.120)$$

The differential equation (4.117) in the x direction becomes, using eq. (3.33)

$$-\frac{dp}{dx}2\pi r\, dr\, \delta x + 2\pi\mu\, \delta x\, dr\, \frac{d}{dr}\left(r\frac{du}{dr}\right) = 0 \qquad (4.121)$$

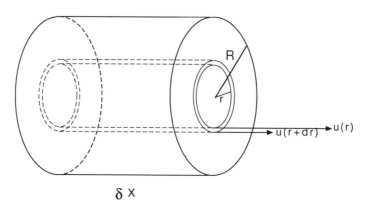

$$\delta x$$

Figure 4.15 Viscous forces in a horizontal cylindrical pipe

or

$$Gr + \mu \frac{d}{dr}\left(r\frac{du}{dr}\right) = 0 \tag{4.122}$$

with the solution

$$u(r) = -\frac{1}{4}\frac{Gr^2}{\mu} + A\ln r + B \tag{4.123}$$

As u must be finite at $r = 0$ it follows that $A = 0$. Requiring $u = 0$ at the radius $r = R$ leads to

$$u(r) = \frac{G}{4\mu}(R^2 - r^2) \tag{4.124}$$

The volume V passing a cross-section per second becomes

$$V = \int_0^R u(r)2\pi r\,dr = \frac{GR^4\pi}{8\mu} \tag{4.125}$$

and the power P required to move the fluid from point 1 to point 2 in a pipe is the force $(p_1 - p_2)$ per unit area times an area $2\pi r\,dr$ times the local velocity $u(r)$ integrated from $r = 0$ to $r = R$. This is for a pipe of length L:

$$P = (p_1 - p_2)V = GLV = \frac{8\mu LV^2}{\pi R^4} \tag{4.126}$$

It follows that increasing the transported volume per second costs a lot of extra power. It would be better to keep V/R constant while increasing R, e.g. by adding pipes. It also becomes clear that decreasing the viscosity μ would help. In fact, crude oil may have a very high viscosity (over $1000\,\mu P$) around the freezing point. Therefore, for transport in Alaska or in Siberia it has to be heated to decrease viscosity; the Alaska pipeline is in fact heated up to $60\,^\circ C$.

Electricity

For electricity the transport is done by high power lines. The energy loss is essentially ohmic, although conduction in the air causes some additional losses. Therefore, when R is the resistance in the wires between the power plant and the consumer the energy lost per second is

$$P_{\text{lost}} = I^2 R \tag{4.127}$$

The total power P_T sent to the consumer is, assuming the current and voltage to be in phase,

$$P_T = IV \tag{4.128}$$

The ratio of the power lost to the power transmitted becomes

$$\frac{P_{\text{lost}}}{P_{\text{T}}} = \frac{I^2 R}{IV} = \frac{IR}{V} = P_{\text{T}} \frac{R}{V^2} \tag{4.129}$$

This ratio should be as small as possible. Decreasing the resistance R would mean thicker wires, which increases costs. For specific applications one could use superconducting wires with $R = 0$. Cooling to the required low temperatures is very expensive as liquid helium is needed. The recent discovery of superconductivity at higher temperatures implies that the cooling might be done with the much cheaper liquid nitrogen. Much development is going on in producing wires for the exotic materials which show high-temperature superconductivity.

The other possibility to diminish losses is to increase the potential V in the denominator of eq. (4.129). The limit here is imposed by the corona discharges of the wires and discharges between the wires and the ground. The higher the value of V, the higher the towers supporting the transmission lines must be and the more expensive the long isolation chains are between wires and towers. Also, one needs to transform the high voltage between the power station and the consumer to acceptable domestic voltages of 120–220 V. Alternating current (a.c.) lines are therefore needed to facilitate transforming.

Despite these advantages of high-voltage a.c. transmission there are some disadvantages compared to direct current (d.c.) lines. First, the towers and insulation of a.c. lines should take account of the amplitude of the voltage which is $\sqrt{2}$ higher than the mean value which determines the energy transport. Second, the time variation of the a.c. more easily induces sparks, perturbing radio transmission, and the current will be concentrated at the surface of the conductor, the so-called *skin effect*. Therefore one often uses a few parallel lines to increase the surface area.

For these reasons d.c. high-voltage connections have also been introduced. The difficulty is to put expensive a.c.–d.c. converters at both ends of the cable, so it is only worth while for long distance transmission. Direct current lines are also in use for particular applications as, for example, the electricity connection between France and Britain under the Channel, by which France exports its surplus nuclear power to the United Kingdom.

Finally it must be mentioned that there are proposals to use electricity from solar energy or from wind energy to produce hydrogen, which is transported, while next this hydrogen can be used in fuel cells to produce electricity again. It appears that in general this option is not competitive compared to transporting electricity directly [2].

4.2.8 REDUCING POLLUTION

Pollutants may be loosely defined as substances that in concentrations much bigger than in the natural environment—a hypothetical environment without human intervention—causes damage to plant, animal and human life. They may occur in air, water and soil and be transported from one place to another. Their sources were summarized in Fig. 1.1: transportation, the production of electricity, the burning of refuse, the burning of fuels for domestic purposes or in industry and all kinds of industrial and agricultural processes.

One should be aware of the fact that primary pollutants, emitted directly, may react with each other or with naturally occurring substances, causing secondary pollutants. This happens on a large scale in the atmosphere, partly influenced by sunlight, and is studied in atmospheric chemistry. The production of photochemical smog by nitrogen oxides and hydrocarbons is an example in case.

One may distinguish pollutants by their chemical composition and will find that the major pollutants are sulphur, nitrogen, carbon and halogen compounds. Or one may distinguish their physical properties and observe them as gas, fluid or solid; the latter may appear as particles in the air of different sizes or as solid waste of any kind. Particles in the air may contribute to an increase of atmospheric albedo and to some cooling, as discussed in Section 3.1.

Finally, one may add radioactive materials to the list and consider thermal pollution as well. Below a brief discussion is given of pollution related to the energy production described in the previous sections; radioactivity will be considered later connected to the sections on nuclear power.

Nitrogen Oxides NO$_x$

The symbol NO$_x$ comprises the nitrogen oxides NO and NO$_2$. Together with hydrocarbons in the air they are one of the major causes of photochemical smog. They result from any type of combustion in which air, consisting of N$_2$ and O$_2$, is brought to a high temperature by burning some fuel. The NO$_x$ is formed in the flame at a temperature of a few thousand K. As the reaction runs endothermic the NO$_x$ formation increases with temperature.

A second source of NO$_x$ is the nitrogen present in the fuel itself. For some coals or oils this is in the order of 1%; in that case some 80% of the NO$_x$ originates from the fuel itself. Natural gas contains little or no nitrogen. Then NO$_x$ production is essentially due to the high combustion temperatures.

The remedy against NO$_x$ emission can be twofold. In the first place one tries to lower the flame temperatures. That may easily lead to lower efficiencies of the cycles concerned. It is a matter of design to avoid local high temperature spots in the flames. Another way of reducing NO$_x$ production is to lower the available oxygen for combustion. It appears that the film of oxides that protect the walls of the combustion chamber against high temperatures then disappear as well, giving the need for new materials. This example serves to show that solving one problem may cause another one.

Much research is being done into *fluidized bed* combustion for power production. Here one has a mixture of crushed coal and some magnesium or calcium carbonate where the air enters from below. The boiler tubes of a steam generator are in direct and very good thermal contact with the bed. The combustion can then be performed at lower temperatures around 1200 K, still producing the same temperature in the steam for the external combustion cycle. This bed has the added advantage that the Ca or Mg additions bind sulphur into sulphates, which may be removed. The fluidized beds are not widely used yet.

For cars and most existing power stations there are no easy ways to remove NO_x after it has been produced, and some production is unavoidable. New designs or operations are being studied to reduce the NO_x production, for example spraying chemicals like NH_3 into the exhaust gases of power stations in an effort to reduce NO_x to N_2.

Sulphur

Coal and oil may have considerable concentrations of sulphur. This must be removed from the emission in order to reduce the occurrence of acid rain. Besides the fluidized beds which are promising for the future, one limits in practice the sulphur concentrations by spraying solutions of calcium carbonate through the flue gases in so-called *wet scrubbers*, again forming calcium sulphates which stay behind. Another way is to use catalysts to remove the sulphur from the exhaust gases.

Carbon Oxides CO and CO$_2$

All fossil fuel burning leads to the production of CO_2, with the consequences for global warming discussed earlier. Technical means of reducing the CO_2 production are hardly imaginable; one will therefore have to resort to removing CO_2 from flue gases and next bind it to calcium carbonate or store it underground.

As to carbon monoxide, CO, one must realize that it will always be produced in fossil fuel burning, for there exists an equilibrium between CO, CO_2 and O_2 according to the equation

$$CO_2 \rightleftharpoons CO + \tfrac{1}{2}O_2 \qquad (4.130)$$

for which the reaction constant K may be written as*

$$K = \frac{[CO][O_2]^{1/2}}{[CO_2]} = e^{-\Delta G^0/RT} = e^{-(\Delta H^0 - T\Delta S^0)/RT}$$

$$= 3 \times 10^4 \, e^{-33770/T} \qquad (4.131)$$

---

*Seinfeld [3, p. 79] uses the tabulated values at standard conditions (a.o. room temperatures), e.g. $\Delta G^0 = 283 \, kJ \, mol^{-1}$. Strictly speaking these values should be adapted for the much higher temperatures for which they are applied.

where R denotes the universal gas constant (cf. Appendix C) and ΔG^0 is the change in Gibbs free energy per mole for standard conditions, ΔH^0 the corresponding enthalpy change and ΔS^0 the entropy change. The sum of the three concentrations may be normalized by $[CO_2] + [CO] + [O_2] = 1$, and the ratio of oxygen to carbon moles may be written as α, leading to

$$\alpha = \frac{2[CO_2] + [CO] + 2[O_2]}{[CO] + [CO_2]} = \frac{\text{moles O}}{\text{moles C}} \qquad (4.132)$$

The relative CO concentration then follows from α and T. When one looks at the complete combustion of octane (representing gasoline) with precisely sufficient oxygen it appears that $\alpha = 3.125$, leading to a relative CO concentration of 0.213 at 3000 K.

Particles

Particulate emissions from power plants are being controlled by mechanical means (e.g. letting the flue gases go round where centrifugal forces bring the particles to the outside), by filters as in a vacuum cleaner, by spraying with water in a scrubber or by having the flue pass electrostatic wires with 30–60 kV, where the particles get a charge and are attracted by grounded plates.

Thermal Pollution

All engines discussed in this section have a thermodynamic efficiency determined by the Carnot efficiency equation (4.74). That implies that a lot of heat must be dumped into the natural environment. For big power plants the steam condensers are cooled either by letting water from lakes and rivers pass the condenser or by big cooling towers. In the latter case hot condenser water is directly sprayed into the tower and is cooled by rising air in which it evaporates, losing latent heat. Fresh water is then needed as input to the cycle (wet towers). Alternatively, cooling takes place by heat exchange between the condenser and tubes in the tower which are cooled by rising air (dry towers).

The cooling in wet towers will not only be more effective but will also lead to a lower temperature T_C of the cool reservoir. The incoming air will absorb water drops as vapour up to its saturation point, effectively cooling itself by providing the evaporation heat (or enthalpy). The resulting temperature is called *wet bulb temperature* in meteorology, measured by thermometers with a wet bulb and used to calculate the relative humidity of the air. The disadvantage of wet towers are the enormous quantities of water required.

An alternative is to use the waste heat for heating houses and buildings or to use it as process heat in industry. In fact, this heat power coupling is not very practical for big power stations where one loses much energy in transport, but is particularly suited for smaller power stations of some 10 MW_e for local industrial complexes or other places where much electricity is used. The point is that one calculates the capacity of the combination such that the heat requirements are fulfilled; one usually has surplus electricity at least during certain hours and too little electricity in other occasions. Therefore a coupling to the main electricity grid is required in order to be certain of its electricity. The question then is whether the utility is prepared to buy surplus electricity for a fair price.

General

Governments are already in the process of making stricter regulations, both for allowable car emissions and for emissions of power plants. This stimulated new designs for engines; it is clear that for external combustion engines where the combustion happens outside, as in the fluidized bed, it is easier to control emissions than for internal combustion engines. The first may therefore have a future.

On the other hand, car manufacturers have succeeded in following the requirements of the environment. The lead that was added to the gasoline to avoid 'pinging' of the engine is being replaced, to mention one example.

The regulations will tend to make fossil fuel combustion both in cars and in power stations more expensive, which may make new methods of energy production, to be discussed in later sections, more competitive.

4.2.9 REFRIGERATION

Cooling can be done by any process where a heat engine runs in the reverse direction, as was shown in Fig. 4.5 and eq. 4.76. A heat engine like the Stirling engine discussed in Section 4.2.3 therefore could do the job. In practice, however, the temperature differences to be maintained in domestic refrigeration are not high. On the one hand one needs a temperature somewhat below the freezing point of water and on the other hand room temperature or atmospheric temperature. The practical problem so far with Stirling-type engines has been to find good heat exchangers between the engine and the place to be cooled.

The Vapour-Compression Cycle

A very effective way of heat exchange is provided by evaporation and condensation. A high amount of heat is absorbed or ejected in this process and when working in a closed cycle the medium itself can transport the heat to the places where it is rejected or used. The usual freezer therefore consists of an evaporator and a condenser, with a cycle as shown in Fig. 4.16.

For the presentation we choose a diagram of pressure p versus enthalpy $H = U + pV$. Just as in the conventional pV diagram, the dotted line denotes where one has a pure liquid (on the left) or a saturated vapour (on the right). In

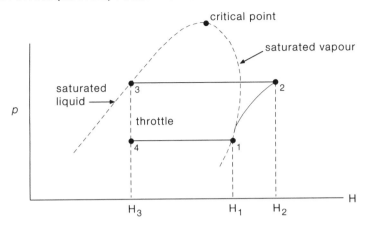

Figure 4.16 The *pH* diagram of a commercial refrigerator cycle. Below the dashed line equilibrium of fluid and vapour exists. The left part of that curve shows saturated liquid; the right part saturated vapour. In between one has a combination of both at a constant temperature

between, going from left to right, the temperature remains constant and the fraction of vapour increases.

Let us follow the steps in the cycle of Fig. 4.16:

(a) From 1 to 2 one compresses adiabatically and a pump is required to deliver a work $W' = H_2 - H_1$. As discussed in Section 4.2.1, the work on the system increases its enthalpy.
(b) At point 2 the fluid enters a condenser; it first cools isobarically until the dotted curve is reached and then condenses at constant temperature. The added heat, according to eq. (4.58), becomes $H_3 - H_2$, which is negative, meaning that heat leaves the system.
(c) From 3 to 4 the fluid expands through a throttle (cf. Fig. 4.4) and $H_4 = H_3$.
(d) From 4 to 1 it expands isobarically and heat $H_1 - H_4 > 0$ is added. Thus heat is extracted from the surroundings, giving cooling.

The coefficient of performance (COP) defined in eq. (4.77) becomes

$$\text{COP} = \frac{\text{cooling achieved}}{\text{work input}} = \frac{H_1 - H_4}{H_2 - H_1} \tag{4.133}$$

As a practical example we consider refrigerant 22 ($CHClF_2$). It operates with a condensing temperature of 35 °C (the temperature of point 3) and an evaporation temperature of -10 °C (the temperature of points 4 and 1). $H_4 = H_3$ and H_1 are easily found from published tables. To find H_2 is more complicated as the fluid is in fact superheated. One knows that the entropy $S_2 = S_1$ and the pressure $p_2 = p_3$. One can then use published graphs of $H(S, p)$ to find H_2. The values are $H_3 = 243.1\,\text{kJ/kg}$, $H_1 = 401.6\,\text{kJ/kg}$ and $H_2 = 435.2\,\text{kJ/kg}$. This gives COP = 4.72 (cf. Ref. 4 from which this example was taken).

As the working fluid for domestic uses one will select a fluid with pressures p_1, p_2 not too far from atmospheric pressure and useful boiling temperatures. In Fig. 4.17 one finds the pT curves for some fluids. One notices that CFCs, NH_3 and C_4H_{10} display reasonable temperatures and pressures. The curve for propane (not shown) is close to the one for HCFC22. In the past C_3H_8 and C_4H_{10} were not chosen as they are inflammable. Also NH_3 was discarded because small leaks caused headaches. Therefore CFCs, which did not react, were chosen as a working fluid in the 1930s. Since it became clear that CFCs destroy the ozone layer in the upper atmosphere one has again returned to the other fluids mentioned. However, it may also be that in the future heat engines with efficient heat exchangers will replace the traditional refrigerators or that one will turn to the absorption refrigerators.

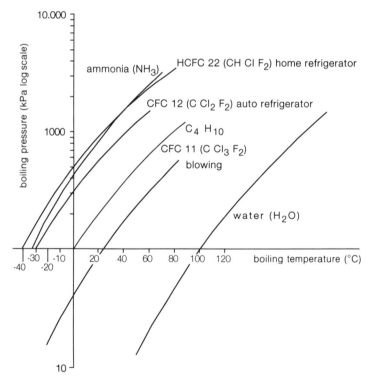

Figure 4.17 Pressures versus boiling points for a few common fluids. It is clear that not many fall in the acceptable range of temperatures between -10 and $60\,^\circ\text{C}$ and reasonable pressures (not too far from the atmospheric pressure of $100\,\text{kPa}$)

Absorption Refrigeration

The essential point of the vapour-compression cycle discussed above is that a large amount of energy is needed for compression to achieve the required high condensation pressure. Another way to achieve this is to absorb vapour in a liquid while removing heat, next increase the pressure with an ordinary pump and then apply heat by which the vapour is released; a high pressure then results. The rest of the cycle is the same as before. The two cycles are compared schematically in Fig. 4.18, where in the dotted rectangle one of the two processes is applied.

The cycle becomes competitive when waste heat from another process is used to heat the mixture, as indicated in step 3 of Fig. 4.18. This would usually come from electric generators. One could in fact use their waste heat in the winter for heating and in the summer for cooling by absorption refrigeration. In practice NH_3–water or LiBr–water combinations are being used.

4.2.10 TRANSPORTATION

When a vehicle (car, truck or train) is moving at a constant velocity u there are two main forces acting to slow down the motion: the air drag F_d and the rolling resistance F_r. The air drag is the resistance which a body with cross-section A_f experiences in a moving fluid; ignoring wind velocities this may be expressed as

$$F_d = (\rho/2)C_d A_f u^2 \tag{4.134}$$

where ρ is the density of air and C_d the *drag coefficient*. One may only calculate this coefficient for a few simple cases such as flow with a low Reynolds number Re around a sphere with radius R (cf. eq. (3.49)). In Section 5.8 we will show that

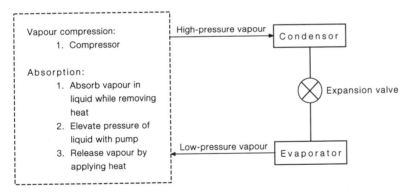

Figure 4.18 Comparison of two refrigerating systems: vapour compression and absorption. (Reproduced by permission of McGraw-Hill from W. F. Stoecker and J. W. Jones, *Refrigeration and Air Conditioning*, McGraw-Hill, New York, 1982, p. 329)

this follows Stokes's law with

$$F_d = 6\pi R\mu u \tag{4.135}$$

where R is the radius of the sphere. When the Reynolds number is defined following eq. (3.49) as

$$Re = \frac{u(2R)}{v} \tag{4.136}$$

one finds for the drag coefficient

$$C_d = \frac{24}{Re} \tag{4.137}$$

As the drag force in this case is proportional to u and not to u^2 as in eq. (4.134) the concept drag coefficient is artificial. The other extreme, however, would be a flat plate. In the simplest approximation the drag force will equal the momentum hitting the plate per second over the area A_f which leads to

$$F_d = (\rho u)uA_f \tag{4.138}$$

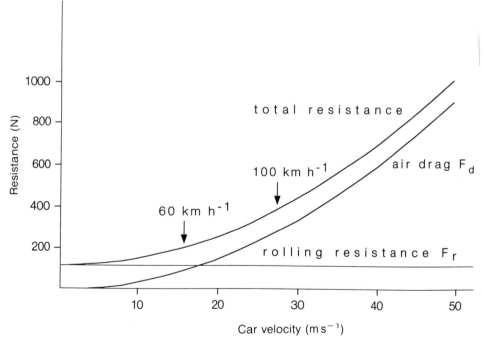

Figure 4.19 Air drag and rolling resistance for a typical light passenger car: $M = 1130\,\mathrm{kg}$, $A_f = 1.88\,\mathrm{m}^2$, $C_d = 0.3$, calculated by using eq. (4.139)

which gives a drag coefficient $C_d = 2$. A more accurate calculation gives a value of around 1.3. For modern passenger cars the drag coefficients vary around 0.3 depending on streamlining and velocity.

The other force acting on a vehicle is the rolling resistance F_r. This certainly will be proportional to the mass M of the vehicle and is mainly due to the internal compression of the part of the tyre touching the road and the depression of the part just leaving the road. An empirical relation is the following:

$$F_r = C_r g M \qquad (4.139)$$

where the coefficient C_r depends on the quality of the road and varies between 0.01 for an asphalt road and 0.06 for a sand road. The coefficient C_r will be smaller for a heavy truck with stiff tyres but in all cases it increases with wear and tear. In Fig. 4.19 we show both forces as a function of velocity for a typical compact car. One notices that for higher speeds the air drag starts to dominate. In fact, the improvement in fuel economy in the 1970s and 1980s is mainly due to a reduction in mass M of the vehicles.

The power consumption corresponding to both forces would be

$$P = (F_d + F_r)u \qquad (4.140)$$

The actual power that needs to be supplied by the fuel is much higher; some 40% would be a good engine efficiency. One then has to subtract internal cylinder losses, frictional losses, accessories (air conditioner) and transmission losses. In the end some 8% might come to support the power needed in eq. (4.140).

4.3 THE PRICE OF ENERGY CONVERSION

Companies that run an installation distinguish between fixed costs, independent of the running of the machine, and variable costs. The fixed costs are essentially the capital costs of paying interest on invested capital and paying back the capital over a number of years. The variable costs are fuel costs and the costs for operation and maintenance, such as wages, reparations, etc. For some installations, such as those that produce electricity from sun or wind, it is not the fuel costs but the saved fuel costs that determine the economics of the installation. To clarify the discussion below it is useful to read Exercise 4.22 as a practical application of the concepts to be introduced.

When a certain amount of money P is invested or put in a bank at an interest rate of i_0 and the interest is calculated n times a year and added to the capital, the total (future) $F(t)$ value of the initial investment after t years would be

$$F(t) = P\left(1 + \frac{i_0}{n}\right)^{nt} \qquad (4.141)$$

This expresses the sum of a geometric series with ratio $(1 + i_0/n)$. For $n \rightarrow \infty$ this approaches the exponential $P\exp(i_0 t)$ so eq. (4.141) represents exponential growth. In order to simplify the discussion it is assumed that interests are compounded at the end of complete years ($n = 1$) and $i_0 = i$. Equation (4.141) then reduces to

$$F(t) = P(1 + i)^t \qquad (4.142)$$

Of course, one may calculate an effective i from i_0 by means of

$$\left(1 + \frac{i_0}{n}\right)^n = 1 + i \qquad (4.143)$$

Consider a uniform payment of A at the end of each year during t years. Then one can reduce each payment back to time $t = 0$ by means of the inverse of eq. (4.142); one obtains a geometric series with a sum

$$A\frac{(1 + i)^t - 1}{i(1 + i)^t} = P \qquad (4.144)$$

which is called the *present value* of the series of uniform payments.

Mathematically, the present value P and the series of payments A are equivalent. In an economic sense this is not completely true. Investors will argue that a single big amount of money P could draw a higher interest rate (corresponding to the long term interest rate) than a large amount of small payments that may be received every month with energy bills; they would at best draw a short term interest rate j. This complication will not be discussed here.

We note that the present value P of a payment $F(t)$, due t years in the future, may be calculated from the inverse of eq. (4.142) as

$$P = \frac{F(t)}{(1 + i)^t} \qquad (4.145)$$

It is important when considering long term investments to remember that prices may rise according to a general inflation rate. Besides, in estimating the economic feasibility of, for example, wind energy one has to estimate costs of fuel saved in the future, which may rise or fall according to the assumptions one makes. Both effects together will result in an apparent *escalation rate e*.

Inflation will mean that a uniform series of payments or receipts has an even lower present value than indicated in eq. (4.144):

$$P = \frac{A[(1 + e)^t(1 + i)^t - 1]}{(1 + e)^t(1 + i)^t(e + i + ei)} \qquad (4.146)$$

In practice, in return for an investment, utilities will expect a series of annual receipts at some escalation rate e, corresponding to a continuous price increase. They will need its present value in order to compare it with the present value of the investment. If the first receipt, after one year, is $A(1 + e)$ the present value of the series over a total of t years becomes

$$P = A(1 + e) \frac{\left(\dfrac{1+e}{1+i}\right)^t - 1}{e - i} \qquad (e \neq i) \qquad (4.147)$$

or

$$P = At \qquad (e = i) \qquad (4.148)$$

The last equation shows that inflation and interest cancel when $e = i$.

Levellized End-of-Year Cost

When comparing alternatives for the production of electric power one has to compare the cost of future payments of interest and the benefit of future receipts or future savings in energy costs. This all should be reduced to the present value P. The parameters of the alternatives such as lifetimes t may be different, however. Therefore, it is convenient to introduce the *levellized end-of-year cost L*, which is the uniform payment to be made at the end of each year up to the end of life of the installation which gives the present value P according to eq. (4.144). This gives

$$L = P \frac{i(i + 1)^t}{(1 + i)^t - 1} \qquad (4.149)$$

The factor multiplying P at the right-hand side is called the *capital recovery factor*. It expresses the factor by which a capital P should be multiplied in order to find the annual payment during t years which precisely recovers the capital P.

Rest Value

Equation (4.144) or (4.147) may be used to calculate the annual payment series starting with A respectively $A(1 + e)$ that one needs in order to recover the capital after t years. It may be the case, however, that at the end of the t years there remains a rest value S. This may be positive when the installation, the building, the land, etc., could be sold, but it may be negative when a big cost of decommissioning of the installation is expected. This is the case for nuclear power plants and for heavily polluted chemical installations.

The rest value S after t years can be reduced to its present value $S(0)$ by eq. (4.142):

$$S(0) = \frac{S}{(1 + i)^t} \qquad (4.150)$$

For long periods of time this results in a small fraction of S. With an interest rate $i = 0.08$ and $t = 25$ years the fraction becomes 0.15 and after $t = 100$ years smaller than 0.000 5. The latter is relevant if one plans to let nuclear power stations cool down for 100 years after the end of operation before dismantling them. Because of the long periods of time involved the cost does not weigh heavy in P or in the levellized costs L during the lifetime of the installation.

Building Times

The capital costs of an installation are influenced by the building times. Usually the commissioning company has to make annual payments that may rise by a factor $(1 + e)$ due to inflation. The company has to pay an interest rate of i over its payments during the building time without getting any money back.

If the first payment A happens after 1 year then the total payments after t years, including interest, are

$$F = \frac{A}{i - e}[(1 + i)^t - (1 + e)^t] \qquad (i \neq e) \tag{4.151}$$

This of course is bigger than the value At for t equal payments A. With an interest rate of $i = 0.08$, an inflation $e = 0.05$ and $t = 6$ years one finds $F = 1.37At$. For a period of $t = 12$ years one finds $F = 2.01At$. Therefore this long construction time doubles the cost. In fact, one of the big causes of cost escalations in building nuclear power plants was the lengthening of the building times due to environmental concerns. It may be noted in passing that eq. (4.151) also represents the future value (after t years) of fuel saved when the price increase of the fuel is e and the money saved is put on an account with interest rate i.

4.4 RENEWABLE ENERGY SOURCES

Renewable energies are defined as those sources that are for all practical purposes in unlimited supply, more specifically that will last as long as the sun is supporting life on earth. Most of them derive from solar radiation which powers the earth with some 120×10^{15} W (eq. (1.1)). Solar energy as discussed below may be used directly as heat or as solar electricity. Indirectly the sun acts as the engine behind atmospheric circulations which give rise to a wind energy of some 1.2×10^{15} W. Part of this energy is transferred to ocean waves which may be used in some applications.

The sun also—and perhaps in the first place—induces photosynthesis on earth which stores energy in biomass. This mass of carbohydrates can be used in several forms as an energy source.

A renewable source that does not originate from the sun is geothermal energy, heat arising from radioactive decay in the earth's crust. The decay times are in the order of billions of years and may safely be regarded as renewable. As the physics involved is mainly heat exchange which is covered in several parts of this text, geothermal energy will not be discussed here.

It should be remarked that renewables, particularly solar energy and wind energy, may be applied in many more ways than described in subsequent sections. They may be used on a small scale, in rural areas and are in many ways suited for developing societies [5].

4.4.1 SOLAR HEAT AND SOLAR ELECTRICITY

The energy received from the sun at a certain place on earth depends on the geographical latitude, the time of the day and the date. This is calculated from Figs. 4.20 and 4.21.

On the left of Fig. 4.20 the earth orbit around the sun is sketched. On the right the annual rotation is shown in a frame fixed on the centre of the earth. It is noticed that the earth axis is making an angle with the orbital plane (the ecliptic). This *inclination* ε amounts to ε = 23.45°. Because of the daily rotation of the earth the sun describes an orbit in the sky. From Fig. 4.20 it follows that at the so-called *equinoxes* the sun is positioned in the equatorial plane of the earth. On these dates day and night are equally long.

The plane through the sun and the earth axis determines the circle on earth where it is noon (facing the sun) or midnight (off the sun). The *declination* δ is defined as the latitude where the sun is in the zenith at noon. It is clear that on 22 December the sun reaches the zenith at 23.45° south of the equator (the Tropic of Cancer) and on 22 June there is the other extreme when the sun is in the zenith at noon at 23.45° north of the equator (the Tropic of Capricorn). Thus the declination varies with the time of the year.

This variation of the declination δ may be found from the right-hand side of Fig. 4.20. Suppose that N days have passed since the sun was in the vernal

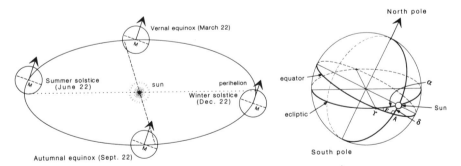

Figure 4.20 On the left the earth's orbit around the sun is shown. The dotted line indicates the major axis of the ellipse; the perihelion is close to the winter solstice (in 1993 on 4 January). The seasons indicated are for the Northern hemisphere. The dates may differ by two days from year to year. On the Southern hemisphere they are shifted by half a year. The earth equatorial plane is dashed. On the right one observes the annual motion of the sun from the centre of the earth

equinox (22 March). The angle along the ecliptic is given by

$$\alpha = N \frac{2\pi}{365.24} \tag{4.152}$$

with the number of days in a year in the denominator. The sine rule on a sphere is applied to the triangle on A, the sun and the vernal point giving

$$\frac{\sin \alpha}{\sin (\pi/2)} = \frac{\sin \delta}{\sin \varepsilon} \tag{4.153}$$

from which it follows that

$$\sin \delta = \sin \varepsilon \sin (2\pi N/365.24) \tag{4.154}$$

This situation is sketched again in Fig. 4.21 where the plane of the sun and the earth axis is indicated for an arbitrary date, determined by declination δ. The *insolation* at D is the solar energy received per second on a given unit surface area in a horizontal plane. To find it, one has to write down the normal to that surface area and calculate $\cos \alpha$ where α is the angle between the normal and the direction of the sun.

The direction of the sun is given by

$$\mathbf{MA} = R \sin \delta \mathbf{e}_1 + R \cos \delta \mathbf{e}_2 \tag{4.155}$$

where R is the earth radius. Point D has a daily rotation with a frequency $\omega = 2\pi/(24 \times 60 \times 60)\,\mathrm{rad\,s^{-1}}$. Including a correction for the annual rotation this

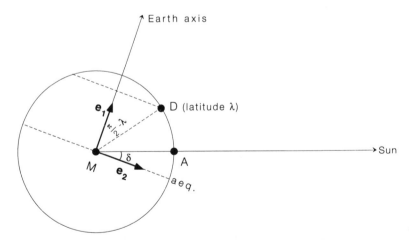

Figure 4.21 The plane through the sun and earth axis, where the circle gives noon (or midnight) on earth. The sun is in the zenith at noon at latitude δ. Point D at latitude λ will perform a daily circle perpendicular on the earth axis. Unit vectors $\mathbf{e}_1$, $\mathbf{e}_2$ and $\mathbf{e}_3$ (not shown) are fixed and do not follow earth's daily rotation

gives $\omega = 7.292 \times 10^{-5}$ rad s^{-1}. The local zenith is given by

$$\mathbf{MD} = R\cos\left(\frac{\pi}{2} - \lambda\right)\mathbf{e}_1 + R\sin\left(\frac{\pi}{2} - \lambda\right)(\mathbf{e}_2\cos\omega t + \mathbf{e}_3\sin\omega t)$$

or (4.156)

$$\mathbf{MD} = R[\sin\lambda\mathbf{e}_1 + \cos\lambda(\mathbf{e}_2\cos\omega t + \mathbf{e}_3\sin\omega t)]$$

where $t = 0$ indicates noon. The required insolation at latitude λ and time t is given by

$$\cos\alpha = \frac{\mathbf{MA}\cdot\mathbf{MD}}{(MA)(MD)} = \sin\lambda\sin\delta + \cos\lambda\cos\delta\cos\omega t \tag{4.157}$$

The times of dawn $(-T)$ or sunset $(+T)$ are given by $\cos\alpha = 0$, leading to

$$\cos\omega T = -\tan\lambda\tan\delta \tag{4.158}$$

For $\lambda = \pi/2 - \delta$ or $\lambda = -(\pi/2) - \delta$ one finds that $T = 12$ hours, so then the sun never sets. For $\delta > 0$ the arctic summer and for $\delta < 0$ the antarctic summer represent this result.

The total amount of sunlight $A(\lambda, \delta)$ received during a day on a horizontal surface is found by integrating eq. (4.157) from dawn to sunset (or in the Arctic or Antarctic regions over 24 hours for the appropriate part of the year). The

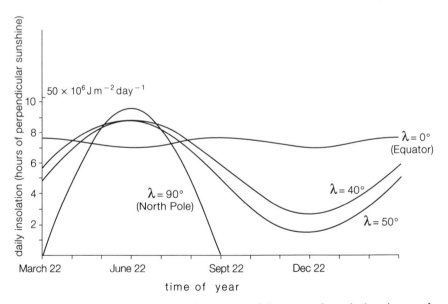

Figure 4.22 The total daily insolation on top of the atmosphere during the year for the equator and 40, 50, 90° north. It is expressed as the number of hours that the sun in the zenith would give the same daily insolation with 4.87×10^6 J m^{-2} as the unit

result is shown in Fig. 4.22 as a function of the day of the year, obtained by varying δ in eq. (4.157). In fact, the results hold for a location on *top* of the atmosphere. It is clear from Fig. 4.21 that at higher latitudes solar light has to penetrate a thicker layer of the atmosphere, leading to more absorption. Therefore, despite the peaks in Fig. 4.22 even in summer the equator surface area will receive more sunlight than the poles.

Let us look at point D of Fig. 4.21. If one wants to receive as much of solar energy as possible on a given surface one should slant it so that its normal faces the sun ($\cos \alpha = 1$). For best results one should change the orientation of the surface during the day, keeping it towards the sun and even correct it from day to day, following the variation of the declination δ with the seasons. Usually this is not worth the cost and one chooses the surface to face the sun at noon, while the angle with the horizon is fixed so as to get the highest results when one needs it most: in winter if home heating is the objective, in summer if one wants solar energy for cooling.*

It must be added that tracking the sun only would influence the amount of *direct* solar radiation received. The indirect, *diffuse* radiation on the earth's surface originates from the scattering of light on particles in the atmosphere. This depends in particular on the humidity in the air. The fraction of diffuse radiation, compared to the total, varies from 20% in desert areas to 60% in cities like London. This is certainly not negligible. When mirrors are used to concentrate solar radiation, they will of course only focus the parallel, direct solar rays and not the diffuse radiation.

Below we shall discuss the use of solar energy in producing hot water for domestic purposes as well as more sophisticated devices to produce heat of higher temperatures to produce electricity. We finish with the direct, photovoltaic, production of electricity by solar radiation.

It should be remarked that much effort is being spent on reducing the need for heating and cooling by the special design of buildings. They are being well insulated, keeping the heat in during wintertime, combined with ways to use direct heating from sunlight that enters the buildings anyway (even in the Netherlands at latitude $\lambda = 52°$ north the average apartment, without special design, receives 12% of its heat by the sun during winter). In summer, special glazing and 'sun curtains' keep the incoming heat low. Use of special low-watt lighting should keep the heat produced inside buildings at a low level.

Solar Collectors

For domestic purposes one uses solar collectors, which absorb solar radiation and heat up to about 80 °C. The heat is transferred to water, passing the collector

---

*Following the daily motion of the sun by some servo-motor mechanism means that one keeps $\cos \omega t = +1$ instead of having it varied from 0 by ($+1$) to 0 again. As the time average of $\cos \omega t = 2/\pi$ one has to make sure that the gain is worth the cost.

reflection

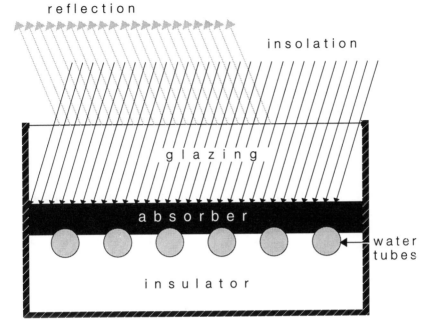

Figure 4.23 A simple solar collector. Solar radiation is absorbed and the resulting heat is transferred to the coolant tubes indicated in grey

in tubes. The simplest case is sketched in Fig. 4.23. One notices the absorber and the coolant tubes. The absorber will be painted black with an absorption coefficient α close to one. The absorber will lose energy by the three processes described in Section 4.1: conduction, convection and radiation. The point is to reduce these losses as much as possible, leaving only conduction to the coolant tubes as a required 'loss' of absorber energy.

Let us start by discussing the radiation losses. For a black body they are described by eq. (4.3):

$$q'' = \sigma T_s^4 \tag{4.159}$$

For a temperature of 80 °C (or $T_s = 353$ K) one finds 880 W m^{-2} which is not far below the maximum insolation $S \approx 1353$ W m^{-2}. Therefore without technical measures there would be no net gain in energy.

The radiation losses may be reduced considerably by using 'black chrome' or similar surfaces. These have a high absorption α_λ in the visible region where the solar spectrum peaks and a small absorption α_λ for long wavelength radiation corresponding with the emission spectrum of a black body of 80 °C (cf. Fig. 1.2). The emissivity ε_λ is high only for those wavelengths where the absorber of Fig. 4.23 will not emit because its temperature is low compared with the solar spectral temperature of about $T = 5800$ K. For the wavelengths λ where the absorber does emit, α_λ is small, which implies a low emissivity ε_λ at those wavelengths.

The losses by conduction to other than the coolant tubes will be reduced by proper insulation, as indicated in Fig. 4.23.

Finally, the losses by convection are reduced by putting glazing (glass or plastic) on top of the collector. This effectively avoids the forced convection with high convection coefficients h, which were indicated in Table 4.2.

It may be interesting to consider the example of a realistic solar collector, following the sketch of Fig. 4.23. It will have an area of $3\,\text{m}^2$ with an incident solar flux of $S = 700\,\text{W m}^{-2}$. The top glass cover will have a transmission of 90%, while the rest is reflected and for a little absorbed in the glass. With an air temperature $T_{air} = 25\,°\text{C}$ the glass will operate at $T_2 = 30\,°\text{C}$. The emissivity of the glass will be $\varepsilon = 0.94$ and the radiation exchange with the surrounding sky will correspond to its temperature $T_{surr} = -10\,°\text{C}$. Convection between glass and air will have a convection coefficient $h = 10\,\text{W m}^{-2}\,\text{K}^{-1}$. The water flow rate will be $m = 0.01\,\text{kg s}^{-1}$ and the specific heat of the water $c_p = 4179\,\text{J kg}^{-1}\,\text{K}^{-1}$. The water enters with a temperature T_i and leaves with a temperature T_o.

The net heat flux entering the collector is given by

$$q'' = 0.9S - \varepsilon\sigma(T_2^4 - T_{surr}^4) - h(T_2 - T_{air}) \tag{4.160}$$

where eqs. (4.2) and (4.5) were used. Inserting the numerical values given above leads to $q'' = 385.7\,\text{W m}^{-2}$. The heat entering should be taken away by the water according to

$$q = 3q'' = mc_p(T_o - T_i) \tag{4.161}$$

leading to $T_o - T_i = 27.7\,°\text{C}$. The efficiency is given by $\eta = 385.7/700 = 55\%$.

One will notice from this example how essential it is to be able to ignore the

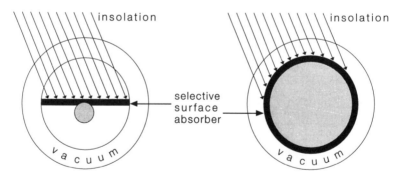

Figure 4.24 Evacuated tube solar collectors. Solar radiation passes a vacuum to reach the absorber, which is blackened with a selective surface. On the left that selective surface is flat, with a traditional tube carrying water. On the right the selective surface is on the outside of a big copper tube with a large water carrying capacity

radiation losses by the absorber and the heat losses through the insulator. The temperature of the glass might be higher in practice by internal convection of the air between the absorber and the glass. One may diminish this effect by evacuating the space between the absorber and the glazing. Stronger and more expensive constructions are then needed, as illustrated in Fig. 4.24, using evacuated tubes.

Electricity by Solar Heat

In producing electricity from heat one needs heat of a high temperature, as discussed in Section 4.2.6. This requires mirrors to focus the direct (!) solar radiation. Long cylindrical mirrors of parabolic shape may be used which have in their focal lines a fluid carrying tube that transports the heat to a central heat engine. In another set-up one uses spherical mirrors and puts a heat engine in its focal point.

As the heat is produced external to the engine, mechanical energy is produced by the heat engines discussed in Section 4.2.3. There are either steam engines* or Stirling engines. In the latter case one often uses a heat pipe to transport the heat from the focal point to the engine itself, a distance of $\approx 10\,\text{cm}$.

The Solar Pond

An amusing application is the solar pond, sketched in Fig. 4.25. A pond of some 2 metres of depth consists of two thin layers, one on top (temperature T_1) and one at the bottom (temperature T_2). Both are well mixed by convection and have a homogeneous temperature. In between there is a central layer with increasing

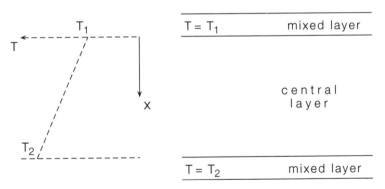

Figure 4.25   The solar pond

---

*The first solar powered steam engine was demonstrated in Paris in 1866. A conically shaped mirror was used to track the sun, and the engine was used to drive a printing press.

salinity going down. Half of the solar radiation is absorbed by the top layer, a quarter by the central layer and a quarter by the bottom layer. The bottom absorption gives that layer a higher temperature than the top $T_2 > T_1$. In the central layer the temperature will increase from T_1 to T_2. This ordinarily would lead to convection as hotter water would be less dense than colder. The increasing salinity compensates for this effect and results in a density increasing with depth. Using a heat exchanger the heat ·at the bottom can be extracted to produce electricity or process heat. When the heat is not needed it will remain in storage in the bottom layer.

Photovoltaics: Direct Conversion of Solar Energy into Electricity

Direct conversion of photons into electrical energy is practised in *solar cells*, which consist of semiconductor materials. Solar cells are connected to modules that may operate independently of each other and are connected in parallel or in series. As Si is in widest use, we take this material as an example in our discussion.

A semiconductor is a (poly-)crystalline material in which only a limited amount of charges can move freely around. The crystal structure is such that there exists an *energy gap* E_g between the valence band in which the outer atomic electrons are bound and the higher lying conduction band in which electrons can move around. For Si the energy gap is $E_g = 1.12\,eV$.

At a temperature T the interaction between the particles in the crystal gives rise to a kinetic energy which on average is kT. At room temperature $T = 293\,K$ and $kT = 0.025\,eV$, which is much too low to cross the energy gap. There is, however, a Boltzmann probability distribution which gives the number of electrons $n(E)\,dE$ with an energy between E and $E + dE$ as

$$n(E)\,dE = ce^{-E/kT}\,dE \tag{4.162}$$

where c is a normalization constant. Consequently there will be some electrons with an energy high enough to cross the gap. For Si at room temperature the density n of these *free electrons* amounts to $n = 1.5 \times 10^{16}\,m^{-3}$. Each free electron leaves a *hole* in the valence band that may be occupied by other valence electrons, of which there are many. Consequently the holes move around freely as well, but behave as electrons with a positive charge. The density n of these particle–hole pairs should be compared with a density for the valence electrons, which is 10^{12} times as big. Therefore, n is very small indeed.

Consider a Si crystal as sketched in Fig. 4.26a for two dimensions. In three dimensions the crystal would have a tetrahedral structure, each atom being positioned on equal distances from its neighbours. In Fig. 4.26a the four valence electrons for each atom that care for the covalent bonds within the crystal are indicated. In Fig. 4.26b one Si atom is replaced by an atom with five valence electrons (e.g. Sb, P, As), resulting in an extra electron; in Fig. 4.26c it is replaced

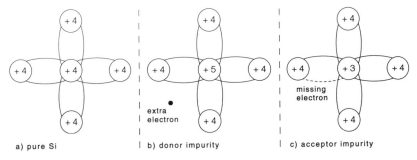

Figure 4.26 A two-dimensional representation of a Si crystal (a) without impurities, (b) with a donor atom replacing a Si atom or (c) with an acceptor atom replacing a Si atom

by an atom with three valence electrons (e.g. B, Al, Ga), resulting in an electron 'missing' in the crystal structure.

In Fig. 4.26b the extra electron is bound with an energy of 0.05 eV* so it can virtually move around as a free electron, hence the name 'donor' impurity. In Fig. 4.26c there is a hole that can accept other valence electrons, leading to a virtually freely moving hole (acceptor impurity). Either can be doped with donors or with acceptors in densities much larger than the above mentioned $n \approx 10^{16}\,\mathrm{m}^{-3}$ (for *n-type* $10^{25}\,\mathrm{m}^{-3}$ and for *p-type* $10^{22}\,\mathrm{m}^{-3}$). Therefore, the conduction in Fig. 4.26b essentially happens by means of donor electrons (n-type Si); in Fig. 4.26c it happens by means of holes (p-type Si).

Consider a single crystal, usually in the form of a round thin cylinder, consisting of a layer of n-type Si on top of a layer of p-type Si. This will act as a *solar cell* if one of the layers is thin ($\approx 1\,\mu\mathrm{m}$) and the other thicker (up to $\approx 100\,\mu\mathrm{m}$). Light can enter the crystal through the thin layer and reach the *junction*, the boundary between both layers.

The situation in the absence of light is sketched in Fig. 4.27. At the n- and p-sides some electrodes have been indicated with an external potential V. Consider first the case with the electrodes disconnected. As we shall discuss in Section 5.1, a concentration difference always gives a *diffusion current* counteracting it. Therefore, the surplus of electrons on the n-side will pass the junction to the p-side and the holes on the p-side will do the reverse. This will lead to a positive charge on the n-side and a negative charge on the p-side (cf. Fig. 4.27) which give a counteracting electric field from n to p.

---

*To make an estimate of the binding of the extra electron one could describe it as a single electron in a Coulomb field $V = -e^2/(4\pi\varepsilon_0\varepsilon_r r^2)$. Following the equations of the hydrogen atom in the Bohr model, taking an effective mass of the electron which is 0.2 times its free mass and using $\varepsilon_r = 11.7$, for Si one finds a binding energy which is 2×10^{-3} times the binding energy in the H atom, about 0.02 eV. The actual value for Si is somewhat higher and for Ge it is around 0.01 eV.

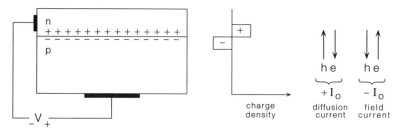

Figure 4.27 A semiconductor junction in the absence of light. On the left one notices the electrodes. Even unconnected diffusion causes a charge distribution (middle). In equilibrium the diffusion current (right) is balanced by the opposite current induced by the field (far right)

Besides the diffusion current there is a current in the opposite direction. Thermal excitation of the Si according to eq. (4.162) will give some electrons on the p-side and holes on the n-side. They cross the junction in the direction of the field, hence the name field current. This is indicated on the right of Fig. 4.27. The charge carriers contributing to the field current are small in number. In practice any field will help them to cross the barrier. The field current is therefore taken to be independent of the field and dependent on the temperature T only. In equilibrium, diffusion and field currents (I_0 respectively $-I_0$) will cancel. In fact the diffusion current is counteracted by the electric field until its magnitude I_0 equals that of the field current. When the electrodes on the left of Fig. 4.27 are connected, no current can run and there would be no potential difference: $V = 0$. We repeat that the field current is produced by electrons from the p-side and holes from the n-side, i.e. from places where these carriers are in short supply.*

Let us now apply a potential V connected with the positive pole to the p-side, as indicated on the left in Fig. 4.27 (a so-called forward bias). This lowers the counteracting potential by V and consequently the diffusion current increases. This does not influence the field current, as this current is much more sensitive to the temperature than to the magnitude of the external field. Therefore, the electrons from the n-side and the holes from the p-side, giving the diffusion current, now need an energy eV less to cross the junction. In fact, these carriers obey a Boltzmann energy distribution (4.162) dominated by the energy kT. It is plausible and can be proven that the increase of the diffusion current follows a Boltzmann exponential as well. This leads to a (diode) current I_D determined by

$$I_D = I_0 e^{eV/kT} - I_0 \qquad (4.163)$$

For $V = 0$ this gives $I_D = 0$, as it should.

---

*The reality is much more complicated than can be summarized here (see, for example, Ref. 6).

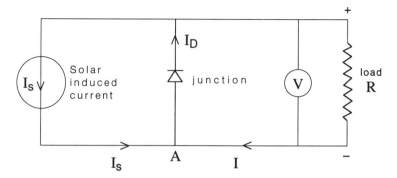

Figure 4.28   Circuit representing the current I_s of the photovoltaic effect, the diode current I_D caused by the potential across the junction and load R

Finally consider a photon entering the solar cell on the thin side. Assume that it has an energy $E > E_g$ so that it creates a particle–hole pair near the junction, e.g. on the n-side. The extra particle (electron) can be readily ignored as there are many of them. The extra hole is added to the small number of holes present and gives an extra current I_s which is positive from n to p across the junction and therefore opposite in sign to the diode current (4.163).

The situation may be characterized by two parallel currents I_s and I_D in opposite directions while the potential decrease V across the junction is caused by an external load R. This is indicated in Fig. 4.28. Conservation of current at A gives

$$I = I_D - I_s = I_0(e^{eV/kT} - 1) - I_s \qquad (4.164)$$

The IV diagram of the solar cell obeys eq. (4.164) and is sketched in Fig. 4.29. For $V \to -\infty$ one finds $I \to -(I_s + I_0)$; for $V = 0$ one finds $I = -I_s$ and for $V \to \infty$ it is clear that $I \to +\infty$ very steeply. The voltage V for which $I = 0$ is called V_{0c}. This value is given by

$$0 = I_0(e^{eV_{oc}/kT} - 1) - I_s$$

$$e^{eV_{oc}/kT} = \frac{I_s}{I_0} + 1$$

$$V_{0c} = \frac{kT}{e} \ln\left(\frac{I_s}{I_0} + 1\right) \qquad (4.165)$$

The power output P for a point M on the diagram of Fig. (4.29) is given by

$$P = IV$$

and is represented by the surface area of the rectangle indicated in the IV diagram.

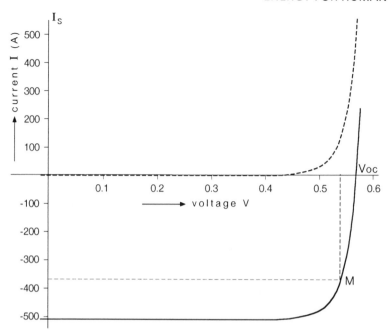

Figure 4.29 *IV* diagram for a solar cell, with realistic Si parameters ($I_0 = 5.9 \times 10^{-8}\,\mathrm{A\,m^{-2}}$; $I_s = 520\,\mathrm{A\,m^{-2}}$; $kT = 0.025\,\mathrm{eV}$). The drawn curve corresponds to insolation, the dashed curve just represents the diode current (4.163)

Efficiency

The efficiency η of a solar cell is defined as

$$\eta = \frac{\text{output power in W m}^{-2}}{\text{incoming radiation in W m}^{-2}} \tag{4.166}$$

The solar spectrum for $T = 5800\,\mathrm{K}$ is reproduced in Fig. 4.30. The energy gap $E_g = 1.12\,\mathrm{eV}$ for Si corresponds with a wavelength $\lambda_g = 1100\,\mathrm{nm}$ by the relations $E = h\nu$ and $c = \nu\lambda$. All photons in the spectrum with $\lambda > \lambda_g$ have too little energy to create a particle–hole pair. In this way 23% of the photons do not contribute to the efficiency η. It may be noted in passing that for materials with a larger energy gap E_g this effect becomes worse.

A second point is that photons with energy $E > E_g$ create a particle–hole pair with energy E. However, only the energy E_g is converted into electric power. The rest, $E - E_g$, is lost as heat in the semiconductor. For Si the two effects together lead to a useful photon energy of only 44% of the solar spectrum, which gives an upper limit for the efficiency (4.166).

Let us make a realistic estimate of the possible efficiency. The current I_s can

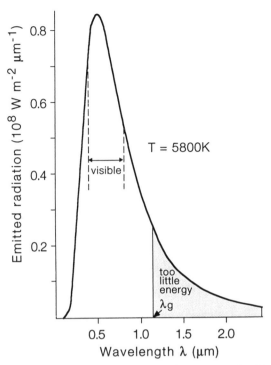

Figure 4.30 The solar spectrum with the energy gap E_g for Si at 1100 nm

be estimated by noting that the average photon energy is 1.48 eV. With an insolation of 1000 W m^{-2} this results in 4.2×10^{21} photons in m^{-2} s^{-1}. Some 77% of them give a photovoltaic effect, resulting in a current of $I_s = 520$ A m^{-2}. The field current I_0 is given in the literature as $I_0 = 5.9 \times 10^{-8}$ A m^{-2}. Equation (4.165) then gives $V_{0c} = 0.57$ volt. These numbers were used to draw Fig. 4.29. The maximum power according to that figure is drawn for a somewhat lower voltage than V_{0c} and a lower current than I_s. The maximum efficiency η of a Si crystal therefore may be about half of 44%, or 22%. It must be mentioned that for Si crystals an efficiency of 23% has been measured.

One could increase the efficiency by using other materials or by increasing the insolation I_s, which would increase V_{0c} according to eq. (4.165). This could be done by focusing sunlight while keeping the temperature constant so as to keep I_0 at its low level.

Cost

Efficiency is not the only factor that counts. Cost is another. Therefore besides Si crystals many varieties of pn junctions are being developed. Using Si one has made polycrystalline materials or even thin films, with a thickness of a few μm. Although their efficiency is usually lower they may be more readily mass produced so the cost per delivered kW h would go down. Besides Si many other atoms or 'compounds' are being studied, sometimes with poisonous materials like As. That will require cautious waste management, if in wide use.

Finally, it should be mentioned that photovoltaic systems deliver d.c. currents. Therefore, when they are coupled to the grid one needs d.c.–a.c. inverters and when used as 'stand-alone' systems one needs a set of batteries as storage. All this adds to the cost.

In the beginning of the 1990s photovoltaic systems were already very competitive in producing electricity when there was no grid connection available. Their maintenance is low as there are no moving parts, their reliability is high, they make no noise (such as with diesel generators) and they do not pollute.

4.4.2 WIND ENERGY

The kinetic energy contained in a unit of volume of air is $\frac{1}{2}\rho u^2$. The amount of energy which passes a unit area perpendicular to the wind velocity **u** in a second is contained in a volume with length u along the direction of the wind velocity. The total energy that passes the unit area per second equals the power that passes the unit area and equals

$$\frac{1}{2}\rho u^3 \tag{4.167}$$

If the wind velocity **u** makes an angle β with the normal **n** on the unit area then the power will be

$$\frac{1}{2}\rho u^3 \cos \beta \tag{4.168}$$

The density ρ is about $1.2\,\mathrm{kg\,m^{-3}}$ but may be put more precisely by using eq. (3.22) with R from Appendix C, which leads to

$$\rho = \frac{p}{287T} \tag{4.169}$$

with p in Pascal and T in Kelvin. Pressure and temperature of the air are usually better to measure than the density itself.

The third power of u in eqs. (4.167) and (4.168) suggests that wind energy should be tapped at places with high wind velocities. As will be shown below this implies that wind turbines are to be placed on towers where the velocity loss due to friction against the ground is small.

The Betz Limit

In order to make an estimate of the power that can be extracted from a wind flow, one may consider a schematically drawn turbine as in Fig. 4.31. It is

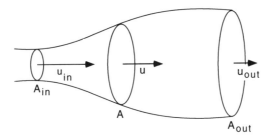

Figure 4.31 Schematic wind turbine. Air enters from the left and leaves at the right losing energy and expanding its cross-section

assumed that air enters the turbine from the left. The undisturbed air far in front of the turbine has a velocity u_{in} passing an area A_{in}. The air leaves at the right and just behind the turbine it has a velocity u_{out} while crossing an area A_{out}. The air flows essentially horizontally; all other components are neglected.

Consider an air mass J_m passing the turbine in a second without change of density ρ. Conservation of mass from A_{in} to A_{out} then leads to

$$J_m = \rho A_{in} u_{in} = \rho A_{out} u_{out} \tag{4.170}$$

Inside the turbine the same mass will pass a cross-section A with velocity u; thus

$$J_m = \rho A u \tag{4.171}$$

Of course the turbine will be constructed so that $u_{out} < u_{in}$ in order to get an energy gain, which means that the area $A_{out} > A_{in}$. For a unit of mass the loss in kinetic energy is written as

$$T_{in} - T_{out} = \tfrac{1}{2}(u_{in}^2 - u_{out}^2) = \tfrac{1}{2}(u_{in} - u_{out})(u_{in} + u_{out}) \tag{4.172}$$

According to Newton's law, the loss of momentum per unit of time of the passing air will be the force exerted on the air by the turbine and this will be equal in magnitude to the force F exerted on the turbine by the air. The loss in momentum per second of the passing air is the mass J_m times its decrease in velocity:

$$F = J_m(u_{in} - u_{out}) \tag{4.173}$$

Following eq. (4.171) a mass $\rho A u$ will pass the turbine in 1 second, so the air will have travelled u metres of length. The air performs work $W = Fu$. This is transferred to the kinetic energy of the turbine. Therefore the energy transferred per unit of mass becomes, using eqs. (4.173) and (4.171),

$$\frac{W}{\rho A u} = \frac{F}{\rho A} = \frac{J_m(u_{in} - u_{out})}{\rho A} = u(u_{in} - u_{out}) \tag{4.174}$$

This should be equal to the loss in kinetic energy of the unit of mass (4.172), which leads to

$$u = \tfrac{1}{2}(u_{in} + u_{out}) \tag{4.175}$$

The velocities u and u_{out} are expressed in terms of u_{in} by a single parameter a as

$$u = u_{in}(1 - a)$$
$$u_{out} = u_{in}(1 - 2a) \tag{4.176}$$

The power P of the turbine equals the energy transferred per unit time. This is expressed in the parameter a by multiplying eq. (4.172) by the mass J_m passing per unit time. This gives

$$P = \tfrac{1}{2}J_m(u_{in}^2 - u_{out}^2) = 2\rho A u_{in}^3 (1 - a)^2 a \tag{4.177}$$

In constructing a turbine, one will consider a certain internal cross-section A and will have a given velocity u_{in}. In the design one then will maximize the ratio

$$C_p = \frac{P}{\tfrac{1}{2}\rho A u_{in}^3} \tag{4.178}$$

which leads to $a = \tfrac{1}{3}$ and $C_p = \tfrac{16}{27}$. This upper limit of the power of a turbine of given cross-section A is called the Betz limit. The *coefficient of performance* C_p of a turbine defined in eq. (4.178) is the ratio of its power output to the input of the wind, shown in the denominator. One observes that the turbine cross-section A is taken instead of A_{in}, as A is known and A_{in} is not. The Betz limit is therefore the upper value of the coefficient of performance of a turbine with which all practical values are compared.

Aerodynamics

The blade of a turbine is shaped somewhat like the wing of an aeroplane, as shown in Fig. 4.32. A flow then produces two forces, a *drag* in the direction of

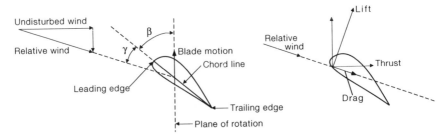

Figure 4.32 The air foil represents a horizontal turbine blade. Shown are the lift and drag forces; horizontal components give the thrust to be absorbed by a tower. As the blade is supposed to move upwards, the relative wind which determines aerodynamics is indicated downwards. The axial symmetry about the undisturbed wind velocity implies that the figure holds for all positions of the blade

the airflow and a *lift* perpendicular to it. These forces are produced with respect to the *relative* airflow; in Fig. 4.32 the turbine blade is assumed horizontal and moving upwards; the relative airflow then has a downwards component as shown.

The air should make an angle γ, the so-called *angle of attack*, with the central cord line of the blade. The air then slows down underneath the blade and speeds up on top. *Bernoulli's law* results in a pressure difference between the top and bottom, the lift force. More precisely, one should say that the shape of the foil produces a vortex around the foil, superimposed on the flow. Along the vortex Bernoulli's law states that $(p + \frac{1}{2}\rho u^2)_{\text{top}} = (p + \frac{1}{2}\rho u^2)_{\text{bottom}}$, which results in the usual expression for the lift:

$$\text{lift} = p_{\text{bottom}} - p_{\text{top}} = \tfrac{1}{2}\rho(u_{\text{top}}^2 - u_{\text{bottom}}^2)$$

The drag force just results from the loss of momentum of air in the horizontal direction. Blades are constructed such that the lift force is 10 to 20 times as big as the drag. In Fig. 4.32 the drag is therefore exaggerated.*

The lift and drag should be decomposed into horizontal and vertical components. The horizontal thrust is absorbed by the tower on which the blades are mounted. The vertical force is the useful force which gives rise to rotation and will eventually drive an electric generator. There is axial symmetry around the unperturbed wind velocity. Therefore the argument holds for all positions of the blade. When one looks at positions on the blade further from the axis the rotational velocity and consequently the relative wind velocity increase; in order to keep the optimal angle of attack one therefore twists the blade a bit or otherwise has to accept loss of performance.

For a high lift force a laminar flow is needed. One therefore tries to avoid turbulence in the wake by choosing a moderate angle of attack.

Wake Effects

In a wind farm many wind turbines are put in an array. The question then is how quickly the velocity reduction due to the lost power is compensated. In the simplest approximation one wind turbine is considered, as in Fig. 4.33 and for the wake it is assumed that the total momentum in a cylinder parallel to the wind velocity is conserved.

The wake is approximated by a cone; at a distance x the velocity $u(x)$ may be found by mass conservation in a cylinder with radius r around the turbine:

$$(r^2 - r_{\text{out}}^2)u_{\text{in}} + r_{\text{out}}^2 u_{\text{out}} = r^2 u(x) \tag{4.179}$$

---

*The maximum ratio between lift and drag may be in the order of 60–150. (See Ref. 7.)

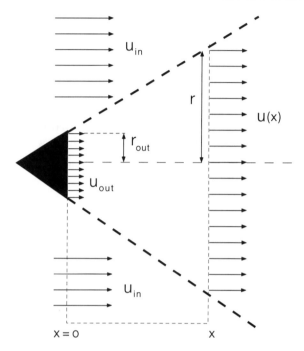

Figure 4.33 Conical approximation of the wake of a wind turbine. Mass in the wake is conserved

With eqs. (4.170), (4.171) and (4.176) one may express r_{out} in the turbine radius R by $r_{out} = R\sqrt{(1-a)/(1-2a)}$. The conical approximation implies $r - r_{out} = \alpha x$ which gives

$$u(x) = u_{in}\left[1 - \frac{2a}{\left(1 + \dfrac{\alpha x}{R\sqrt{(1-a)/(1-2a)}}\right)^2}\right] \qquad (4.180)$$

A semi-empirical expression for α is $\alpha = 1/[2\ln(h/z_0)]$, where h is the height of the turbine and z_0 a roughness parameter to be discussed below. For an altitude of 11 m and a surface roughness of 2.5 cm one finds $\alpha = 0.082$. Using the Betz value $a = 1/3$ one finds at 10 turbine radii a velocity $u = 0.73u_{in}$ and at 20 radii a velocity $u = 0.86u_{in}$. In practice one has to take into account friction with outer layers that give some extra velocities and turbulence which may reduce the wake effect. Equation (4.180) seems a reasonable approximation. Of course, in a real wind park one has to consider a superposition of wakes, which may be calculated in a similar way.

Dependence on Space and Time

A difficulty in using wind energy is that the wind velocity field is not uniform and homogeneous as assumed above, but that it is dependent on position and time. As to the last, the wind velocity is rapidly fluctuating, which causes even bigger fluctuations in the power absorbed by the turbine. This poses strains on the construction of the turbine which should be able to operate under adverse conditions. It also means that the supply of wind energy cannot be planned in accordance with power demands, so if wind energy is to provide more than some 10% of the electricity demand, ways of energy storage should be found.

As to the location of wind turbines, positions will be selected where it is found empirically that high wind velocities exist. The vertical dependence of the wind velocity is one of the properties that is open to some physical analysis.

At the earth's surface the air velocity will be zero as the air molecules stick to the ground. Here one has the laminar boundary layer where friction is the dominant force. The thickness δ of this layer is found by demanding that the Reynolds number

$$Re = \frac{UL}{v} = \frac{u_* \delta}{v} \tag{4.181}$$

defined in eq. (3.50), should be in the order of 10. The velocity u_* in this relation is the *friction velocity* or *shear velocity*; it is related to the tangential stress τ along the surface, which is defined as the force per unit area along the surface (and may be seen as the vertical transport of horizontal momentum per unit time and unit area). As another relevant quantity for the boundary layer will be the air density ρ, dimensional analysis (to be discussed in Section 5.5) suggests the following definition for the friction velocity:

$$u_* = \sqrt{\frac{\tau}{\rho}} \tag{4.182}$$

Above the laminar boundary layer experience shows an increase of velocity with height. The simplest expression would be

$$\frac{\partial u}{\partial z} = \frac{u_*}{kz} \tag{4.183}$$

where u is the dominant horizontal velocity and k is the so-called Von Karman constant, which is in the order of 0.4. Equation (4.183) leads to the logarithmic dependence

$$u = \frac{u_*}{k} \ln z + B = \frac{u_*}{k} \ln \frac{z}{z_0} \tag{4.184}$$

Here, z_0 is a measure for the surface roughness. For $z = z_0$ the equation suggests that u will vanish. In practice z_0 is around 10% of the height of the grass stems that determine the surface roughness. Equation (4.184) holds only approximately for z around z_0. It holds well for heights $z > z_0$ up to some hundred metres. Engineers are usually interested only in the changes of u with height and use an empirical formula

$$\frac{u_1}{u_2} = \left(\frac{z_1}{z_2}\right)^{0.17} \tag{4.185}$$

It is clear that for turbines with blades of 20 metres, the velocity change with height is considerable. Construction problems, however, arise particularly from the turbulent variation of wind velocity in time over the surface area of the circle drawn out by the blades.

4.4.3 WAVES

Ocean waves are caused by winds that blow the irregularities of the ocean surface into running waves. In order to get an estimate of the power connected with ocean waves we consider very deep oceans with gravity as the only active force. A water particle is defined by its equilibrium position (x, y, z, t). In general, the water particle will have a displacement $\mathbf{s}$ from this position. The local pressure $p(x, y, z, t)$ is defined as the deviation from the equilibrium pressure. In the neighbourhood of the surface $(z = 0)$ this will be due to the extra water above, so

$$p = \rho g s_z \tag{4.186}$$

We now calculate the kinetic energy in the wave motion with the simplest assumptions. First, assume that everywhere rot $\mathbf{s} = 0$; then

$$\mathbf{s} = -\nabla\psi \tag{4.187}$$

In the equation of motion (3.32) we only keep the pressure term as the gravity and equilibrium pressure will cancel and the others are ignored:

$$\rho\frac{d\mathbf{u}}{dt} = -\operatorname{grad} p \tag{4.188}$$

The next simplification is the equalling of the time derivative with the local derivative, ignoring the non-linear terms in eq. (3.51). With $\mathbf{u} = d\mathbf{s}/dt \approx \partial\mathbf{s}/\partial t$ this leads to

$$\frac{\partial^2\mathbf{s}}{\partial t^2} = -\frac{1}{\rho}\nabla p \tag{4.189}$$

and with eq. (4.187)

$$\frac{\partial^2}{\partial t^2}(-\nabla\psi) = -\frac{1}{\rho}\nabla p \tag{4.190}$$

or

$$\rho \frac{\partial^2 \psi}{\partial t^2} = p \qquad (4.191)$$

With eqs. (4.186) and (4.187) this gives

$$g \frac{\partial \psi}{\partial z} = - \frac{\partial^2 \psi}{\partial t^2} \qquad (4.192)$$

Assume that there is no net outflow of mass, so div s = 0 and with eq. (4.187) it follows that div grad $\psi = 0$. If ψ only depends on one horizontal direction x, the vertical z and time t, this gives

$$\frac{\partial^2 \psi}{\partial x^2} + \frac{\partial^2 \psi}{\partial z^2} = 0 \qquad (4.193)$$

A solution of this equation for a wave with propagation velocity v is

$$\psi = \frac{a}{k} \sin k(x - vt) e^{kz} \qquad (4.194)$$

For $z \to -\infty$ this vanishes, as it should. It is now easy to find the displacement s from eq. (4.187) and the particle velocity $ds/dt \approx \partial s/\partial t$.

The kinetic energy T of a unit volume is found by substitution:

$$T = \tfrac{1}{2}\rho \left(\frac{\partial s}{\partial t} \right)^2 = \tfrac{1}{2}\rho a^2 k^2 v^2 e^{2kz} \qquad (4.195)$$

For a vertical column of a unit surface area the total kinetic energy T_{col} is found by integration of eq. (4.195):

$$T_{col} = \tfrac{1}{4}\rho a^2 k v^2 \qquad (4.196)$$

Combining eqs. (4.192) and (4.194) gives, for the wave velocity v,

$$v^2 = \frac{g}{k} \qquad (4.197)$$

Finally, one has to multiply T_{col} by the velocity v to find the power P, the energy passing a metre perpendicular to the propagation per second:

$$P = \tfrac{1}{4}\rho g v a^2 \qquad (4.198)$$

In practice one can only use a fraction of this power, as it is not possible to use the whole column.

Converters

In order to get an idea of a wave energy conversion device we show in Fig. 4.34 an apparatus designed and applied by Masuda. The sea water performs a vertical

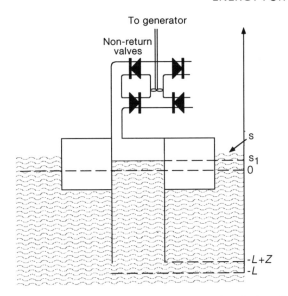

Figure 4.34 Masuda's pneumatic wave energy conversion device. (Reproduced by permission of Academic Press Limited from Bent Sørensen, *Renewable Energy*, Academic Press, London, 1979, Fig. 257, p. 450)

motion $s(x, t)$. A buoy floating on the water has an open tube in its middle through which the sea water may enter. The waves (4.194) passing the buoy will cause a harmonic vertical movement $Z(x, t)$ of the buoy and a similar vertical movement $s_1(x, t)$ inside the tube. The three movements will be out of phase and will have different amplitudes:

$$s = a \sin k(x - vt)$$
$$Z = Z_0 \sin k(x - vt - \delta_z) \qquad (4.199)$$
$$s_1 = s_0 \sin k(x - vt - \delta_1)$$

This means that with respect to the buoy the tube water will perform an up-and-down movement. One may observe this effect in old-fashioned fishing-boats with a well in the middle; the water goes up and down with respect to the boat. The valves in Fig. 4.34 are positioned in such a way that with both up and down movement the air passes the propeller near the turbine in the same, upward, direction.

4.4.4 BIOENERGY

In eq. (1.2) we gave the basic equation for photosynthesis by which carbohydrates ($C_6H_{12}O_6$) are formed:

$$6\,H_2O + 6\,CO_2 + 4.66 \times 10^{-18}\,J \rightleftharpoons C_6H_{12}O_6 + 6\,O_2 \qquad (4.200)$$

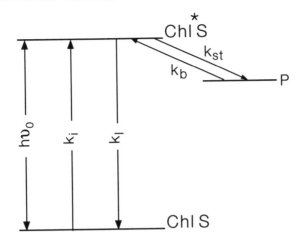

Figure 4.35 The basic mechanism of photosynthesis. A molecule indicated by S is present with chlorophyll molecules Chl, together represented as Chl S. It is excited with a rate k_i to a state Chl*S. The excited chlorophyll state either decays with (loss) rate k_l or turns molecule S into a molecule P with (storage) rate k_{st}. Energy is stored in P, but it may decay back with rate k_b

This reaction equation immediately shows the advantage of using photosynthesis to absorb sunlight. It not only uses solar energy, but contrary to solar cells and wind turbines the mechanism also stores energy in the form of chemical free energy. This is summarized by the simplified scheme of Fig. 4.35.

A ground state Chl S consisting of a so-called substrate molecule S and chlorophyll molecules Chl may absorb an energy $h\nu_0$ and gets into an excited state Chl*S. This state either returns to the ground state with loss rate k_l (which includes loss due to conversion into heat and triplet formation) or uses its energy to change S into a product molecule P with storage rate k_{st}. Here the energy is stored, but it may also return to state Chl*S with rate k_b. It should be remembered that a rate is defined as the number of transitions per molecule per second. Typical numbers are given in Table 4.4.

Table 4.4 Typical rates in photo-synthesis (s^{-1})

k_l	k_{st}	k_i (high)	k_i (low)
10^9	2×10^{10}	10	10^{-4}

Efficiency

Figure 4.35 presents a very schematic picture of reaction (4.200). In reality one has oxidation of H_2O:

$$2 H_2O \rightarrow O_2 + 4 H^+ + 4 e^- \qquad (4.201)$$

and reduction of CO_2:

$$CO_2 + 4 H^+ + 4 e^- \rightarrow (H_2CO) + H_2O \qquad (4.202)$$

of which the net effect produces reaction (4.200). One needs four photons to drive reaction (4.201) and four other photons to drive reaction (4.202). So at least eight photons are needed for process (4.200) to run. The wavelength of these photons are in the range $400 < \lambda < 700$ nm and this is called the photosynthetically active region. Let us take the lowest energy photon with $\lambda = 700$ nm. Its energy is $E = h\nu = hc/\lambda = 0.284 \times 10^{-18}$ J. Eight quanta have energy 2.27×10^{-18} J and store an energy per H_2O molecule of 4.66×10^{-18} J$/6 = 0.78 \times 10^{-18}$ J. The efficiency of the storage process in light of 700 nm is therefore $0.78/2.27 = 0.34$.

In real sunlight most of the photons absorbed have higher energy which reduces the efficiency by a factor of two. There is some 20% loss by reflection and 40% loss due to respiration. Therefore the maximum efficiency of photosynthesis would be $0.34 \times 0.5 \times 0.8 \times 0.6 = 8\%$. A more accurate calculation gives 6.7% or—if one assumes that one needs ten photons instead of eight—it becomes 5.5%.* Farming in fertile regions gives an efficiency of about 1% if one divides the total annual output by the annual solar input. It should be noted that the energy content of fertilizers should be subtracted from the biomass energy output in order to calculate the net energy gain.

Storage

In order to calculate the maximum attainable storage efficiencies of photosynthesis at a given light intensity one has to study Fig. 4.35 in some detail. The photosynthetic process is described as a single-step conversion of light into Gibbs free energy, defined in eqs. (4.68) and (4.69). Per molecule the Gibbs free energy is called the chemical potential μ. It depends on the absolute temperature by

$$\mu(T) = \mu^0 + kT \ln [\text{Chl}] \qquad (4.203)$$

where we used the ground state of chlorophyll with concentration [Chl] as an example. The difference $\Delta\mu$ in chemical potential for the excited state and ground

---

*For eq. (4.202) to run, one needs ATP, of which the formation also requires two photons. Hence it takes ten photons to run (4.200). (See Ref. 8.)

state may be written as

$$\Delta\mu = \mu^0(Chl^*) + kT \ln [Chl^*] - \mu^0(Chl) - kT \ln [Chl] \qquad (4.204)$$

$$\Delta\mu = h\nu_0 + kT \ln \frac{[Chl^*]}{[Chl]} \qquad (4.205)$$

The difference in chemical potential at $T = 0$ is just the excitation energy $h\nu_0$ for the Chl* level. We note in passing that, in the dark, at thermal equilibrium $\Delta\mu = 0$, and we find the Boltzmann distribution (4.162):

$$[Chl^*] = [Chl]e^{-h\nu_0/kT} \qquad (4.206)$$

Let us first ignore the back reaction $P \rightarrow S$ in Fig. 4.35. In *steady state* the formation of Chl* equals its decay:

$$k_i[Chl] = (k_1 + k_{st})[Chl^*] \qquad (4.207)$$

The yield ϕ_{st}^0 of the storage process is given by

$$\phi_{st}^0 = \frac{k_{st}}{k_1 + k_{st}} \qquad (4.208)$$

For photosynthesis to work one should have $k_{st} \gg k_1$ or $\phi_{st}^0 \approx 1$, which is indeed the case in nature (cf. Table 4.4).

Let us now switch on the sunlight and take into account the unavoidable back reaction with the rate k_b. When the light is switched off again we will have a steady state if the formation of equals its decay:

$$k_b[P] = (k_1 + k_{st})[Chl^*S]_{dark} \qquad (4.209)$$

as the formation from Chl S vanishes. Since $k_{st} \gg k_1$, eq. (4.209) looks like an equilibrium between P and Chl^*S_{dark}, which implies (roughly) the same chemical potential:

$$\mu(Chl^*S)_{dark} = \mu(P) \qquad (4.210)$$

With the light switched on, the left-hand side of eq. (4.209) will also contain the formation of excited states by light absorption of Chl S and we find for steady state

$$k_b[P] + k_i[Chl\,S] = (k_{st} + k_1)[Chl^*S]_{light} \qquad (4.211)$$

The efficiency for storage ϕ_{st}^{ss} in steady state with light becomes

$$\phi_{st}^{ss} = \frac{k_i[Chl\,S] - k_1[Chl^*S]_{light}}{k_i[Chl\,S]} \qquad (4.212)$$

Using eqs. (4.211), (4.209) and (4.208) this becomes

$$\phi_{st}^{ss} = \phi_{st}^0 - \frac{k_1}{k_i}\frac{[Chl^*S]_{dark}}{[Chl\,S]} \qquad (4.213)$$

Let us write down eq. (4.203) for $Chl*S_{dark}$ and $Chl S$:

$$\mu(Chl*S)_{dark} = \mu^0(Chl*S)_{dark} + kT \ln [Chl*S]_{dark}$$

$$= \mu^0(Chl S) + hv_0 + kT \ln [Chl*S]_{dark} \qquad (4.214)$$

$$\mu(Chl S) = \mu^0(Chl S) + kT \ln [Chl S] \qquad (4.215)$$

In eq. (4.214) we used μ^0 (Chl S) to refer to the ground state and found that it is independent of the presence or absence of light. We now use eq. (4.210) to put $\mu(P)$ equal to eq. (4.214) and subtract (4.215) and find that

$$\ln \frac{[Chl*S]_{dark}}{[Chl S]} = \frac{\mu(P) - \mu(Chl S) - hv_0}{kT} = \frac{\Delta\mu_{st} - hv_0}{kT} \qquad (4.216)$$

where $\Delta\mu_{st}$ is the free energy stored. From eqs. (4.216) and (4.213) one can determine the steady state storage rate ϕ_{st}^{ss} for realistic cases.

However, we can also argue the other way round. We obviously require a high value of storage quantum efficiency ϕ_{st}^{ss}, say $\phi_{st}^{ss} = 0.9$. We find from eq. (4.206) and Table 4.4 that $\phi_{st}^0 = 0.95$. With the data from Table 4.4 we find that for strong sunlight

$$\Delta\mu_{st} - hv_0 = -0.54 \, eV \qquad (4.217)$$

while for very weak sunlight

$$\Delta\mu_{st} - hv_0 = -0.82 \, eV \qquad (4.218)$$

Obviously, there is only one value of $\Delta\mu_{st}$. However, it should be considerably lower than the energy of the incoming photons (which ranges around 2 eV) to have a good storage efficiency also in the lower parts of the plant. Remember that photosynthesis has to have both a high energy conversion efficiency and a high storage efficiency. Therefore, nature had to compromise.

Stability

In a period of time Δt with the light switched on the amount of Gibbs free energy stored will be

$$G_{st} = k_i \Delta t [Chl S] \phi_{st}^{ss} [\mu(P) - \mu(Chl S)] \qquad (4.219)$$

In the dark the loss of free energy per second through the unavoidable back reaction is given by

$$R_1 = k_1 [Chl*S]_{dark} [\mu(P) - \mu(Chl S)] \qquad (4.220)$$

This represents night or winter and a lower limit for loss after harvesting. It gives a time

$$t_1 = \frac{G_{st}}{R_1} = \frac{k_i \Delta t [Chl S] \phi_{st}^{ss}}{k_1 [Chl*S]_{dark}} \qquad (4.221)$$

in which the stored energy is lost again (Exercise 4.28).

Use of Biomass

If one does not have the time to wait for fossilization there are other means to harvest the bioenergy. The easiest way is drying biomaterials and burning them; another way is composting. More sophisticated methods include the production of biogas, essentially CH_4, from carbohydrates. This is still done in a traditional way at some farms that are not connected to the national grid (they have a pressure tank in a ditch where rotting plants produce methane). Gas, of course, is easier to use for cooking and heating than raw biomaterials. Other possibilities are the use of manure to produce biogas. A cow, for example, is able to produce $1 \, m^3$ or 26 MJ of biogas per day.

Finally ethanol and other liquid fuels can be produced from biological raw materials. They may then be used in cars for combustion.

An interesting way to produce energy, in which you can kill two birds with one stone, is to gasify biomass, such as sugar cane, and use the gas to fuel a gas turbine, which produces electricity and heat very efficiently. Part of the heat is used as input for the gasification process. In this way an equilibrium could be maintained: plant as many sugar canes as are being gasified. There is thus no net CO_2 production as for each carbohydrate molecule dissociating into CO_2 another is formed, binding CO_2. Of course care should be taken that no other greenhouse gases are allowed to escape. The main net effect is that solar energy is converted into electricity.

It should be mentioned here that CH_4 may be used to produce H_2 gas which is stored then or transported for use in fuel cells to produce electricity (cf. Section 4.4.5).

Still in the realm of speculation and research is the idea of a bio-photocell. In the process of photosynthesis a potential difference is built up between the two sides of a membrane. This Coulomb energy is eventually converted into chemical energy. Perhaps it is possible to stop the process once the charge separation has been performed and to construct a battery with high stability: a bio-photocell. Another option would be to use photosynthesis to produce marketable C_xH_y compounds without any intermediary steps.

4.4.5 HYDROPOWER AND FUEL CELLS

Although not much space will be devoted to these topics, they should be mentioned as they are important in producing renewable energy.

Hydropower

Hydropower stations are usually sited at a dam in a river where the water passes out of dam by a tube and a turbine. The maximum power output may be calculated by assuming that all potential energy mgh is converted into kinetic energy and all of this into electric energy. With the density of water ρ and a flow of $Q\,m^3\,s^{-1}$ this leads to

$$P = \rho g h Q \quad J\,s^{-1} \approx 10 h Q \quad kW \tag{4.222}$$

On a much smaller scale one could convert part of the kinetic energy of a river to rotational motion and hence into electricity. For a velocity u and a flow of $Q\,m^3\,s^{-1}$ that is tapped one finds

$$P = \tfrac{1}{2} Q \rho u^2 \approx \tfrac{1}{2} Q u^2 \quad kW \tag{4.223}$$

The turbines to be used must be tailor-made to the application. For large power stations their efficiency may be higher than 90%.

The Fuel Cell

In a fuel cell chemical energy is converted directly into electrical energy just like in a battery. The advantage over a battery is that the fuel can be supplied continuously. The fuel cells most discussed and under development are those based on oxidation of hydrogen, methane or natural gas. All are gases, so are easily transportable. Hydrogen may be produced by hydrolysis of water by electricity, while methane is produced in nature by rotting biomass, a process that of course can be stimulated. Natural gas existing essentially of methane is widely found in nature.

In order to explain the principle, a hydrogen–oxygen fuel cell is sketched in Fig. 4.36 with a acidic liquid electrolyte between the anode and cathode. Hydrogen gas enters a porous anode where it comes into contact with an electrolyte and a catalyst and produces positive ions (protons) according to the reaction

$$2\,H_2 \to 4\,e^- + 4\,H^+ \tag{4.224}$$

The protons move through the electrolyte to the cathode where the reaction

$$4\,e^- + 4\,H^+ + O_2 \to 2\,H_2O \tag{4.225}$$

discharges the ions, so the net effect is the oxidation of hydrogen to water according to

$$2\,H_2 + O_2 \to 2\,H_2O \tag{4.226}$$

The electrons move from the anode to the cathode by an outer load, producing a potential difference. Thus the chemical energy is converted into an electrical energy W_e. As explained in Section 4.2.1, the stored chemical energy is represented by the enthalpy H of the system and in the oxidation (4.226) the enthalpy increase ΔH is negative, $-\Delta H$ being the energy that is liberated. This energy is used to perform electric work ΔW_e; besides heat $-T\,\Delta S$ will be rejected. The consequence is that

$$-\Delta H \geqslant \Delta W_e - T\,\Delta S \tag{4.227}$$

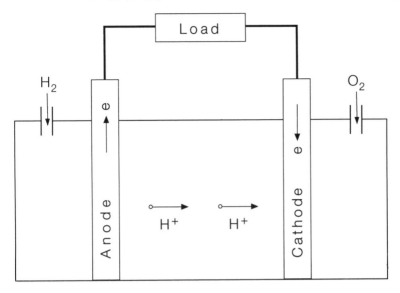

Figure 4.36 Example of a fuel cell. Hydrogen enters at the anode where it produces positive ions transported in an acidic electrolyte to the cathode, where it reacts with entering oxygen to produce water

where the inequality sign shows that chemical energy may be lost due to irreversibility etc. as well. It follows that

$$\Delta W_e \leqslant -(\Delta H - T\Delta S) = -\Delta G \qquad (4.228)$$

similar to relation (4.67). The maximum efficiency of the fuel cell becomes

$$\eta_{max} = \frac{output}{input} = \frac{\Delta W_e}{-\Delta H} = \frac{-\Delta G}{-\Delta H} \qquad (4.229)$$

From tables one may deduce that the maximum efficiency for the hydrogen fuel cell would be 0.83, whereas for the methane fuel cell it would be 0.92. For amusement one may note that in some cases $-\Delta G > -\Delta H$, which would mean that the fuel cell would cool the environment and would use the heat to produce electricity.

The voltage V of a fuel cell may be deduced from the tabulated Gibbs free energy by using eq. (4.228). Consider n moles of hydrogen as in eq. (4.225); the Gibbs free energy ΔG can then be written as

$$\Delta G = n\mu = nN_A e\Phi \qquad (4.230)$$

where N_A is the Avogadro number and $e\Phi$ the Gibbs free energy per ion. The electrical energy ΔW_e may be written as

$$\Delta W_e = n_e N_A e V \tag{4.231}$$

where n_e is the number of moles of electrons. As $n = n_e$ it follows from eqs. (4.230), (4.231) and (4.228) with the equality sign that

$$V = \Phi = \frac{\Delta G}{n N_A e} \tag{4.232}$$

From the tables one may read that the formation Gibbs free energy of water equals $237 \times 10^3 \, \text{J mol}^{-1}$; for one molecule of water one has two electrons exchanged, so $n_e = 2$. Substitution gives

$$V = 1.23 \, \text{volt} \tag{4.233}$$

4.5 NUCLEAR ENERGY

The energy to be gained by controlled nuclear reactions originates from the beginning of the solar system some 10^{10} years ago. The idea is explained in its simplest way by looking at the graph of binding energy per nucleon versus the mass number A for the stable or very long lived isotopes as in Fig. 4.37. The curve exhibits a maximum around $A = 60$, in the vicinity of ^{56}Fe. Even a superficial look at Fig. 4.37 shows that fissioning of one nucleus of $A = 235$ into two of $A = 118$ would liberate about 1 MeV per nucleon, which would be 236 MeV in total. Although, as we shall see below, fissioning in practice does not

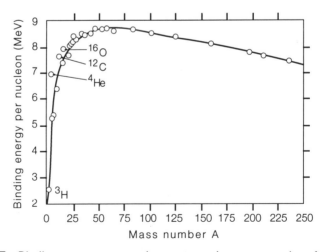

Figure 4.37 Binding energy per nucleon versus the mass number A for stable nuclei

give two equal nuclei, the order of magnitude is correct. In fact, for the fission of $A = 235$ it averages at about 200 MeV. Similarly fusion of light nuclei could give an energy gain as shown on the very left of the graph.

In order to appreciate the order of magnitude of the energy released one should remember that combustion of fossil fuels takes place by chemical reactions, i.e. by rearranging electrons in atomic or molecular orbits with energies in the order of eV. As the number of electrons is in the order of the number of nucleons, the use of nuclear power gives a gain of energy with a factor 10^6, which in principle would be reflected in a reduction of the volume of fuel required for a given amount of energy by the same factor and a reduction in the volume of waste by the same factor as well. The traditional argument in favour of nuclear energy is that this factor of a million buys a lot of safety facilities.

The factor of a million makes the difference between a nuclear society and a traditional industrial society. It is interesting to compare this with the energy used in a pre-industrial society, which would use gravity. The free fall of a nucleon from a height of 10 m representing hydropower or human labour produces an energy of 10^{-6} eV, which is again a factor of a million smaller than the energy released by chemical reactions.

In this section we first discuss in Section 4.5.1 how a nuclear power station works. In the next section a short discussion of nuclear fusion follows. Then, in Section 4.5.3 there is a discussion of the health aspects and finally in Section 4.5.4 some points about the nuclear fuel cycle including the management of radioactive waste.

4.5.1 POWER FROM NUCLEAR FISSION

The essentials of the nuclear fission process are described in textbooks on nuclear physics. Operation of a nuclear reactor in order to produce electricity belongs to the field of reactor physics. Below the essentials of how a reactor works will be discussed.

A fission process initiated by slow neutrons on ^{235}U may be written as

$$^{235}U + n \text{ (slow)} \rightarrow {}^{236}U \rightarrow X + Y + \nu n \text{ (fast)} \tag{4.234}$$

A slow neutron is captured by the ^{235}U nucleus; a compound nucleus ^{236}U is formed with an energy high enough to fission spontaneously into two reaction products X and Y and a number of fast neutrons ν. For ^{235}U this number averages at $\nu \approx 2.43$; for other fissionable materials it may be slightly different (for ^{239}Pu it amounts to 2.87 and for ^{233}U to 2.48). The fast neutrons have energies of on average 2 MeV. The slow neutrons initiating the fission have an energy corresponding to thermal equilibrium with the surroundings; for $T = 293$ K the maximum of the Maxwell distribution is found at $kT = 0.025$ eV.

The fission products X and Y, for which many pairs occur in practice, usually decay further. In some instances this gives rise to *delayed neutron emission*. The fission product ^{87}Br, for example, has a half-life against β decay of 55.6 s. Part of that decay feeds an excited state of ^{87}Kr, which then rapidly emits a neutron. About 0.65% of the neutrons emitted by the compound nucleus ^{236}U fission are delayed by a time $t_d \approx 9$ s.

The gain in binding energy for the process (4.234) is around 200 MeV. The bulk (165 ± 5 MeV) goes into the kinetic energy of the large fission fragments; these have a very small mean free path ($\leqslant 1$ mm), so this energy is rapidly converted into heat. In a fission reactor this heat is harnessed into electric power by means of a heat engine (Section 4.2.3). It obviously pays to operate the power plant at a high temperature. In Fig. 4.38 a nuclear reactor system is represented schematically. One notices a reactor core with a moderator to slow down the fast neutrons until they are so slow that process (4.234) can occur again. One observes the coolant of the reactor core, which generates steam for a turbine by means of a heat exchanger. This coolant sometimes acts as a moderator as well. The reflector and the control rods shown in the figure will be discussed later. Some typical parameters for operating reactors are shown in Table 4.5, from which one can read the type of fuel used, its enrichment, the moderator and the coolant.

In order to discuss the fission reactor we need a few definitions. We remind the reader that the decay constant λ is defined by

$$\mathrm{d}N = -\lambda N \, \mathrm{d}t \tag{4.235}$$

as the number of nuclei $\mathrm{d}N$ decayed in a time $\mathrm{d}t$ is proportional to the number N. This equation leads to

$$N = N_0 \mathrm{e}^{-\lambda t} \tag{4.236}$$

and the half-life when $N = N_0/2$ is given by

$$T_{1/2} = \frac{\ln 2}{\lambda} \tag{4.237}$$

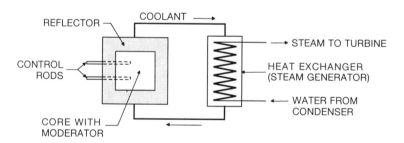

Figure 4.38 Scheme of a nuclear reactor system. The heat from the reactor core is transported by a coolant. A reflector tries to keep as many neutrons in the core as possible, whereas the control rods can absorb superfluous neutrons, if necessary

Table 4.5 Typical data on current nuclear reactors.[a] (Summarized and reproduced by permission of John Wiley & Sons from James J. Duderstadt and Louis J. Hamilton, *Nuclear Reactor Analysis*, Wiley, New York, 1992, pp. 634–5).

	PWR	BWR	CANDU	HTGR
Fuel	UO_2	UO_2	UO_2	UC, ThO_2
Enrichment (% ^{235}U)	≈ 2.6	≈ 2.9	Natural	93.5
Moderator	H_2O	H_2O	D_2O	Graphite
Coolant	H_2O	H_2O	D_2O	He
Electric power (MW_e)	1150	1200	500	1170
Coolant out (°C)	332	286	293	755
Maximum fuel temperature (°C)	1788	1829	1500	1410
Net efficiency (%)	34	34	31	39
Pressure in reactor vessel (bar = 10^5 Pa)	155	72	89	50
Conversion ratio[b]	≈ 0.5	≈ 0.5	≈ 0.45	≈ 0.7
Specific power (MW_{th}/ton fuel)	37.8	25.9	20.4	77

[a]PWR = pressurized water reactor where the pressure prevents steam formation. Steam is produced in a second loop by a heat exchanger. BWR = boiling water reactor, where the steam is generated in the reactor itself. CANDU = Canadian deuterium–uranium reactor. HTGR = high-temperature gas-cooled reactor.
[b]The conversion ratio shown is the number of fissible Pu nuclei formed per fission from the present ^{238}U and Pu nuclei in the reactor.

The mean lifetime is defined as

$$\frac{1}{N_0}\int t(-\,\mathrm{d}N) = -\frac{1}{N_0}\int_0^\infty t\,\frac{\mathrm{d}N}{\mathrm{d}t}\,\mathrm{d}t = \frac{1}{\lambda} \qquad (4.238)$$

The cross-section σ of a nuclear reaction like (4.234) is measured in barns (1 barn = 10^{-28} m^2); it expresses the area by which an incident particle 'sees' the nucleus. For the process (4.234) the geometric cross-section would be 1.7 barns; the measured cross-section for thermal neutrons amounts to 582 barns, which indicates that the reaction is very much favoured.

Process (4.234) is an example of fission; the corresponding cross-section is indicated as σ_f. For absorption one uses σ_a, for scattering σ_s, etc. In a reactor one has to deal with many nuclei, say N per unit of volume. We introduce the *macroscopic cross-section* by the upper case letters Σ, Σ_f, Σ_a, Σ_s, etc., defined by

$$\Sigma = N\sigma \qquad (4.239)$$

and correspondingly for the other Σ values.

These quantities are useful, as they are directly related to the mean free path for a certain process. Consider a half-infinite piece of material with its boundary in the yz plane. A beam of neutrons with intensity I_0 (the number of neutrons

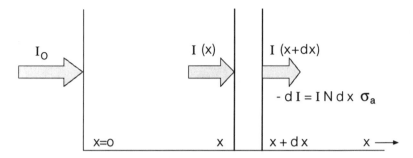

Figure 4.39 The macroscopic cross section for absorption Σ_a is defined as the relative decrease of intensity $-dI/dx$ at position x, due to absorption only. So $\Sigma_a = N\sigma_a$

passing 1 m² per second) is entering from the negative x direction, as is illustrated in Fig. 4.39. Consider only absorption of neutrons with cross-section σ_a. At a certain position x the intensity of the beam is decreased to $I(x)$. Over the next slab dx with surface area 1 m² one finds $N\,dx$ absorption centres. Therefore, the relative intensity decreases by the total area the beam 'sees':

$$-dI = IN\,dx\sigma_a = I\Sigma_a\,dx \tag{4.240}$$

$$I = I_0 e^{-\Sigma_a x} \tag{4.241}$$

The number of neutrons absorbed in the layer dx from eq. (4.240) covered a path of length x. The mean free path λ_a of the incident neutrons therefore becomes

$$\lambda_a = \frac{1}{I_0}\int x(-dI) = \frac{1}{I_0}\int_0^\infty I\Sigma_a x\,dx = \frac{1}{\Sigma_a} \tag{4.242}$$

Let us now take a closer look at a fission reactor based on uranium; for natural uranium one has 99.3% of ^{238}U and 0.7% of the lighter isotope ^{235}U. For reasons that will become clear soon, the fraction of ^{235}U is often enriched up to a few %.

We start by considering n slow neutrons to be absorbed by the fuel, in our case uranium.* Some of them will be captured by ^{235}U without fission (cross-section σ_c) and some will be captured by ^{238}U without fission. Therefore, the actual number of fast neutrons produced is smaller than vn. Assume that there are $N(235)$ atoms of ^{235}U and $N(238)$ of ^{238}U. Only the former will be fissioned by thermal neutrons. The number η of fast neutrons per slow neutron

---

*In actual life ^{238}U also captures neutrons, leading by means of ^{239}U and ^{239}Np to ^{239}Pu (cf. Fig. 4.49). This ^{239}Pu is in itself fissionable, and at the end of the fuel cycle it produces 50% of the fissions. The plutonium has properties slightly different to those of ^{235}U, which implies that one must adjust the parameters of the reactor in the course of the fuel lifetime. This complication does not influence the derivation given here.

absorbed is the fission yield

$$\eta = v \frac{N(235)\sigma_f(235)}{N(235)[\sigma_f(235) + \sigma_c(235)] + N(238)\sigma_c(238)} \qquad (4.243)$$

These ηn fast neutrons should be forced to slow down by collisions with a so-called *moderator*. The cross-section for fission of ^{235}U is very small for neutron energies of MeV; with decreasing energy it increases until it reaches the huge thermal cross-sections. In elementary mechanics courses it is shown that a mass of one unit, which collides with a mass of A units, loses a fraction $(A + 1)^{-1}$ of energy. Therefore the mass A of the moderator should be as small as possible The neutrons should slow down quickly, otherwise there is a chance that they will be lost in processes other than (4.234), as we will see shortly. In practice one uses H_2O (easily available, but it may absorb neutrons itself), heavy water D_2O (expensive, but it has only a small cross-section σ_a for neutron absorption; if it does absorb it forms radioactive tritium which might leak away) and carbon in the form of graphite (easily available, a small σ_a, but with $A = 12$ on the edge). From Table 4.5 it is seen that water used as a moderator also acts as a coolant. For safety reasons the pressure in the reactor should not be above some 100 bar and therefore Fig. 4.17 shows that water-moderated reactors do not operate above 300 °C.

Let us follow the adventures of the ηn fast neutrons. Before slowdown some of them will induce fission in ^{238}U and ^{235}U, leading to a *fast fission factor* ε which is slightly larger than 1. This results in $\varepsilon\eta n$ fast neutrons. For a finite size reactor a fraction l_f will leak out, leading to

$$\eta\varepsilon(1 - l_f)n$$

neutrons available for slowdown. The reflector shown in Fig. 4.38 should reduce the leakage l_f.

During the slowdown the neutrons face obstacles. In particular there are some sharp resonances in the absorption cross-section $n + ^{238}U$ not leading to fission. If during slowdown a neutron happens to have a resonance energy it may be readily lost for the reaction process. This is taken into account by introducing a *resonance escape probability* p. Obviously, a light moderator would do better here than a heavy one, as the resonance energy may be jumped.

After slowdown some of the slow neutrons may escape the reactor before doing their duty; this is taken care of by a leakage factor l_s for slow neutrons. This results in

$$\eta\varepsilon p(1 - l_f)(1 - l_s)n$$

slow neutrons. Finally, only a fraction f, the so-called *thermal utilization factor*, will be absorbed in the fuel, the rest being absorbed in the moderator and the cladding of the fuel elements. So, the number of neutrons kn available for another round of fission becomes

$$kn = \eta \varepsilon p f (1 - l_f)(1 - l_s) n \tag{4.244}$$

Here k is called the *multiplication factor* of the reactor. For a very large reactor, theoretically infinitely big, the leakage factors vanish, leading to the four-factor formula

$$k_\infty = \varepsilon \eta p f \tag{4.245}$$

For a typical reactor this will be in the order of 1.10 or 1.20. Taking into account the leakage factors the factor k could obtain the value $k = 1$, required for stable operation.

Let us look in more detail at eq. (4.245). The quantities ε and η are both larger than 1, whereas p and f are smaller than one. As to η, by virtue of (4.243) it will increase with the enrichment of the uranium. For natural uranium it amounts to 1.33, for 5% enriched uranium it becomes $\eta = 2.0$, close to the value for pure ^{235}U of 2.08.

The fissioning material is traditionally prepared in long rods, set in a moderating medium. Let us assume that the moderator/fuel ratio increases. Then the thermal utilization factor f decreases, but the chance p to escape resonance absorption increases (there are fewer absorbing nuclei and the better moderation gives a greater chance to jump the resonances). The result is that the product pf has a maximum at a certain ratio of moderator to the fuel. This is shown in Fig. 4.40.*

The peak property of k_∞ leads to a straightforward stability consideration. In Fig. 4.40 a sketch is given for natural uranium with D_2O as the moderator. One observes that the peak value of k_∞ is larger than one. When leakage is taken into account the resulting multiplication factor k will exhibit a similar peak-like behaviour. Assume that the reactor of Fig. 4.40 operates at a certain value of N_{mod}/N_U and $k = 1$ and that the temperature rises. In cases where water acts as the moderator the water will expand or boil, resulting in a smaller ratio of N_{mod}/N_U. If the reactor operated to the left of the maximum, the value of k would go down, becoming smaller than 1, and the reactor would slow down. If the reactor operated at the right of the peak, the value of k would rise as well, resulting in $k > 1$ and in an increase in the reactor power, leading to a runaway effect.

---

*The magnitude of this ratio depends on the diameter d. The resonance absorption determining p is so strong that it happens at the surface of the fuel elements when a slow neutron escapes from the moderator. For a constant amount of fuel, increasing d means decreasing the relative surface area and increasing p. It turns out that for a combination of natural uranium and graphite as the moderator the peak in k_∞ as a function of the moderator/fuel ratio lies at $k_\infty = 1.05$ for $d = 3$ cm while the mean free path of the neutrons in the moderator is in the order of 54 cm. This was used to build such a pile in Chicago in 1942. With some leakage this could give $k = 1.0$. Of course, in designing reactors the first task is to look at k_∞.

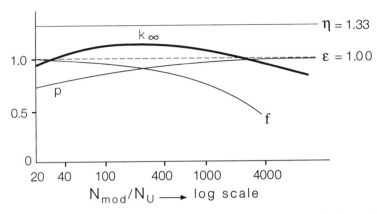

Figure 4.40 Reactor physical parameters for natural uranium as a fuel and D_2O as a moderator. Horizontally one finds the ratio of their atoms or molecules. (Reproduced by permission of H. van Dam/J. E. Hoogenboom from Dr J. E. Hoogenboom lecture notes, IRI, Delft, Netherlands)

The Reactor Equations

For a finite reactor the leakage of fast and slow neutrons out of the reactor has to be taken into account. It will depend on the neutron fluxes at the boundaries of the reactor which have to be calculated by means of a diffusion equation. In practice one writes down the equations for a finite reactor and requires the reactor to run stationary. The time dependence of the fluxes of neutrons then should disappear and the so-called critical size of the reactor is found. In principle one should distinguish neutrons of different energies. For simplicity we assume that only one energy is present. Below, we indicate the differential equations and their solutions. As variable one may choose either the neutron flux ϕ or the density n. As for a constant velocity u, they are proportional by $\phi = nu$, which amounts to the same. Below we shall use the neutron density n.

Consider a volume element. Conservation of the number of neutrons may be expressed as

$$\partial n/\partial t = \text{sources} - \text{absorption} - \text{leakage} \qquad (4.246)$$

Let us write down the terms on the right and start with the absorption term. It may be written as $\Sigma_a nu$ for according to (4.242) $1/\Sigma_a$ is the mean free path. The velocity u divided by this number would be the number of absorptions per neutron per second; multiplied by n it gives the number of absorptions per m^3 per second. The sources are easily written as k_∞ times the absorption, as leakage terms are considered separately.

The leakage out of the volume element is written as div $\mathbf{J}$, where $\mathbf{J}$ is the neutron flux, the net number of neutrons passing a m^2 per second. This process is a diffusion effect, just as in eq. (4.1) and discussed more extensively in Section 5.1. Therefore, $\mathbf{J}$ will be proportional to the gradient of neutron density n. For historic reasons one usually expresses it in terms of $\phi = nu$, with

$$\mathbf{J} = -D \operatorname{grad}(nu) \tag{4.247}$$

The *diffusion coefficient* D has the dimensions of a length. When it is taken as constant the leakage term $-\operatorname{div} \mathbf{J}$ in eq. (4.246) may be written as

$$-\operatorname{div} \mathbf{J} = D \operatorname{div} \operatorname{grad}(nu) = D \Delta(nu) \tag{4.248}$$

Taking all terms together one finds

$$dn/dt = k_\infty \Sigma_a nu - \Sigma_a nu + D \Delta(nu) \tag{4.249}$$

It is useful to look in passing at the last two terms only, representing a diminishing flux. One then obtains the equation

$$-\Sigma_a nu + D \Delta(nu) = 0 \tag{4.250}$$

When this is solved in one dimension, one will find an exponential decay

$$nu = (nu)_0 e^{-x/L} \tag{4.251}$$

which shows that

$$L^2 = D/\Sigma_a \tag{4.252}$$

is the square of a *diffusion length* L. For a few moderators the values of L are given in Table 4.6. Within the fuel rods the figures will be different.

Returning to eq. (4.249), it may be simplified as

$$dn/dt = (k_\infty - 1)\Sigma_a nu + D \Delta(nu) \tag{4.253}$$

Table 4.6 Properties of thermal neutrons in several moderators at $T = 293$ K. (Reproduced by permission of Chapman and Hall, London, from Samuel Glasstone and Alexander Sesonske, *Nuclear Reactor Engineering*, 3rd ed., Van Nostrand, New York, 1981, pp. 148 and 233.)

Moderator	ρ (10^3 kg/m^3)	L (m)	D (mm)	Lifetime l (s)[a]
Water	1.00	0.0275	1.6	2.1×10^{-4}
D_2O	1.10	1.00	8.5	0.14
Be	1.85	0.21	5.4	3.9×10^{-3}
Graphite	1.70	0.54	8.6	1.6×10^{-2}

[a]The lifetime applies to an infinite medium.

The Stationary Reactor

For a stationary state of a reactor the time dependence in eq. (4.253) should vanish. When we assume the reactor to be homogeneous (fuel mixed with moderator) there is no variation of the parameters in eq. (4.253) with position. This leads to

$$\Delta n + B^2 n = 0 \tag{4.254}$$

with

$$B^2 = (k_\infty - 1)\frac{\Sigma_a}{D} = \frac{k_\infty - 1}{L^2} \tag{4.255}$$

Consider a rectangular reactor with sizes a, b and c in the x, y and z directions respectively. The differential eq. (4.254) can then be separated with $n = n_x(x)n_y(y)n_z(z)$ and $B^2 = B_x^2 + B_y^2 + B_z^2$. In the x direction this gives

$$\frac{d^2 n_x}{dx^2} + B_x^2 n_x = 0 \tag{4.256}$$

$$n_x = n_{x0} \cos\frac{\pi x}{a} \tag{4.257}$$

$$B_x^2 = \left(\frac{\pi}{a}\right)^2 \tag{4.258}$$

with similar results in the y and z directions. Here we assumed that the density is maximal in the centre $(x, y, z) = (0, 0, 0)$ of the reactor and has no nodes inside. The density is assumed to vanish at $x = \pm a/2$, $y = \pm b/2$ and at $z = \pm c/2$.*

From eqs. (4.255) and (4.258) one determines the critical size of the reactor by

$$\frac{k_\infty - 1}{L^2} = \left(\frac{\pi}{a}\right)^2 + \left(\frac{\pi}{b}\right)^2 + \left(\frac{\pi}{c}\right)^2 \tag{4.259}$$

A bigger reactor would be supercritical as there are fewer neutrons leaking out.

Connection between k and k_∞

The difference between k and k_∞ is the leakage of neutrons out of the reactor. This may be estimated by defining a non-leakage probability P as

$$P = \frac{\text{neutrons absorbed}}{\text{neutrons absorbed} + \text{leaked out}} = \frac{\Sigma_a nu}{\Sigma_a nu - D\Delta(nu)} \tag{4.260}$$

---

*The values a, b and c are a little bit bigger than the actual size of the reactor as there is an outward flux. This small difference will be further ignored.

The numerator has been discussed in connection with eq. (4.249). The second term in the denominator, which represents leakage, will be recognised as div **J** defined in eq. (4.248). With eq. (4.254) one may replace Δn by $-B^2 n$, which with the help of eq. (4.252) leads to

$$P = \frac{1}{1 + L^2 B^2} \tag{4.261}$$

Because of its definition

$$k = P k_\infty \tag{4.262}$$

Time Dependence of a Reactor

The next step would be to solve the time-dependent equation (4.253). This is complicated and we therefore will take a short-cut. As stated before, the fission process (4.234) leads to prompt neutrons and a small fraction β of delayed neutrons. Ignoring the latter for the time being, one may say that N fissioning neutrons lead to kN slow neutrons after a time determined by two effects: slowdown of the fast neutrons and diffusion in the moderator before being absorbed. The diffusion time l dominates and is given in Table 4.6.

One may conclude that in a time $dt = l$ the number of neutrons increases by $dN = (k-1)N$. Note that here the factor k is given with leakage taken into account. Dividing gives

$$\frac{dN}{dt} = (k-1)\frac{N}{l} \tag{4.263}$$

For $k = 1$ the reactor is stable with $N = N_0$. Assume, however, that at $t = 0$ the multiplication factor jumps to $k = 1 + \rho$. Then

$$\frac{dN}{dt} = \frac{\rho N}{l} \tag{4.264}$$

leading to

$$N = N_0 e^{\rho t/l} \tag{4.265}$$

which implies a time constant for a nuclear power station of

$$\tau = \frac{l}{\rho} \tag{4.266}$$

This is the time after which the neutron density increases with a factor e. For a reactor using water as the moderator Table 4.6 shows that with l at around 10^{-4} s and $\rho = 10^{-3}$ the time constant would become 0.1 s, which is much too small to activate countermeasures, for after 10 seconds the flux would have increased by a factor of 10^{43}.

It must be mentioned that for a reactor operating on fast neutrons alone, such as a fast breeder reactor, the time l is determined by the short time in which the fast neutron crosses its mean free path, which is about 10^{-7} s, leading to $\tau = 0.1$ ms for the same small increase in reactivity.

Although a time of 0.1 ms is much too short for human interference or for the use of automatic mechanical devices, the same holds for times of 0.1 s. It is fortunate, however, that the delayed neutrons cause a much larger time constant for small increases in reactivity, as we shall show.

Assume that at $t = 0$ there are N_0 neutrons available for fission. Again at $t = 0$ the multiplication factor k jumps from $k = 1$ to a value $k = 1 + \rho$. Eventually this will give rise to $kN_0 = (1 + \rho)N_0$ new neutrons. However, we now have to distinguish between the prompt neutrons that reproduce after a time l and the delayed neutrons that reproduce after an average time t_d (or $t_d + l$, but this essentially is t_d by virtue of the numbers given earlier).

Take a time $t > 0$ with N slow neutrons around, which all lead to fission within the next instant. After a time interval $dt = l$ the resulting prompt neutrons give rise to $(1 - \beta)kN$ new slow neutrons and after a time interval $dt = t_d$ the delayed neutrons give rise to another βkN slow neutrons. Let us calculate the reproduction time of slow neutrons from the moment that they cause fission up to the moment that their 'children' cause renewed fission.

After fission, the prompt neutrons produce

$$(1 - \beta)kN = (1 - \beta)(1 + \rho)N \approx (1 + \rho - \beta)N \qquad (4.267)$$

fissioning neutrons after $dt = l$. If $\rho < \beta$ there is no net increase in neutrons after this short time. Therefore, one has to take into account the delayed neutrons as well. The reproduction time for the delayed neutrons is essentially t_d, so the average reproduction time dt will be (as $\beta t_d \approx 0.05$ s)

$$dt = (1 - \beta)l + \beta t_d \approx \beta t_d \qquad (4.268)$$

In this time dt the increase in the number of neutrons will be

$$dN = (k - 1)N = \rho N \qquad (4.269)$$

which leads to a similar differential equation (4.264) and a time constant for an increase with a factor e of $\tau = \beta t_d/\rho$. With $\rho = 10^{-3}$ and a water moderated reactor the time constant becomes 50 s, long enough to control the reactor.

For larger ρ, and thus larger increases in the multiplication factor k, the time constant becomes smaller. When $\rho > \beta$ eq. (4.267) shows an increase in number of neutrons, even from the prompt neutrons alone. Starting at $t = 0$ one will first have an exponential increase with a time constant $l/(\rho - \beta)$, but after $t = t_d$ the time constant will become l/ρ.*

---

*A more comprehensive derivation, starting from the differential eq. (4.253) and taking into account six groups of delayed neutrons, is given in Ref. 9.

Passive Safety and Inherent Safety

Following the definitions of the International Atomic Energy Agency (IAEA), which among other things has been established to stimulate the 'peaceful use of nuclear energy', we distinguish between three types of components for safety measures. A *passive* component does not need any external component to operate, such as a sprinkler system used against fires. An *active* component are all other components of the safety system, such as the fire brigade.

Inherent safety is achieved by the elimination of a specified hazard by means of the choice of material and concept. The example here would be a concrete building filled with glassware and pottery where there is nothing to burn. The concept of inherent safety is useful for the chemical industry as well. If one has to move petrol by a pipe system with a pressure of 7×10^5 Pa at a temperature of 100 °C a small leak will lead to an explosion; with a temperature of 10 °C this is very unlikely. Obviously the pipe system should be designed to operate in a virtually inherently safe way [10].

Following this discussion, it is clear that passive safety measures will depend on the design of the reactor. The worst thing that could happen during the operation of a reactor is that the temperature rises, for example by loss of cooling. A passive safety measure would be to design a reactor such that it has a negative temperature coefficient dk/dT around the operational parameters, for then the reactivity would go down with a rise of temperature and the temperature would presumably fall. The temperature coefficient should, on the other hand, not be so negative that a cold reactor would have a multiplication factor $k \gg 1$, which shows that the design is complicated.

A negative temperature coefficient was discussed in connection with Fig. 4.40 regarding moderation. Another example of passive safety is the use of the Doppler effect. If the temperature of a reactor rises, the absorption lines for neutrons in ^{238}U broaden. This means that the neutrons will be absorbed at a broader range of energies, which implies that the chance p to escape resonance decreases. Thus, with rising temperature the multiplication factors k_∞ and k decrease, which implies a negative temperature coefficient. Note that this argument holds only when the resonance absorption in ^{238}U is a significant effect, which is only the case for a reactor with uranium that is not too highly enriched.

The points mentioned above are taken into account in the design of virtually all reactors except the Chernobyl types. The cause of the Chernobyl accident in the then Soviet Union in 1986 was an experiment with the reactor in which the engineers cut off certain safety signals. During the experiment the k value of the reactor became bigger than $k = 1$, which was not corrected in time. This caused a high temperature, the fuel was destroyed and injected into the coolant, reactions between high-temperature water and Zr of the fuel cladding produced hydrogen, which initiated a chemical explosion, with the result that the reactor containment (weaker than in the West) blew up [11].

The other well-known accident with a nuclear reactor happened in the United States in Harrisburg in 1979. In this case a valve was not closed properly, the operator misinterpreted the signals and made decisions by which the reactor lost its coolant. The decay heat was not taken away and the reactor largely melted. This so-called core melt-down was known as a possibility and the containment of Western reactors was designed to survive such an accident. Still, in Harrisburg, radioactive materials escaped into the environment, although to a much smaller extent than in the Chernobyl case, as will be related in Section 4.5.3.

The nuclear industry has reacted to these accidents by improving the existing safety measures, and also in envisaging new designs that essentially depend only on passive systems and inherently safe constructions. The latter should essentially have two aspects. First, when the temperature of the reactor core rises significantly, the reactor should shut down because of physical principles. Second, when the reactor has been shut down, the decay heat of the many radioactive nuclides in the reactor should be transported away by virtue of physical principles alone.

As an example we mention the PIUS design, which has not been put into practice yet.* The coolant of the reactor is water under high pressure, which is pumped around as indicated in Fig. 4.38. The water acts as a moderator as well so 3.5% enriched uranium is used as a fuel. The high pressure of 5 MPa enables the water to be heated up to 290 °C without boiling. The reactor core is embedded in a pool of cold water with a high concentration of the element boron. This element has a very high cross-section for neutron absorption over a wide range of neutron energies ($\sigma_a = 759$ barn at thermal energies). Therefore, if this water were to enter the reactor core the reactor would shut down immediately.

The construction of the reactor is such that the pool is at two points in open connection with the hot coolant water being pumped around. The pressure of the pump and the density difference between the two kinds of water prevent the boron-containing water from entering the reactor core. There exists a delicate hydraulic balance that would break down if the coolant fluid were to boil, so in the case of a temperature rise the absorbing fluid would enter the core, shutting down the reactor. The same water is so abundant that its heat capacity is sufficient to absorb the heat of the decaying nuclides for a reasonably long time.

Another promising development is the high-temperature gas cooled reactor (HTGR). The fuel is in the form of particles individually coated with multiple layers of ceramic material. It can withstand high temperatures. Some authors are doubtful, however, because the moderator, graphite, is combustible. Therefore, the discussion has not reached a final stage yet.

To finish this discussion it should be mentioned that there will always be a need for active possibilities to change the multiplication factor k, for example to start or stop a reactor. When fuel rods age the fission products will capture

---

*For PIUS and other examples of passive safety measures, consult Ref. 12.

neutrons without fission, effectively reducing k. In addition the ^{235}U will burn up during the reactor operation. Although the latter effect is in some way compensated by the production of fissile Pu (illustrated later in Fig. 4.49) the ageing of the fuel requires adaptation of the reactor parameters during the process. In Fig. 4.38 therefore so-called control rods are indicated containing strongly neutron absorbing materials like the boron mentioned earlier. When the reactor starts these rods are placed deeply into the reactor; when the fuel ages the rods are drawn out. A recent development is to put Gd in the fuel. The odd isotopes have a large neutron absorption cross-section. When they subsequently become even in the course of time, the absorption is much smaller. This effect therefore compensates for the ageing of the fuel.

Nuclear Explosives

Nuclear bombs work according to the same principles explained above. An explosive operates on fast neutrons only as there is no time for slowdown. Fissions of ^{235}U or ^{239}Pu are initiated by respectively putting together two pieces each below the critical size or by compression with a conventional detonation until a critical density is reached. Note that for a higher density the macroscopic cross-section (4.239) increases and by eqs. (4.252) and (4.259) the critical size decreases. Depending on how much the actual density exceeds the critical one a multiplication factor of $k = 1.5$ or $k = 1.6$ is reached.

Fusion bombs work according to the principles explained in Section 4.5.2 where the high temperatures needed for fusion are caused by a classic fission explosion.

4.5.2 POWER BY NUCLEAR FUSION

From Fig. 4.37 it was concluded that fusion of the lightest nuclei could lead to energy gains. In practice the following reactions are possible:

$$^2D + {}^3T \rightarrow {}^4He + {}^1n + 17.6\,\text{MeV} \tag{4.270}$$

$$^2D + {}^2D \rightarrow {}^3He + {}^1n + 3.27\,\text{MeV} \tag{4.271}$$

$$^2D + {}^2D \rightarrow {}^3T + {}^1H + 4.03\,\text{MeV} \tag{4.272}$$

$$^2D + {}^3He \rightarrow {}^4He + {}^1H + 18.3\,\text{MeV} \tag{4.273}$$

In these equations deuterium and tritium are indicated by ^2D and ^3T, although the more proper notation would have been ^2H and ^3H.

From eqs. (4.270) to (4.273) it is clear that the positively charged nuclei need to overcome an appreciable Coulomb barrier before the attractive nuclear forces can lead to the energy gains indicated. The Coulomb barrier amounts to 500 or 600 keV, but because of the quantum mechanical tunnelling through the barrier the maximum cross-section is reached at 100 keV for the $^2D + {}^3T$ reaction and at a few hundred keV for the other reactions. For the near future attention is focused on the first reaction (4.270).

In principle, a plasma of deuterium and tritium is heated up to energies of at least some 10 keV. These kinetic energies correspond by means of the Maxwellian distribution with a temperature; in fact at temperature T the peak of the energy distribution is at kT. For 10 keV this would correspond with $T = 10^8$ K. It is custom in this field to express the temperatures in units of keV. As the ionization energy of the hydrogen atoms concerned is around 13.5 eV all atoms are ionized at the high kinetic energies and an electrically neutral mixture of nuclei and electrons is obtained: a *plasma*. We note in passing that the temperatures of ions and electrons may be different.

The deuterium required to fuel the plasma is in ample supply. In fact, on 6700 hydrogen atoms in water there is one deuterium atom. The hydrolysis efficiency of water depends on the weight of the atoms. Therefore by (repeated) hydrolysis one obtains pure deuterium. The (electric) energy required should count as a (small) cost for fusion. The tritium with a half-life of 12.3 years is produced by neutron bombardment of lithium, which is in ample supply as well. It is planned to have the plasma surrounded by a lithium blanket; the neutrons from reaction (4.270) will then breed the required tritium.

The ions and electrons in a plasma must be prevented from hitting the walls for otherwise so many alien nuclei would be sputtered from the walls that a fusion reaction would stop immediately. The integrity of the plasma is kept by magnetic confinement. In fact, about 80% of all research is done on the so-called Tokamak design (Fig. 4.41), a Russian word that means toroidal magnetic chamber.

In the following we will discuss the energy balance (losses and gains) in a plasma and consider a unit of volume. The plasma will be confined for a finite time τ_E which is determined by instabilities, by the fraction of ions that hit the walls after all, by collisions between ions and electrons and by the synchrotron radiation originating from the helix-circular shape of the particle orbits. When n is the number of ions per m^3, the number of electrons will be the same. The energy per m^3 is therefore $2n \times 3kT/2 = 3nkT$. The loss per second from the processes mentioned will therefore be

$$3nkT/\tau_E \qquad (4.274)$$

Another energy loss is caused by Bremsstrahlung occurring when charged particles meet. The number of encounters will be proportional to $n(n-1)$ which is essentially n^2 and the number per unit of time will be proportional to their relative velocity u as well. This velocity will on average be determined by the square root of the kinetic energy, and thus by $(kT)^{1/2}$. The Bremsstrahlung loss may therefore be written as $\alpha n^2 T^{1/2}$, where α is a proportionality constant. The total power leaving the plasma per m^3 will be

$$P_L = \alpha n^2 (kT)^{1/2} + 3n \frac{kT}{\tau_E} \qquad (4.275)$$

TOROIDAL-FIELD
MAGNET

CENTRAL
TRANSFORMER
MAGNET

VERTICAL-FIELD
MAGNET

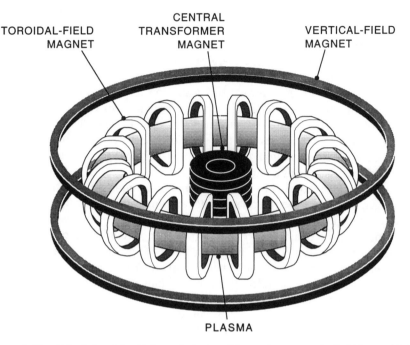

PLASMA

Figure 4.41   Principle of the Tokamak design. The main magnetic field is toroidal; the vertical field provides a centripetal Lorentz force by which the ions follow the main toroid. Other magnets produce induction currents that heat the plasma as far as required.

The production of energy is determined by the reaction rate R, which will be proportional to the density of deuterium nuclei n_d, the density of tritium nuclei n_t and the cross-section* of the process multiplied by the number of collisions. The latter is given by the relative velocity u, so we have $\langle \sigma u \rangle$. The averaging $\langle \ \rangle$ is done by writing a Maxwellian distribution for both nuclei and separating centre-of-mass and relative energies. Therefore, the number of reactions R in $m^{-3} s^{-1}$ is written as

$$R = n_d n_t \langle \sigma u \rangle \qquad (4.276)$$

Both the cross-section σ and the averaging process depend on the energy kT. The resulting behaviour is displayed in Fig. 4.42. Between 10 and 20 keV this double logarithmic scale looks like a straight line, to be represented by

$$\langle \sigma u \rangle = 1.1 \times 10^{-24} (kT)^2 \qquad m^3 s^{-1} \qquad \text{if } (kT) \text{ is in keV} \qquad (4.277)$$

---

*The cross-section is defined as the number of reactions divided by the number of incident particles per m^2 per second. Therefore eq. (4.276) gives precisely the reaction rate R.

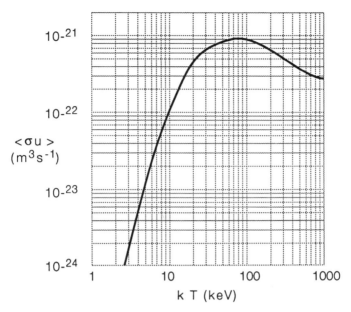

Figure 4.42 The rate $\langle \sigma u \rangle$ for the $^2\text{D} + {}^3\text{T}$ reaction as a function of energy. (Reproduced by permission by Oxford University Press from John Wesson, *Tokamaks*, Clarendon Press, Oxford 1987, Fig. 1.3.1, p. 7)

For simplicity we take $n_\text{d} = n_\text{t} = n/2$, which leads to a production rate of thermonuclear energy of

$$P_\text{th} = \langle \sigma u \rangle E \frac{n^2}{4} \tag{4.278}$$

where E is the energy of 17.6 MeV produced in the reaction. The break-even point for the production of fusion power, ignoring the power required to organize the system, would be reached when the energy production becomes larger than the energy loss. This statement can be made more precise, leading to the so-called *Lawson criterion*. Its derivation assumes a set-up as sketched in Fig. 4.43.

The external heating power P_H is provided to counter the losses P_L of eq. (4.275). The total amount of power P_T leaving the plasma is the sum of P_L and P_th:

$$P_\text{T} = P_\text{L} + P_\text{th} = n^2 \left[\frac{\langle \sigma u \rangle E}{4} + \alpha (kT)^{1/2} \right] + \frac{3nkT}{\tau_\text{E}} \tag{4.279}$$

It is assumed that a generator is able to convert a fraction η into useful energy, so the criterion for energy gain is that

$$\eta P_\text{T} = \eta (P_\text{L} + P_\text{th}) > P_\text{L} \tag{4.280}$$

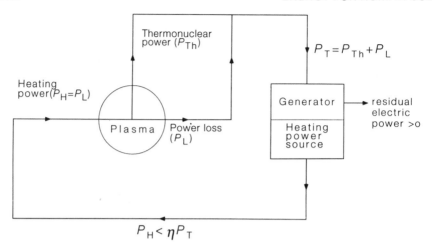

Figure 4.43 Power flows in deriving the Lawson criterion. (Reproduced by permission by Oxford University Press, from John Wesson, *Tokamaks*, Clarendon Press, Oxford, 1987, Fig. 1.4.1, p. 9)

or

$$P_{th} > (1 - \eta)\frac{P_L}{\eta} \tag{4.281}$$

This leads to

$$n\tau_E > \frac{3kT}{\left(\dfrac{\eta}{1-\eta}\right)\dfrac{1}{4}\langle \sigma u \rangle E - \alpha(kT)^{1/2}} \tag{4.282}$$

The right-hand side of this inequality is pictured in Fig. 4.44. In drawing it, Lawson's parameter values $\eta = 1/3$ and $\alpha = 3.8 \times 10^{-29}\,\mathrm{J}^{1/2}\,\mathrm{m}^3\,\mathrm{s}^{-1}$ were used as well as $\langle \sigma u \rangle$ of Fig. 4.42. From the figure Lawson's criterion can be read as

$$n\tau_E > 0.6 \times 10^{20}\,\mathrm{m}^{-3}\,\mathrm{s} \tag{4.283}$$

Another way of looking at the energy balance would be to require that the plasma, once ignited, would maintain its own energy. This could be achieved when the energy of the ^{4}He nuclei in eq. (4.270) could be kept within the plasma, using their relatively short range and high charge. This condition is written as

$$P_\alpha > P_L \tag{4.284}$$

where

$$P_\alpha = \frac{n^2}{4}\langle \sigma u \rangle E_\alpha \tag{4.285}$$

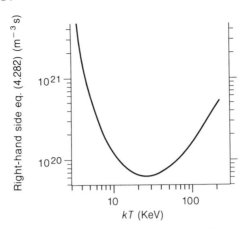

Figure 4.44   The right-hand side of eq. (4.282) as a function of energy kT

in the same way as the thermal power of eq. (4.278). Here $E_\alpha = E/5$ because of the mass ratios in reaction (4.270). Combining eqs. (4.284), (4.285) and (4.275) while ignoring the Bremsstrahlung contribution to P_L leads to

$$n\tau_E > \frac{12(kT)}{E_\alpha \langle \sigma u \rangle} \qquad (4.286)$$

Substituting the approximation for $\langle \sigma u \rangle$ given in eq. (4.277) for the energy region 10–20 keV leads to

$$n\tau_E(kT) > 30 \times 10^{20} \, \text{m}^{-3} \, \text{s keV} \qquad (4.287)$$

The left-hand side of eq. (4.287) is called the *fusion product*. Its value together with the temperature achieved indicates how far a certain machine is from fusion conditions. A fusion diagram that combines the data for several machines is given in Fig. 4.45. One notices the good status of the JET, the Joint European Torus, in the United Kingdom, a combined effort of the countries of the European Community, Sweden and Switzerland.

All major industrialized countries are now working together on a project called ITER (international thermonuclear experimental reactor). This should be finished around the year 2005 and produce more energy than it consumes. Commercial production of electric power would take another 30 years.

4.5.3 RADIATION AND SAFETY

It is well known that working or living with radioactive materials poses an inherent health risk. Let us start with defining the units.

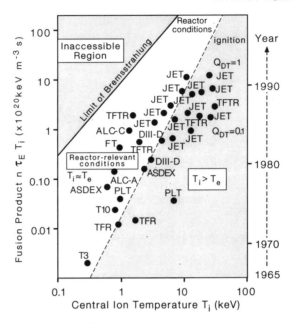

Figure 4.45 Fusion quality diagram as a function of the plasma temperature for several experiments. (Reproduced by permission of JET Joint Undertaking from JET, *Nuclear Fusion, Energy for Centuries to Come*, JET office for Public Relations, p. 12) Note that the inequality (4.287) refers to the top of the graph. The dotted line is a sophisticated version of eq. (4.283)

Radioactivity

The radioactivity of a sample is one *becquerel* (1 Bq) when per second one nucleus of that sample is decaying: $[Bq] = [s]^{-1}$. An old-fashioned unit to measure this is the curie, corresponding with the radioactivity of 1 gram of radium: $1\,Ci = 37 \times 10^9$ Bq, which is taken as an exact definition of the curie.

Dose and Dose Equivalent

A living creature receives a dose of 1 *gray* (1 Gy) when it absorbs 1 joule of energy per kg of its body. $[Gy] = [J\,kg^{-1}]$. An old-fashioned unit to measure this is the rad (radiation absorbed dose) with 1 rad = 0.01 Gy.

The radiation damage in human tissue not only depends on the dose but also on the kind of radiation received. Therefore the dose-equivalent has been introduced, being the dose multiplied with a quality factor Q. The unit is the *sievert* (Sv) with the dimension $[J\,kg^{-1}]$. The old-fashioned unit is the rem (radiation equivalent man) with 1 rem = 0.01 sievert. The quality factor Q has the following values:

(a) $Q = 1$ for X-rays, γ-rays and electrons.
(b) $Q = 2.3$ for neutrons with energies $\leqslant 0.025\,\text{eV}$ (thermal).
(c) $Q = 10$ for other neutrons, protons and similar particles with charge = 1. The quality factor for neutrons may, for specific purposes, be taken more accurately as a function of the neutron energy.
(d) $Q = 20$ for α-particles and similar particles with a charge larger than one.

From natural sources, such as the radioactivity in the atmosphere and soil, the average human being is receiving a dose equivalent of approximately 2 mSv a year. The precise value depends on the geographic location. In present-day life one receives radiation from medical applications and from materials used in buildings (especially concrete). For the average person in industrialized countries this amounted to 2.5 mSv in the year 1987.

There are official norms for the amount of dose-equivalent that is taken as acceptable. The International Commission on Radiological Protection (ICRP) has given norms, in which natural and medical causes were not included. In simplified form they are given in Table 4.7. The norms are higher for radiological workers, people who come into contact with radiation in their profession, as it can be seen as a professional risk. Consideration is being given to reducing the norms by a factor of 2.5. The simplest way to look at radiation from non-natural sources is that they should not exceed the natural dose of 2 mSv/yr for the population at large, or to look at the geographic variation of the natural dose. From that point of view the numbers of Table 4.7 seem reasonable.

Even when one is able to connect a quantitative risk with a radiation dose, to be discussed below, there are three fundamentally different points of view to deal with risks from radiation:

(a) Avoid any risk.
(b) The risk should be related to the possible benefit.
(c) Any dose below the politically agreed limits is acceptable.

Table 4.7 Norms for upper limit of radiation as recommended by ICRP (1966) and under consideration for the year 2000. (The actual rules describe doses for several organs separately).

	ICRP (mSv/yr)	Proposed (year 2000) (mSv/yr)
Population at large	1.7	0.4
Members of the public	5	2
Radiologic workers	50	20

The norms of Table 4.7 are interpreted nowadays in the sense of point (c); formerly the point of view of (b) was taken up.

In Chapter 8 the matter of risk and how to deal with it will be discussed in general. Some points which are rather specific for radioactivity will be discussed below.

A sophisticated point of view is to relate radiation risks to other risks in society and to quantify it. For the latter purpose one would need a dose–effect relationship for several conceivable ill-effects. Here, the fate of the victims of the nuclear explosions at Hiroshima and Nagasaki of 1945 and experience with people irradiated for medical purposes have given some information. One usually distinguishes between *non-stochastic* or *deterministic* effects, in which radiation above a certain threshold value destroys cells or tissues and in which the seriousness of the effect depends on the dose, and, on the other hand, *stochastic* processes. In the latter case there is no threshold; the probability of the effect depends on the dose, but the seriousness does not. Examples are cancers and genetic effects. Figure 4.46 shows dose–effect relations in a qualitative way. Data are available for doses on the right of D_2. For smaller doses the effects of radiation are difficult to distinguish from other causes of harm and the curve shown is based on model calculations. The curve in Fig. 4.46 exhibits a maximum because for higher doses the irradiated cells are weakened or killed, so they cannot multiply and cannot cause cancers or genetic effects.

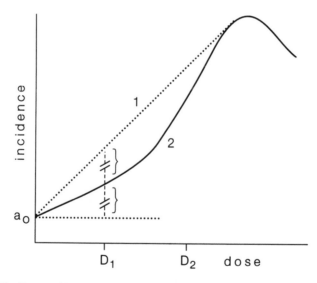

Figure 4.46 Dose–effect relation for stochastic processes. The curve on the right of D_2 is based on data and on the left on models. For a dose D_1 the ICRP norms are derived from the linear extrapolation. They are assumed to be on the safe side. The vertical number a_0 denotes a natural incidence of the ill-effect

It is clear that for very small doses one has to extrapolate the effect. A linear extrapolation overestimates the biological effect, as an organism is sometimes capable of repairing its cells in cases of small damage. The difficulties of interpretation explain why in Table 4.7 the norms are reduced: the Japan data were incorrectly interpreted. For the Hiroshima explosion an explosion yield was used which was stronger than in reality (15 kton instead of 12.5 kton TNT) and the humidity in the air (reducing the radiation intensity) was not properly taken into account.

Normal Use of Nuclear Power

Governments have ordered strict rules as to the radioactive emissions of their nuclear installations. The present collective radiation dose in the Netherlands due to nuclear power (i.e. summed over the population) is estimated as 0.1 milli-man Sv/yr; in countries with more nuclear power stations and other nuclear installations like France it may be higher. The individual dose in the Netherlands is calculated as 10^{-10} Sv/yr. The doses are mainly due to consumption of food, in particular fish, because of the biological concentration of elements like ^{60}Co. These data are very much below the limits of Table 4.7. One warning is in order here: it may well be that locally the doses are much higher than the average numbers.

Accidents

As normal use produces very small emissions only it is more useful to look at what might happen in the case of accidents. During recent years we have experienced two major accidents with nuclear power stations: Harrisburg, USA in 1979 and Chernobyl, USSR (now Ukraine), in 1986.

The reactor core in Chernobyl contained 40×10^{18} Bq of radioactive materials. Virtually all activity present in inert gases was released (1.7×10^{18} Bq) and some 1 to 2×10^{18} Bq of other materials. A considerable amount of the radioactivity was long lived: 4×10^{16} Bq with a half-life in the order of 10^4 days and 6×10^{13} Bq with a half-life in the order of 10^6 days. They will stay in the environment for some time. It has been estimated that in Europe the accident will result in 39 000 cancer deaths over 50 years after the accident. It is a matter of opinion whether such a risk should prohibit the further development of nuclear power (cf. Section 8.1).

Other accidents happened in the Urals near Sverdlovsk, where both in 1957 and in 1967 a vessel with nuclear waste exploded, because of the temperature rise from its decay heat. It contaminated a region of tens of kilometres with ^{90}Sr (half-life 28 yr) and ^{137}Cs (half-life 30 yr). Being β emitters they do most harm inside the body, when they become built into the tissues. The contamination was in the order of 10^{10} Bq km^{-2} over many square kilometres.

From Harrisburg the consequences were much less serious: 10^{17} Bq of inert gases were released and 6.5×10^{11} Bq of ^{131}I. A single individual may have received a dose of 0.8 mSv and the average dose for the population within a distance of 80 km has been a factor of 100 less.

Nuclear Fusion

From a safety point of view fusion of deuterium and tritium in a plasma has advantages over nuclear fission. This originates from the fact that at any moment only 1 gram of D + T is really present in the plasma. The radioactivity of 1 gram of T amounts to 370×10^{12} Bq. In the walls of the reactor vessel a kg of T may be present with another 1 or 2 kg in storage around the reactor. Tritium is the most dangerous of the radioactive materials connected with fusion. As it is gaseous it is easily released by an accident and moreover the atom is so small that it easily leaks through container walls. The total activity of the T present would be about 10^{16} Bq, a few orders of magnitude lower than the activity in a nuclear fission power station.

There is of course radioactivity in the fusion reactor walls, arising from the long irradiation with neutrons, some 10^{20} Bq for a 1000 MW$_e$ stainless steel power station. This is in the same order of magnitude as in a fission reactor (10^{21} Bq) but by contrast a fusion reactor cannot melt down. If anything goes wrong the plasma will collapse and fusion will stop. Note that reactions (4.272) and (4.273) do not produce neutrons and could be of some advantage from the irradiation point of view.

As T poses the worst danger, studies have been made of what the consequences would be of the daily routine discharges. For people living around the reactor this would amount to 0.015 mSv/yr, well below the limits of Table 4.7. It is believed that the worst conceivable accident would release less than 200 g of T. This could lead to a dose of 60–80 mSv at a distance of 1 km. As this is comparable to the admissible dose for a radiological worker it would not pose an unacceptable risk, or so the argument goes.*

4.5.4 MANAGING THE FUEL CYCLE; WASTE

The nuclear fuel cycle is a concept that is fundamental to the exploitation of nuclear fission. The scheme is given in Fig. 4.47. Below we describe the different elements of the fuel cycle bearing in mind three aspects:

---

*The numbers quoted for irradiation only refer to the total doses. For a thorough discussion one should also study the locations in the human body in which elements like T, Sr, Cs and I would concentrate, and the damage they would do there.

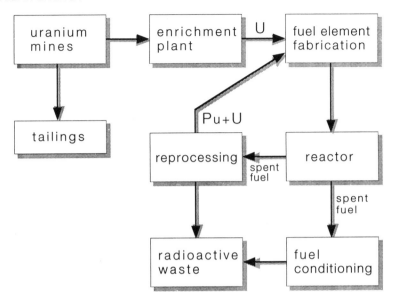

Figure 4.47 The nuclear fuel cycle. From the reactor one either uses the residual U and Pu again by reprocessing or prepares the fuel elements for storage as radioactive waste (no real cycle)

(a) Protection against production of nuclear warheads by governments or terrorist groups,
(b) Health hazards during the operation of the system,
(c) Long term health hazards, (long) after the power plants have been shut down.

Uranium Mines

The required uranium is mined in the form of uranium oxides, often chemically bound to other oxides; the richer ores contain some 1–4 kg uranium per 1000 kg of ore. This is processed locally producing yellow cake with some 80% of U_3O_8. The tailings consist of residues with still enough uranium to produce ^{222}Rn gas escaping into the surroundings; in fact 85% of the radioactive materials of the ore are still in the tailings. In nature they would have been there as well, of course, but now they are kept close to the open environment, usually in ponds underneath water at the processing site.

Enrichment

Only 0.7% of the uranium nuclei are of the isotope ^{235}U, the rest being ^{238}U. Although it is possible to run nuclear reactors with natural uranium (cf.

Table 4.5) many reactor builders prefer uranium containing some 3–4% of ^{235}U. As explained below eq. (4.245), for power production a higher enrichment does not make sense.

For a nuclear explosive higher enrichments of ^{235}U are required. Table 4.8 gives the critical mass (rather than the critical size) of a core of enriched uranium, surrounded by a layer of natural uranium that is to reflect back neutrons to the fissioning core. From the table it is clear that isotope separation may be applied to obtain nuclear explosives. The separation must use physical principles as the atomic structure of the uranium atoms and accordingly the chemistry will be essentially the same. Therefore chemical separation methods would not work. We now describe the physics of three separation methods:

(a) *Gaseous diffusion.* Uranium is processed into the gas UF_6. At a certain temperature all molecules will have the same average kinetic energy $mu^2/2$; hence the molecules of both uranium isotopes will have slightly different velocities. This results in slightly different penetration probabilities through a porous medium. In fact, the lightest molecule will penetrate a little more easily. By repeating the process in a cascade the enrichment will increase.

(b) *Gas centrifuge.* A particle with mass m rotating with an angular momentum L has a rotational kinetic energy of $L^2/(2mr^2)$, where L is the angular momentum and r the distance to the centre. When one puts a gas of UF_6 in a rotating centrifuge, this device should create a force field counteracting the centrifugal forces of the particles. One may say that the centrifuge exerts an external potential field of magnitude $V = -L^2/(2mr^2) = -mu^2/2$. In such an external field the particles will arrange according to a Boltzmann distribution

$$n = n_0 e^{-V/(kT)} = n_0 e^{(mu^2)/(2kT)} \tag{4.288}$$

where n_0 is the number of particles where the field vanishes, i.e. on the axis of the centrifuge. It is seen that the mass enters into the equation, just as with gas diffusion; so a cascade of centrifuges will enrich uranium to the desired degree.

Table 4.8 Critical mass of uranium versus enrichment with a 15 cm thick reflector of natural uranium. (From Taylor, quoted in an APS report.)

Enrichment % ^{235}U	Critical mass (kg)	^{235}U content (kg)
100	15	15
80	21	17
60	37	22
40	75	30
20	250	50
10	1300	130

(c) *Laser separation.* The mass difference of the two uranium nuclei causes a slightly different size and shape of both nuclei which results in a slightly different Coulomb field in which the atomic electrons move. Added to this the position of the centre of mass will be slightly different in both cases. This together results in an isotope shift, a small difference in energy of the atomic levels of the nuclei.

In uranium the major configuration of the outer atomic electrons is $(5f)^6(6d)(7s)^2$ (so there is an odd number of electrons in inner shells). The wave functions of the s-electrons extend to the atomic nucleus and 'feel' the changes in the Coulomb field. The resulting isotope shift is larger for certain levels than the line widths due to Doppler and hyperfine splittings. Also, in ^{238}U hyperfine splitting is absent as the nuclear spin is zero. Consequently, the energy levels for the electrons are different enough to separate vaporized atoms by exciting them with laser beams of very well defined frequencies.

The principle is shown in Fig. 4.48, where in the ^{238}U spectrum the positions of corresponding ^{235}U levels are dashed. It appears that because of auto-ionization effects in a three-step process between certain ^{235}U levels rather high cross-sections for ionization can be reached. Subsequently, the ^{235}U ions are separated from the ^{238}U atoms by an electric field. The pressure may not be too high, otherwise charge exchange would give ^{238}U ions as

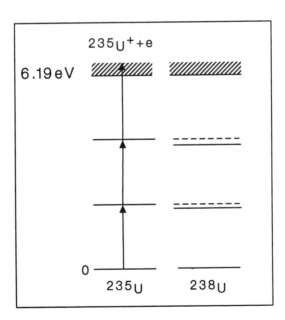

Figure 4.48   Laser separation of ^{235}U and ^{238}U. Shown is a three-step process with a high cross-section. Not shown is the hyperfine splitting of ^{235}U. (From Diakov, personal communication)

well. Thus, one still needs several successive separations to obtain high enough enrichment in sufficient high quantities to produce nuclear warheads. Contrary to the two enrichment methods mentioned before, this one does not require high investments and large facilities. One only needs to know the (still secret) frequencies of the excitations. As there are some 500 000 lines in the spectrum this 'pincode' will be difficult to find. Moreover, the technology of evaporation of metallic uranium at high temperatures (3000 K) is extremely difficult because of its chemically aggressive nature.

The first two processes are long established. The third technology is likely to mature in the 1990s, although commercialization remains to be seen.

Nuclear Reactor

After fabrication of the fuel (see below) the fuel elements enter the fission process described in Section 4.5.1. Assume that one starts with enriched uranium only. Because of the high neutron density many neutrons will be absorbed, resulting in heavier nuclei, where from time to time a neutron is exchanged for a proton, accompanied by β-decay as shown in Fig. 4.49.

One notices a buildup of transuranic nuclei of which the amounts will increase with time. Eventually this will lead to increased absorption of neutrons and a decrease of the multiplication factors η and k. The fuel has then to be replaced. Typically, when one starts with one kg of uranium which contains some 30 g of ^{235}U it is replaced it when it has still some 8 grams of unused ^{235}U and some 6 grams of Pu, more than half of which is fissile ^{239}Pu. This isotope has a half-life of 24 000 years so that its decay during the fuel cycle is negligible.

Reprocessing

The fact that there is still so much fissionable material in the spent fuel is the rationale behind reprocessing. In a chemical plant uranium and plutonium are separated from each other and from all the rest of the materials: not only the actinides of Fig. 4.49, but the fission products, indicated by X and Y in eq. (4.234), and their daughter nuclei as well. One then produces so-called MOX, mixed oxide fuel, consisting of the Pu mixture and either natural uranium or the depleted uranium residue from the enrichment process. MOX is used to fuel up to one third of the reactor, where the Pu essentially replaces ^{235}U as the fissile component.

Fuel Conditioning

If the spent fuel is not reprocessed, it will be prepared for storage as waste. In this case one does not use a 'fuel cycle', but rather a once-through process.

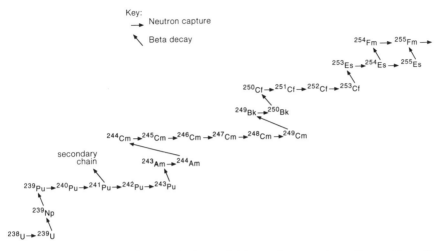

Figure 4.49 Buildup of transuranic nuclei in nuclear reactors. One notices the subsequent neutron captures and occasional β-decay

Radioactive Waste

One cannot avoid producing radioactive waste in the production of nuclear electricity. In fact, operation of nuclear power plants in the past has already produced waste that has to be kept from entering the natural environment. Table 4.9 gives an estimate of the radioactivity of the fuel that is annually taken out of a typical light water reactor with an electric power of $1000\,MW_e$. It must be mentioned that the much smaller amount of radioactivity of the reactor components is not taken into account in these estimates.

From the numbers given in Table 4.9 it can be seen that the spent fuel is usually

Table 4.9 Radioactivity in the one third of the fuel elements taken annually out of a typical light water reactor of $1000\,MW_e$ after a time t in 10^{16} Bq. (Based on data from Jack J. Kraushaar and Robert A. Ristinen, *Energy and the Problems of a Technical Society*, Wiley, New York, 1984, p. 144; A. V. Nero, *A Guidebook to Nuclear Reactors*, Berkeley, University of California Press, 1979; and H. Küsters, *Atomwirtschaft*, June 1990, p. 287.)

	$t = 0$	$t = 6$ months	$t = 10$ years	$t = 300$ years
Fission products	15 000	480	37	$\approx 4 \times 10^{-4}$
Actinides	4 500	16.5	9.5	≈ 2
Fuel cladding	16	3.7	0.4	Negligible
Total	19 500	500	47	≈ 2

stored for some time at the reactor facility. When it is 'cooled down', i.e. has lost the activity of the shorter lived radioactive nuclides, it can be either processed or put away. Before discussing waste management, however, we make some remarks on the nuclear weapons aspects of the fuel cycle.

Proliferation

It is generally recognized that a world in which many countries have nuclear weapons (the explosives combined with missiles or aircraft) is not a safe place. Even worse is the prospect that subgovernmental or terrorist groups may have these weapons at their disposal. This is the rationale behind the non-proliferation treaty and the safeguards precautions.

It should be explained at this stage that most countries have become parties to the nuclear non-proliferation treaty (NPT) where besides the recognized nuclear weapon states (the United States, the United Kingdom, France, Russia and China) all participants pledge not to use nuclear reactions for weapon production; in exchange they get all possible help in the civilian use of nuclear energy. An organization situated in Vienna, the IAEA, International Atomic Energy Agency, is called on to verify that state parties comply (by inspection teams). The IAEA strongly promotes civilian use of nuclear power. Technologies which may be used for weapon production are called sensitive technologies.

The major concerns in this respect are the following:

(a) *Enrichment.* Although most countries have put their enrichment facilities under international safeguards, for the laser isotope separation this seems difficult to enforce. In general all three technologies move towards smaller size with less energy consumption and the facilities are therefore more difficult to detect.

(b) *Nuclear reactors.* The easiest way to produce fissile material is to put an extra sample of natural uranium in a working reactor. When one does this for a short time only, the series of Fig. 4.48 does not have time to build up. From the Pu isotopes one chiefly has the ^{239}Pu of which the critical mass with a thick reflector of natural uranium is around 4.5 kg with a spherical volume of about 0.28 litres (without the reflector it would be 10.2 kg). Although the critical mass does not increase very much with the addition of other Pu isotopes the design of a reliable weapon becomes more difficult. A chemical separation of the Pu from the other atoms in the irradiated sample is a process well within the reach of most countries.

(c) *Reprocessing.* As the objective here is to separate Pu from other elements much Pu will be around. The unavoidable small inaccuracies in Pu book-keeping could be used to divert Pu from the process.

The problem of safeguarding the elements of the nuclear fuel cycle against military misuse will obviously increase with the amount of Pu being around. The 'best' way to handle this problem, if the use of nuclear power continues, is to have as many of the parts of the fuel cycle at the same place under international protection, inspection and safeguards. Only power stations have to be dispersed because of the cost of transporting electricity (cf. Section 4.2.7); their fuel can be mixed so as to discourage misuse.

Radioactive Waste Management

In discussing radioactive waste it should be recognized that one has several categories of waste, originating not only from nuclear power generation (or from nuclear weapons production) but from medical uses as well. The official definitions are given in Table 4.10. In Table 4.10 units of mGy are used (milligray, cf. Section 4.5.3) for solid waste which is the dose a human being would receive when handling the waste unprotected. For solid waste the categories mentioned are also denoted as low-level waste (LLW), medium-level waste (MLW) and high-level waste (HLW).

Waste that is not very long lived may be stored and cooled by air or water until the activity rate is so low as to be acceptable. The longer lived nuclei pose the biggest problem. As may be seen from Table 4.9, the decay of the mixture of actinides happens slowly because of the long half-lives present.

One tries to reduce the volume of the waste by solidification, which is evaporation of the fluid components; next one would encapsulate it in glass and find geological formations where it is expected to be kept out of the natural environment for geological lifetimes.

Granite

It is easy to make mistakes. On some american test site of deep granite, for example, it was assumed that migrating Pu compounds would diffuse into little

Table 4.10 Categories of waste following IAEA definitions

		Radiation
Solid		
Category	1 (LLW)	$<2\,mGy/h$
	2 (MLW)	2–$20\,mGy/h$
	3 (HLW)	$>20\,mGy/h$
Fluid		
Category	1	$<4 \times 10^{-5}\,GBq/m^3$
	2	4×10^{-5}–$4 \times 10^{-2}\,GBq/m^3$
	3	4×10^{-2}–$4\,GBq/m^3$
	4	$>4\,GBq/m^3$

haircracks in granite and then would be absorbed. It was found, however, that the net Pu velocity was a few orders of magnitude larger than calculated [13]. The reason was to be that the Pu was picked up by colloids which were too big to enter the haircracks and instead were only going by the bigger cracks in which the groundwater velocity was much higher. It will be discussed in Chapter 8 how far the possibility of mistakes should let a technology be discarded altogether or whether a 'learning' period should be accepted.

Salt Domes

Consider the storage of high level waste from nuclear power stations in a deep underground formation and take salt as an example. The following aspects then be considered:

(a) the temperature rise in the salt as a function of position and time;
(b) the consequent changes in its mechanical properties, movements and creep;
(c) the possible changes in crystal structure of the salt in which potential energy could be stored and suddenly released causing damage to the protecting layer;
(d) leakage of radioactivity from the glass or other containers and their migration to the natural environment.

In order to get an impression of the order of magnitude concerned, we consider the storage of the HLW from a few power stations together with 3600 MW_e of power during their lifetime, when the waste has first been stored for 10 years to 'cool' it down.* The initial heat production of the HLW (13 years after removal from the nuclear reactor) will be 2.4 MW_{th}. The HLW is divided over 19 positions, the salt parameters are taken as $k = 4.885\,W\,m^{-1}\,K^{-1}$ and $\rho c_p = 1.88 \times 10^6\,J\,m^{-3}\,K^{-1}$. This would lead to a temperature increase of 206 K at a distance of 10 m from the source. More realistic calculations take into account the fact that the source is distributed vertically over some tens of metres and gives a temperature rise of 90 K occurring after some 15 years, close to the source.

Exercises

4.1   An uninsulated hot water pipe passes through a room with temperature 20 °C. The outside diameter of the pipe is 30 mm; its surface temperature is 90 °C and its emissivity 0.7. The heat convection coefficient may be taken as $h = 15\,W\,m^{-2}\,K^{-1}$. Calculate the heat loss from the surface of the pipe per metre per second. Find 136 $W\,m^{-1}$. (Adapted by permission of Wiley

---

*Data from a study done for the Netherlands by Dr Jan Prij (thesis). His results, referred to in the text, are found on p. 158.

from F. P. Incropera and D. P. DeWitt, *Introduction to Heat Transfer*, Wiley, New York, 1990, example 1.2, pp. 11–2.)

4.2 A furnace brick wall with a thermal conductivity $k = 0.72\,W\,m^{-1}\,K^{-1}$ and a surface emissivity of $\varepsilon = 0.8$ is 0.15 m thick. Its outer surface temperature is 110 °C; the free convection coefficient with the air is $h = 20\,W\,m^{-2}\,K^{-1}$ and the temperature of the surroundings is 20 °C. Calculate the inner brick temperature as 609 °C. (Adapted by permission of Wiley from F. P. Incropera and D. P. DeWitt, *Introduction to Heat Transfer*, Wiley, New York, 1990, example 1.5, p. 20.)

4.3 A glass window is 5 mm thick and has a thermal conductivity of $k = 1.4\,W\,m^{-1}\,K^{-1}$. Its inside surface temperature is 20 °C; its outer surface temperature is 5 °C. Calculate the heat loss by conduction for a window of $3\,m^2$.

4.4 A thermopane glass window consists of two layers of glass of thickness 2 mm with 4 mm of air in between. Use the values of Table 4.1 to estimate the thickness of a glass window with the same heat resistance. Calculate the heat loss by conduction for a window of $3\,m^2$ with an inside temperature of 5 °C and an outside temperature of 20 °C. Compare with the results of Exercise 4.3.

4.5 A square chip of width 5 mm is insulated on the bottom and its sides. On the top it is cooled by an air stream with a temperature of 15 °C and a convection coefficient $h = 200\,W\,m^{-2}\,K^{-1}$. If the chip temperature may not exceed 90 °C calculate the maximum allowable chip power, taking into account convection only. (Adapted with permission from F. P. Incropera and D. P. DeWitt, *Introduction to Heat Transfer*, Wiley, New York, 1990, example 1.13, p. 31.)

4.6 Consider a semi-infinite concrete wall, undergoing daily temperature variations. Calculate the damping depth as 0.14 m and the time delay as 27.7 hours. Perform the same calculation for the annual cycle.

4.7 At what depth should a water pipe be buried so that it will never be frozen? Calculate the value for soil with an average temperature T_{av} of 10 °C and an amplitude of annual variations of 15 °C. Find 0.48 m. Repeat the calculation for numbers referring to your own country, if it ever freezes.

4.8 Take a concrete floor with a temperature of $T_0 = 0$ °C. At $t = 0$ a temperature T_1 is applied to the surface. Calculate the time at which the temperature change becomes measurable at a depth of 0.5 m, i.e. when 1% of the temperature difference is realized. Find this at 7.6 hours.

4.9 Study the heat current density q'' for the problem of a sudden change of temperature as given by eq. (4.27) at the surface $x = 0$. Study what happens there as the time t goes to infinity. Conclude that there is no reason to heat the floor of a factory which is in constant use.

4.10 Apply eq. (4.29) to find the temperatures of several materials for floor covering giving the required contact temperature with the human foot of

25.4 °C. In particular, check the numbers given in the text.

4.11 The human tongue has a b value of 1400 units and a temperature of 37 °C. Compare the contact temperatures of an iron railing with that of a soft wooden one when they both have the same temperature of -5 °C. In which case would a child's tongue freeze to the railing. (If this ever happens pour hot water on the tongue.)

4.12 The rear window of a car is 4 mm thick. It is being defogged at the inside by passing air of 40 °C with a heat convection coefficient $h = 30$ W m^{-2} K^{-1}; the outside air has a temperature of -10 °C with a heat convection coefficient of 65 W m^{-2} K^{-1}. Calculate the inner and outer surface temperatures of the window. (Adapted by permission of Wiley from F. P. Incropera and D. P. DeWitt, *Introduction to Heat Transfer*, Wiley, New York, 1990, example 3.2, p. 143.)

4.13 The same window as in the previous exercise with the same outside air is now heated by a fine electric element which gives a temperature of 15 °C at the window inside. The interior air temperature is 25 °C with a heat convection coefficient of 10 W m^{-2} K^{-1}. What power must be provided per unit area? (Adapted by permission of Wiley from F. P. Incropera and D. P. DeWitt, *Introduction to Heat Transfer*, Wiley, New York, 1990, example 3.3, p. 143.)

4.14 Urethane foam is used to insulate all sides of a cylindrical hot water tank, which is electrically heated. The tank is 2 m high with a diameter of 80 cm; the insulation around is 40 mm thick. The tank is in ambient air with a temperature of 10 °C and $h = 10$ W m^{-2} K^{-1}. How many kW h a day are required to maintain the inner surface of the tank at a constant temperature of 55 °C? Calculate the cylindrical wall by using eq. (B.2) and finding a logarithmic temperature variation. Add top and bottom losses and find 4.15 kWh a day. (Adapted by permission of Wiley from F. P. Incropera and D. P. DeWitt, *Introduction to Heat Transfer*, Wiley, New York, 1990, example 3.30, p. 150.)

4.15 Radioactive waste with $k = 20$ W m^{-1} K^{-1} is stored in a very long spherical steel container with $k = 15$ W m^{-1} K^{-1}, an inner radius of 50 cm and an outer radius of 60 cm. Heat is generated uniformly at a rate of $\dot{q} = 10^5$ W m^{-3} and the outer surface is exposed to a water flow for which $h = 1000$ W m^{-2} K^{-1} and $T_\infty = 25$ °C. Write down the radial heat equation and calculate the outer and inner steady state surface temperatures. Derive an expression for the inner temperature $T(r)$ and evaluate at $r = 0$. (Adapted by permission of Wiley from F. P. Incropera and D. P. DeWitt, *Introduction to Heat Transfer*, Wiley, New York, 1990, example 3.74, p. 160.)

4.16 A brick wall of 30 cm thickness is covered with TIM materials. The black surface experiences a temperature fluctuation with an amplitude of 20 °C. Find the time after which the peak of the heat flux reaches the inside of the wall.

4.17 Show that for isothermal processes the free energy F represents the maximum work that a system can deliver. Show similarly that the Gibbs free energy G represents the maximum non-pV work for isothermal, isobaric processes. Use the Clausius inequality.

4.18 Reproduce the calculations in eqs. (4.130) to (4.132). Make a table for some values of α around 3.125 and of a temperature around 3000 K. Check that the mole fraction of CO increases with temperature and with the fuel/air ratio α^{-1}.

4.19 Take the example discussed around Fig. 4.16 of a vapour-compression refrigerator. With the values of enthalpies given there for refrigerant 22 and the knowledge that the installation delivers 50 kW of refrigeration capacity, find the refrigerating effect, the flow rate (the amount of refrigerant circulating in kg/s) and the compressor power delivering the work for step $1 \rightarrow 2$. Find the COP and compare with the value mentioned in the text.

4.20 A compact car has a frontal area of $A = 1.94\,\mathrm{m}^2$ and $C_\mathrm{d} = 3.20$ with a mass of $M = 1160\,\mathrm{kg}$. It runs for 16.7 km on 1 litre of gasoline at a speed of $90\,\mathrm{km\,h}^{-1}$. Calculate the power against friction and the total power. Calculate the fraction of the gasoline used to combat friction. Compare with the figure given below eq. (4.140).

4.21 For the mortgage on a house one borrows 100 000 ECU (or $, if you wish). This has to be paid back in 25 years with an interest rate of $i = 10\%$. Calculate using eq. (4.144) the annual payment that is due. What is the capital recovery factor from the point of view of the lender?

4.22 A person considers buying a solar hot water installation with an expected lifetime of 20 years and a cost of 3000 ECU, for which he has to take a loan with interest rate $i_1 = 0.10$. The rest value is estimated as 500 ECU. The fuel saved is expected to increase in price with $e = 0.06$. The money saved (A_1 after 1 year) could be put into a savings account with an interest rate $i_2 = 0.05$. Find the final value of the fuel saved after 20 years in terms of A_1. Calculate the present value of this money and of the rest value in order to find the 'break-even' point in the fuel bill A_1. Also calculate the present bill A_0 for the break-even point.

4.23 Determine the latitude λ where you live and the date. Find the length of the day by using eq. (4.158). Compare with the date from your newspaper (or from direct measurement). Comment on the deviations, which should not be more than up to 20 minutes.

4.24 Determine the constant c in eq. (4.162) by requiring $\int n(E)\mathrm{d}E = 1$. Check that the average energy is $\bar{E} = kT$.

4.25 The energy spectrum of black body radiation at temperature T may be written as

$$u(E) = \frac{aE^3}{e^{E/(kT)} - 1}$$

with $kT = 0.50$ eV for $T = 5800$ K and with a constant a which gives the normalization. As explained in the text, for a solid state solar cell with a gap energy E_g photons with $E < E_g$ do not contribute; photons with $E > E_g$ give electric energy E_g. Write down the requirement for the maximum efficiency E_g. Calculate E_g with a simple program on your PC. The answer should be $E_g \approx 1.5$ eV. Calculate the (maximum) efficiency for the Si photocell with $E_g = 1.12$ eV.

4.26 From the wind velocity data measured over some period of time, one may deduce the probability $f(U)$ that a certain wind velocity occurs. The Weibull distribution

$$f(U) = \frac{k}{a}\left(\frac{U}{a}\right)^{k-1} e^{-(U/a)^k}$$

is a good empirical approximation of the measured data in many cases. One usually finds the parameters k and a by plotting

$$G(U) = 1 - \int_0^U f(U)\,dU$$

on a double logarithmic scale.
(a) Give an interpretation of $G(U)$.
(b) Collect the curve $G(U)$ from a local meteorological station and find a and k. For Amsterdam airport $k = 1.85$ and $a = 6.9$ m s^{-1}.
(c) Express the average value of U^3 as a function of a and k.
(d) Calculate the energy content passing a m^2 in a year at your local meteo station. (For Amsterdam airport one finds 6×10^9 J m^{-2} per year).

4.27 Equation (4.194) was derived for oceans of infinite depth. Write down a wave type solution of eq. (4.193) for an ocean with depth D and a boundary condition such that $s_z = 0$ at $z = -D$. Find the propagation velocity v for this general case. Check that in the limit $D \to \infty$ you will get back eq. (4.197).)

4.28 Calculate the amount of time it takes to lose the energy stored during 8 hours of photosynthesis, assuming $k_i = 10$ s^{-1} and $k_i = 10^{-4}$ s^{-1} for various values of $\Delta\mu_{st}$ in the range determined by eqs. (4.217) and (4.218). If one wishes to have 90% of the stored energy still available after one month, what would be the required value of $\Delta\mu_{st}$?

4.29 Use some tables on chemical thermodynamics to check the maximum efficiency for the hydrogen fuel cell and the methane fuel cell.

4.30 Show that fission of 1 g of ^{235}U a day in a reactor will produce power of approximately 1 MW$_{th}$.

4.31 Write down the equations of a spherical nuclear reactor and derive a formula for the radius R in which it will be just critical.

4.32 Fission of ^{235}U produces neutrons with an energy of about 1 MeV. The fission cross-section of ^{235}U at this energy is about 2.1 barn and its density is 19.0×10^3 kg m^{-3}. Calculate the macroscopic cross-section Σ_f for fission of a sample of pure ^{235}U at these energies and the corresponding mean free path. The diffusion coefficient D at this energy may be calculated from the scattering cross-section as $D = 1.72 \times 10^{-2}$ m. Calculate B^2 from reinterpretation of eq. (4.255) and by adapting eq. (4.245). Calculate the critical size of a pure ^{235}U sphere and its critical mass using the result of Exercise 4.31.*

4.33 Plutonium nitrate $PuO_2(NO_3)_2$ is dissolved in water. Show that the minimum concentration for which the solution may become critical is 9 g/kg of ^{239}Pu. What do you conclude for the storage of solutions? You may use the following data: for ^{239}Pu one has $v = 3.0$; fission cross-section $\sigma_f = 664$ barn; capture cross-section $\sigma_c = 361$ barn; for N a capture cross-section $\sigma_c = 1.78$ barn and for water molecules $\sigma_c = 0.66$ barn. Ignore all surface effects. (Reproduced by permission of R. Stephenson from R. Stephenson, *Introduction to Nuclear Engineering*, McGraw-Hill, New York, 1954, example 4.13, p. 160.) We now double the concentration of ^{239}Pu and take a finite size cubic reactor with leakage. Deduce from eq. (4.259) the critical size of the reactor. As leakage is caused by fast neutrons one may use a higher value of the diffusion length than given in Table 4.6, say $L = 0.06$ m.

4.34 In the derivation of eq. (4.287) the Bremsstrahlung contribution was ignored. Check with the parameters given in the text whether this is justified for energies $kT = 10$ keV.

4.35 Discuss the heat diffusion of a point source at the origin which produces $\Phi(t)dt$ between t and $t + dt$. Solve the heat eq. (4.15), first for an instantaneous point source at $t = 0$ and next for the continuous point source. The resulting integrals require a specification of $\Phi(t)$. For $\Phi(t) = \Phi_0$ find the solution analytically; for a decaying $\Phi(t)$ you may do it numerically. It may be added that in a different context this exercise is solved in Section 5.1.

References

1. Frank P. Incropera and David P. DeWitt, *Introduction to Heat Transfer*, John Wiley, New York, 1990. Use for Section 4.1. A standard text with many exercises, some of which were used in the present text.
2. Gerald W. Braun, Alexandra Suchard and Jennifer Martin, Hydrogen and electricity as carriers of solar and wind energy for the 1990s and beyond, *Solar Energy Materials*, **24** (1991) 62–75.
3. John H. Seinfeld, *Atmospheric Chemistry and Physics of Air Pollution*, John Wiley, New York, 1986. Use for Section 4.2. This book is written on a more advanced level than the present text.

*In a more precise calculation one should take into account the fact that the neutron density vanishes somewhat outside the sphere, reducing the mass by approximately a factor 3. Putting a ^{238}U reflector at the outer surface of the sphere would reflect the outgoing flux, reducing the critical mass again by a factor of 3–15 kg (cf. Table 4.8).

4. W. F. Stoecker and J. W. Jones, *Refrigeration and Air Conditioning*, McGraw-Hill, New York, 1982.

5. P. D. Dunn, *Renewable Energies: Sources, Conversion and Application*, Peter Peregrinus, London, 1986. Use for Sections 4.2 and 4.4. Due attention is paid to the needs of developing countries; the mathematics is kept simple, the physics is stimulating.

6. Neil W. Ashcroft and N. David Mermin, *Solid State Physics*, Holt, Rinehart and Winston, New Work, 1976, Ch. 29.

7. J. B. Dragt, Wind energy conversion, *Europhys. News*, **24** (1993) 27–30.

8. David O. Hall, Solar energy conversion through biology—could it be a practical energy source? *Fuel*, **57** (June 1978) 322–333.

9. Samuel Glasstone and Alexander Sesonske, *Nuclear Reactor Engineering*, 3rd ed., Van Nostrand, New York, 1981. Use for Section 4.5. A classical text on reactor engineering.

10. Trevor A. Kletz, *Cheaper, Safer Plants, or Wealth and Safety at Work*, Institute of Chemical Engineers, Rugby, Warwickshire, 1985.

11. H. van Dam, *Rep. Prog. Phys.*, **11** (1992) 2025–77; cf. p. 2073.

12. Charles W. Forsberg and William J. Reich, Worldwide advanced nuclear power reactors with passive and inherent safety: what, why, how and who, Oak Ridge Report ORNL/TM-11907, 1991. Use for Section 4.5.

13. R. W. Buddemeier and J. R. Hunt, Transport of colloidal contaminants in groundwater: radionuclide migration at the Nevada test site. *Applied Geochemistry*, **3** (1988) 535–48.

Bibliography

Atkins, P. W., *Physical Chemistry*, Oxford University Press, Oxford, 1990. A beautiful text, well suited as a background for the chemical thermodynamics of Section 4.2.

Culp Jr, Archie W., *Principles of Energy Conversion*, McGraw-Hill, New York, 1991. Although written from an engineer's point of view, this book contains a lot of useful information for physicists as well. Use for Sections 4.2 and 4.3.

Devins, Delbert W., *Energy, Its Physical Impact on the Environment*, John Wiley, New York, 1982. A book with much physics explained from first principles; useful for most sections of Chapter 4.

Duderstadt, James J. and Louis J. Hamilton, *Nuclear Reactor Analysis*, John Wiley, New York, 1992. A standard reference for Section 4.5.1.

Efficient Use of Energy, American Institute of Physics Conference Proceedings No. 25, American Institute of Physics, New York, 1978. Use for Section 4.2. Much attention is paid to second-law efficiency and ways of energy economizing.

Johnson, Gary L., *Wind Energy Systems*, Prentice Hall, Englewood Cliffs, New Jersey, 1985. Use for Section 4.4. The text is rather engineering oriented.

Krenz, Jerrold H., *Energy, Conversion and Utilization*, Allyn and Bacon, Boston, 1976. A book on a similar physics level as the present one with more attention to policy. Most of the topics of Chapter 4 are discussed here, usually at more length.

Palz, Wolfgang, *Solar Electricity*, UNESCO, 1978. A well-written introductory text for Section 4.4.1. It contains many details on what happens in a solar cell.

Report to the APS by the study group on nuclear fuel cycles and waste management, *Rev. Mod. Phys.*, **50**, no. 1, part II, January 1978. Use for Section 4.5. This report is very detailed on physics and on costs; it includes proliferation aspects.

Wesson, John, *Tokamaks*, Clarendon Press, Oxford, 1987. Use for Section 4.5.

Zemansky, Mark W. and Richard H. Dittman, *Heat and Thermodynamics*, McGraw-Hill 1981. Use for Section 4.2. The standard thermodynamics text.

5

Transport of Pollutants

The best way to deal with pollutants is to prevent their coming into existence. This, however, is not always possible, socially or technically. In that case three ways of conduct are possible:

(a) To change pollutants into harmless matter by chemical or biological methods or by irradiation.
(b) To dilute pollutants to concentrations that are considered harmless. Here, however, the fact should be borne in mind that there are biological mechanisms that concentrate the pollutants again. Fish, for example, filter enormous quantities of water from which they keep not only the food but also poisonous elements like mercury, Hg.
(c) To put away pollutants in a safe place where they will not enter the environment.

With reference to dilution, the physicist could use turbulent flows that occur everywhere in nature. A plume of smoke originating from a combustion process is seen to diverge and disappear because of eddies in the air motion.

With reference to putting away pollutants, one should study the ways in which pollutants could bypass their enclosures, for example by ground water motion.

More generally, transport physics is the relevant field where one may distinguish:

(a) transport of momenta in flowing media, discussed in fluid dynamics,
(b) transport of energy, particularly of heat, already discussed in Section 4.1,
(c) transport of matter.

Similar equations govern all of these processes. Below, in Section 5.4, the fundamental *Navier–Stokes equations* are derived and their limitations pointed out. As an introduction, in Section 5.1 diffusion is discussed, in Section 5.2 flow in rivers and in Section 5.3 groundwater flow. These topics may be understood from first principles.

The Navier–Stokes equations appear to be non-linear and in almost all cases must be tackled numerically with the help of the most sophisticated computers.

These methods almost always involve putting a grid over a region of space that one wants to describe and calculating spatial and temporal derivatives at the points of the grid. This implies that any computation is made for a specific geometry and has to be repeated for other geometries, and moreover that physical phenomena smaller than the grid spacing have to be described by approximations and models anyway.

It is therefore important to get a feeling for the physics involved. The old-fashioned analytical methods give a frame of reference for any more precise calculation; they are usually quickly done on PCs and they are a source of inspiration for any approximations that one might want to make. In Section 5.2 it will become clear that turbulent diffusion is the major mechanism for dispersion of pollutants. Therefore, in Section 5.5 the traditional discussion of turbulence is summarized in order to appreciate the fact that as a first approximation diffusion type equations may be taken with adapted parameters. A simplified approach is again taken in Section 5.6 where the average behaviour of a plume in the air is described by Gaussian formulas, the so-called Gaussian plume model, which is much better in practice than one would expect.

In Section 5.7 turbulent jets and plumes in water are discussed in its simplest way, making extensive use of dimensional analysis to find the essential equations in a zero-order approximation. Finally, in Section 5.8 the behaviour of small particles in the atmosphere is summarized.

The material is organized in such a way that a reader may skip the more theoretical oriented Sections 5.4 and 5.5 without loss of coherence.

5.1 DIFFUSION

If there is a concentration of a pollutant at a certain position in a gas or liquid it will distribute in time over the entire expansion of the gas or liquid, even when macroscopically the host is at rest. The physical origin of this process concerns the collisions between atoms and molecules. If the size of the suspended molecules and the host molecules is comparable, one speaks of *molecular diffusion*; if the size of the suspended molecules is appreciably larger, this phenomenon is called *Brownian motion*. Molecular diffusion is observed when one looks at the dispersion of a drop of ink in water at rest without any temperature differences or other disturbing influences.

It should be mentioned that, usually, molecular diffusion is a minor effect in distributing pollutants. More important are flows of liquids (rivers, oceans) or gases (air) and the turbulence associated with them. This will be discussed in later sections. There are, however, examples of molecular diffusion that are important from an environmental point of view, e.g. diffusion of highly radioactive materials in clays or in ground water at rest. Another reason to discuss molecular diffusion is that the techniques used may be applied to large scale turbulent diffusion as well.

The concentration $C(x, y, z) = C(\mathbf{r})$ of a diffusing substance is defined as its mass ΔM, divided by the sample volume ΔV in which it is dispersed:

$$C = \frac{\Delta M}{\Delta V} \tag{5.1}$$

in which it is assumed that ΔV is large compared to a^3, where a is the mean free path between the diffusing molecules or particles. It is further assumed that $C(\mathbf{r})$ is a continuous and differentiable function, as usual. Finally, the concentration C is assumed to be so small that the change in mass of volume elements due to diffusion may be ignored.

The flux $\mathbf{F}$ is pointing into the direction in which the suspending particles flow; its magnitude F is the mass of diffusing particles crossing a unit area (m^2) in a unit of time (s) in the direction of $\mathbf{F}$. The relation between flux $\mathbf{F}$ and concentration C is known as *Fick's law*:

$$\mathbf{F} = -D\nabla C \tag{5.2}$$

The constant D is called the *diffusivity*, *diffusion coefficient* or *diffusion constant*. Its dependence on temperature, molecular weight, etc., can be derived from the kinetic theory of gases. Fick's law may be derived rather rigorously from that theory. Some values of the diffusivity are given in Table 5.1. One observes that Fick's law (5.2) is exactly analogous to eq. (4.1) which describes the heat flow $\mathbf{q}$ as proportional to the gradient of the temperature T. Therefore eq. (4.1) is often called the law of heat diffusion. Another point to be mentioned is that a form of Fick's law was discussed in eq. (4.247) where the neutron flux $\mathbf{J}$ was taken as proportional to the gradient of the neutron density n.

Table 5.1 Diffusion coefficients D in $m^2 s^{-1}$ at 25 °C and atmospheric pressure. (From L. P. B. M. Janssen and M. M. C. G. Warmoeskerken, *Transport Phenomena Data Companion*, Edward Arnold, London, 1987, p. 143.)

	$D(m^2 s^{-1})$
CO_2 in air	16.4×10^{-6}
H_2O vapour in air	25.6×10^{-6}
C_6H_6 = benzene in air	8.8×10^{-6}
CO_2 in water (20 °C)	1.60×10^{-9}
N_2 in water	2.34×10^{-9}
H_2S in water	1.36×10^{-9}
NaCl in water (20 °C)	1.3×10^{-9}

The concentration C may be a function of the time t, resulting in $C = C(\mathbf{r}, t)$. The equation of continuity, or conservation of mass of the dispersing substance, gives

$$\frac{\partial C}{\partial t} + \nabla \cdot \mathbf{F} = 0 \tag{5.3}$$

Consider the case of dispersion within a laminar flow with a velocity $\mathbf{u}$ independent of the position $\mathbf{r}$. Then the flux $\mathbf{F}$ also has a component due to the velocity $\mathbf{u}$:

$$\mathbf{F} = \mathbf{u}C - D\nabla C \tag{5.4}$$

The first term on the right-hand side describes the flux of a concentration C which has a velocity $\mathbf{u}$. This effect is called *advection* of the particles with the flow. Conservation of mass then gives, by combining eqs. (5.3) and (5.4),

$$\frac{\partial C}{\partial t} + \mathbf{u} \cdot \nabla C - D\nabla^2 C = 0$$

assuming that $\mathbf{u}$ and D are independent of the position. Rearranging terms, one finds

$$\frac{\partial C}{\partial t} + \mathbf{u} \cdot \nabla C = D\nabla^2 C \tag{5.5}$$

The left-hand side appears to be the total derivative of the function $C(x(t), y(t), z(t), t)$:

$$\frac{\mathrm{d}C}{\mathrm{d}t} = \frac{\partial C}{\partial t} + \frac{\partial C}{\partial x}\frac{\mathrm{d}x}{\mathrm{d}t} + \frac{\partial C}{\partial y}\frac{\mathrm{d}y}{\mathrm{d}t} + \frac{\partial C}{\partial z}\frac{\mathrm{d}z}{\mathrm{d}t} = \frac{\partial C}{\partial t} + \nabla C \cdot \mathbf{u} = \frac{\partial C}{\partial t} + \mathbf{u} \cdot \nabla C \tag{5.6}$$

Here, in $C(x, y, z, t)$ the coordinates are functions of time $x(t), y(t), z(t)$, such that one follows the flow. Their time derivatives therefore give the flow velocity $\mathbf{u}$.

It may be helpful to look at eq. (5.6) in a simple one-dimensional way. Ignoring diffusion and restricting oneself to advection the local concentration increase ∂C at point (x, t) will be written as

$$\partial C = C(x, t + \mathrm{d}t) - C(x, t) = C(x - \mathrm{d}x, t) - C(x, t) \tag{5.7}$$

for the concentration C at $(x, t + \mathrm{d}t)$ originates from $(x - \mathrm{d}x, t)$ where the velocity is given by $u = \mathrm{d}x/\mathrm{d}t$. Taylor expansion then gives

$$\partial C = C(x, t) - \frac{\partial C}{\partial x}\mathrm{d}x - C(x, t) = -\frac{\partial C}{\partial x}\frac{\mathrm{d}x}{\mathrm{d}t}\mathrm{d}t = -u\frac{\partial C}{\partial x}\mathrm{d}t$$

$$\frac{\partial C}{\partial t} = -u\frac{\partial C}{\partial x} \tag{5.8}$$

or

$$\frac{\partial C}{\partial t} + u\frac{\partial C}{\partial x} = 0$$

which is precisely the one-dimensional form of eq. (5.5) while ignoring diffusion. Indeed, in eq. (5.7) we used $dC/dt = 0$ when following the flow; therefore eq. (5.8) is a special case of eq. (5.6).

Going back to eq. (5.6), it must be mentioned that this equation holds for any quantity which is a function of position and time. Therefore one could write the operator identity

$$\frac{d}{dt} = \frac{\partial}{\partial t} + \mathbf{u}\cdot\nabla \tag{5.9}$$

It should be noted that in this equation $\mathbf{u}$ may be a function of time and position. In deriving eq. (5.5) from eq. (5.4) it was assumed that $\nabla\cdot\mathbf{u} = 0$ and with that condition one may rewrite (5.5) as

$$\frac{dC}{dt} = D\nabla^2 C \tag{5.10}$$

This equation holds both for media in rest and for moving media with $\nabla\cdot\mathbf{u} = 0$. In both cases it was assumed that D = constant. The diffusion equation (5.10) is identical to the one encountered with heat diffusion, eq. (4.18). Therefore the same examples could be given as in Section 4.1, albeit with a different physical meaning. Instead some slightly different examples will be discussed.

Instantaneous Plane Source in Three Dimensions

Consider a homogeneous medium at rest, i.e. $\mathbf{u} = 0$ and D = constant. Look at the case in which the concentration C is a function of x only and independent of both y and z. The differential equation (5.5) simplifies itself as

$$\frac{\partial C}{\partial t} = D\frac{\partial^2 C}{\partial x^2} \tag{5.11}$$

A solution of this equation is the Gaussian function

$$C(x,t) = \frac{Q}{2\sqrt{\pi Dt}}e^{-x^2/(4Dt)} \tag{5.12}$$

as may be checked by substitution (cf. Exercise 5.3). The factor in front is chosen such that for $t \to 0$ this becomes the delta function, discussed in Appendix A:

$$C(x, t \to 0) = Q\delta(x) \tag{5.13}$$

One may, therefore, interpret the solution (5.12) as the situation where at $t = 0$

an amount Q kg m^{-2} is released in the plane $x = 0$ (an instantaneous plane source). Alternatively, one might interpret eq. (5.12) as the solution of a one-dimensional problem where the amount Q kg is released at $x = 0, t = 0$ (an instantaneous point source in one dimension). This will be used extensively in Section 5.2. For times $t > 0$ the concentration disperses symmetrically, keeping the same integral

$$\int_{-\infty}^{\infty} C \, dx = Q \tag{5.14}$$

Another way of looking at the Gaussian distribution (5.12) is by means of the mean square distance σ to which the particles have diffused:

$$\sigma^2 = \frac{1}{Q} \int_{-\infty}^{\infty} C x^2 \, dx = 2Dt \tag{5.15}$$

or

$$C = \frac{Q}{\sigma \sqrt{2\pi}} e^{-x^2/(2\sigma^2)} \tag{5.16}$$

in which the time dependence is hidden in σ.

A Finite Size Cloud

Consider a 'cloud' released at $t = 0$. The initial conditions for a simple case may be expressed as

$$C = C_0 \text{ at } -\frac{b}{2} < x < \frac{b}{2}, \qquad t = 0$$
$$C = 0 \text{ elsewhere}, \qquad\qquad t = 0 \tag{5.17}$$

The cloud will diffuse perpendicular to the $x = 0$ plane. In order to solve eq. (5.11) with initial conditions (5.17) one could interpret this cloud as a superposition of a big number of delta functions

$$C_0 \delta(x - x') dx'$$

Each has as a solution eq. (5.16) with the origin moved to $x = x'$:

$$\frac{C_0 dx'}{\sigma \sqrt{2\pi}} e^{-(x-x')^2/(2\sigma^2)} \tag{5.18}$$

The complete solution therefore becomes

$$C(x) = \frac{C_0}{\sigma \sqrt{2\pi}} \int_{-b/2}^{b/2} e^{-(x-x')^2/(2\sigma^2)} dx' \tag{5.19}$$

in which one should remember that $\sigma^2 = 2Dt$, independent of x'. As the problem is similar, albeit not identical to the one discussed below eq. (4.23), it comes as no surprise that the error function again appears in the solution of eq. (5.19). Following the definitions given in Appendix A, eq. (A.5), one finds

$$C(x) = \frac{C_0}{2} \left[\text{erf}\left(\frac{b/2 + x}{\sigma\sqrt{2}} \right) + \text{erf}\left(\frac{b/2 - x}{\sigma\sqrt{2}} \right) \right] \qquad (5.20)$$

Instantaneous Line and Point Sources in Three Dimensions

Equation (5.16) gave the solution of the diffusion equation (5.10) for an instantaneous plane source at $x = 0$ at the time $t = 0$. By analogy one would write down the solutions for an instantaneous line source or point source as

$$C = \frac{Q}{(\sigma\sqrt{2\pi})^n} e^{-r^2/(2\sigma^2)} \qquad (5.21)$$

where r is the distance to the source and n is the number of Cartesian coordinates (x, y, z) required to define r. For $n = 1$ and $r^2 = x^2$ this reduces to the solution (5.16). There is one significant dimension x and the problem is equivalent to a one-dimensional problem, as was noted before. For an instantaneous line source there are two essential dimensions and we would put $n = 2$ in eq. (5.21) and $r^2 = x^2 + y^2$. For an instantaneous point source there are all three dimensions $n = 3$ and $r^2 = x^2 + y^2 + z^2$. It can be shown by substitution that eq. (5.21) is correct, cf. Exercise 5.3. In all cases

$$\sigma^2 = 2Dt \qquad (5.22)$$

It is clear that σ is expressed in metres and the concentration C in kg m^{-3}. Thus, for a plane source Q is expressed in kg m^{-2}, for a line source in kg m^{-1} and for a point source in kg. For a point source in one dimension C is expressed in kg m^{-1} and Q therefore again in kg.

Continuous Point Source in Three Dimensions

Consider a point source at the origin which, starting at $t = 0$, emits continuously at a rate of $q \text{ kg s}^{-1}$. In a time interval dt' it therefore emits $q \, dt'$. At a certain time t and position $\mathbf{r}$ one may find the concentration $C(\mathbf{r}, t)$ by integrating (5.21), for $n = 3$, i.e. adding all individual and instantaneous 'puffs' at times t' earlier than t. One should realize that $\sigma^2 = 2Dt$ in eq. (5.21) should be written down explicitly, for at time t the contribution from a puff at t' will be found by putting $\sigma^2 = 2D(t - t')$. One finds

$$C = \frac{q}{8(\pi D)^{3/2}} \int_0^t e^{-r^2/[4D(t - t')]} \frac{dt'}{(t - t')^{3/2}} \qquad (5.23)$$

One should try to bring this into a form resembling the error function eq. (A.5) or its complement eq. (A.7). By substitution of $\beta^2 = r^2/4D(t-t')$ one finds

$$C(r) = \frac{q}{4\pi Dr} \operatorname{erfc}\left(\frac{r}{2\sqrt{Dt}}\right) \tag{5.24}$$

Point Source in Uniform Wind

Let us consider a point source in a flow of uniform velocity **u** (either air or water). The x axis is taken in the direction of **u**. For an *instantaneous* point source at $t = 0$ at the origin one would use an x', y', z' system moving with the fluid. Then eq. (5.21) gives

$$C = \frac{Q}{(\sigma\sqrt{2\pi})^3} e^{-(x'^2+y'^2+z'^2)/(2\sigma^2)} \tag{5.25}$$

Using a Galilei transformation to return to the fixed coordinate system one finds

$$C = \frac{Q}{(\sigma\sqrt{2\pi})^3} e^{-[(x-ut)^2+y^2+z^2]/(2\sigma^2)} \tag{5.26}$$

Here one should remember that $\sigma^2 = 2Dt$. One may verify that the concentration C from eq. (5.26) obeys eq. (5.5) which applies to a uniform flow with velocity **u** (Exercise 5.5). We note in passing that for a one-dimensional case one should take solution (5.12) with x replaced by $x - ut$.

In order to find the concentration C resulting from a *continuous* point source at the fixed origin in a uniform flow **u** one combines the procedures which led to eqs. (5.23) and (5.26). As in eq. (5.23) one considers a delta puff at time t' with a magnitude $q \, dt'$. The integrand of eq. (5.23) contains the factor r^2 which now should read as

$$r^2 = (x')^2 + (y')^2 + (z')^2$$

In order to return to the fixed coordinate frame one should write

$$x' = x - u(t - t')$$
$$y' = y$$
$$z' = z$$

as for $t = t'$ both frames coincide. This argument leads to the integral

$$C = \frac{q}{8(\pi D)^{3/2}} \int_0^t e^{-\{[x-u(t-t')]^2+y^2+z^2\}/[4D(t-t')]} \frac{dt'}{(t-t')^{3/2}} \tag{5.27}$$

Simplifying this by substitution of $\beta^2 = r^2/4D(t-t')$ leads to an integral with a lower boundary $r/2\sqrt{Dt}$ and upper boundary ∞. As this is the only time dependence left, one takes the limit $t \to \infty$ in order to put the lower boundary at zero and finds, using integral tables,

$$C = \frac{q}{4\pi rD} e^{-u(r-x)/(2D)} \tag{5.28}$$

in which $r^2 = x^2 + y^2 + z^2$. Relation (5.28) describes a plume of dispersing material in a flow with velocity u after a very long period of time. In Fig. 5.1 we give in a $z = 0$ plane the lines of constant concentration. As dimensionless variables one may take $\bar{x} = ux/D$ and $\bar{y} = uy/D$. In terms of these variables eq. (5.28) becomes

$$C = \frac{qu}{4\pi \bar{r} D^2} e^{-(\bar{r} - \bar{x})/2} \tag{5.29}$$

where $\bar{r}^2 = \bar{x}^2 + \bar{y}^2$. Taking $\tilde{C} = 4\pi D^2 C/(qu)$ leads to

$$\tilde{C} = \frac{1}{\bar{r}} e^{-(\bar{r} - \bar{x})/2} \tag{5.30}$$

In Fig. 5.1 one observes how 'slender' the plume is. For big x (downstream) one therefore approximates, in the usual variables x, y and z,

$$r = x + (y^2 + z^2)/(2x)$$

and finds for large x (and after a long time t)

$$C = \frac{q}{2\pi u \sigma^2} e^{-(y^2 + z^2)/(2\sigma^2)} \tag{5.31}$$

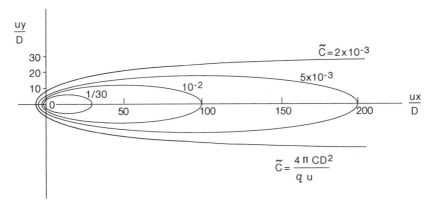

Figure 5.1 Lines of constant concentration in the $z = 0$ plane of a continuous point source at the origin and a uniform velocity field in the x direction after a long period of time. (Reproduced by permission of Kluwer Academic Publishers from G. T. Csanady, *Turbulent Diffusion in the Environment*, Reidel Publishing, Dordrecht, Netherlands, 1973, Fig. 1.9a, p. 18)

with $\sigma^2 = 2Dx/u$. The exponential function describes the dispersal perpendicular to the flow. Its standard deviation σ is apparently proportional to the distance x from which the dispersal originates, or equivalently proportional to the time t that the dispersal needed to arrive at x. It does not depend on puffs emitted earlier or later. Therefore, individual puffs do not interfere and diffusion along the x axis is negligible in the limit of large x and t.

From eq. (5.31) one may write down directly the expression for an infinitely long line source which emits $q\,\text{kg m}^{-1}\text{s}^{-1}$ perpendicular to the uniform flow. The line is supposed to be in the z direction; therefore the z dependence disappears and in virtue of eq. (5.21) one has to multiply by $\sigma\sqrt{2\pi}$. This gives

$$C = \frac{q}{\sqrt{4\pi Dxu}}\,e^{-y^2 u/(4Dx)} \tag{5.32}$$

The Effect of Boundaries

In the presence of walls the differential equations will still apply, if the proper boundary conditions are used. Consider the one-dimensional problem of an instantaneous plane source without external wind and with a cloud propagating in the x direction. If there is a wall at $x = -L$ the flow F at that point should be zero, giving

$$F = -D\frac{\partial C}{\partial x} = 0 \qquad (x = -L) \tag{5.33}$$

It is clear that eq. (5.12) does not obey the boundary condition, although it does solve the diffusion equation (5.11). Mathematically one could add an imaginary source at $x = -2L$ that would give the same flow at $x = -L$, but in the other direction. The sum of both solutions is again a solution of the linear equation and obeys the boundary condition; so the final solution becomes

$$C = \frac{Q}{2\sqrt{\pi Dt}}[e^{-x^2/(4Dt)} + e^{-(x+2L)^2/(4Dt)}] \qquad (x > -L) \tag{5.34}$$

This would be sufficient to solve diffusion type equations in the air where the ground acts as a wall. For a river there would be two boundaries, at $x = L$ and $x = -L$ respectively. Adding an imaginary source at $x = -2L$ gives the right condition at $x = -L$, but not at $x = L$. In fact one needs a whole series of mirror sources $x = -2L, -4L, -6L, \ldots$ and also at $x = 2L, 4L, 6L, \ldots$ to get both conditions right at the same time. This gives

$$C = \frac{Q}{2\sqrt{\pi Dt}}\sum_{n=-\infty}^{\infty} e^{-(x+2nL)^2/(4Dt)} \tag{5.35}$$

5.2 FLOW IN RIVERS

The question to be tackled is how a concentration of pollutants that enter a river at a factory is dispersed by the flow somewhere downstream. Of course, one expects the pollutant to be dispersed so that downstream a smaller concentration spreads over a larger region. The direction of the flow is always called the x direction; the average velocity $\mathbf{u}$ points in this direction. The vertical coordinate is called z and y represents the remaining direction, horizontally perpendicular to the average flow.

The three-dimensional diffusion equation in a moving flow with velocity $\mathbf{u}$ was derived in Section 5.1. The flux $\mathbf{F}$ was written in eq. (5.4) as the sum of advection $\mathbf{u}C$ and diffusion:

$$\mathbf{F} = \mathbf{u}C - D\nabla C \qquad (5.36)$$

Next the equation of continuity, eq. (5.3), was used, leading to

$$\frac{\partial C}{\partial t} + \nabla(\mathbf{u}C) - \nabla(D\nabla C) = 0 \qquad (5.37)$$

For a constant velocity field $\mathbf{u}$ and a diffusion constant D that is also independent of position the resulting equation could be written as

$$\frac{\partial C}{\partial t} + \mathbf{u}\cdot\nabla C - D\nabla^2 C = 0 \qquad (5.38)$$

$$\frac{dC}{dt} - D\nabla^2 C = 0 \qquad (5.39)$$

where again the total derivative dC/dt has been written. It is important to realize the assumptions made in writing down these equations as in real flows the velocity may change with position and the diffusion coefficients may be different in the three x, y and z directions. This would lead to a more complicated expression (5.36).

A One-Dimensional Approach

The simplest approach to the problem is to reduce it to one dimension. One assumes that the pollutant is inserted over the total width and depth of the flow; the only relevant coordinate then would be x, along the flow, and we should find a one-dimensional form of eq. (5.38).

In a river such as sketched in Fig. 5.2, there will be a velocity distribution $u(y)$ where the velocity will be largest near the middle of the stream. Consider two particles close to one another. Assume that one diffuses a little into the y direction

TRANSPORT OF POLLUTANTS

while the other one does not. The first particle would get another velocity in the x direction while the second would not and so the first would obtain a certain distance from the second one more quickly than with diffusion in the y direction only. This implies that the effective coefficient D must be larger than the diffusion constant. We will first discuss molecular diffusion only and then include the effect of eddies.

In order to derive the relevant equations it is convenient to introduce quantities averaged over the width W of the river and weighted with the depth $h(y)$. We assume that the variables do not change with the vertical coordinate z. So, besides $u(y)$ we define a mean $\bar{u}$ by

$$\bar{u} = \frac{1}{A} \int_0^W u(y)h(y)\,\mathrm{d}y \qquad (5.40)$$

where A is the cross-sectional surface area of the river. We note that $A\bar{u}$ is the volume of water passing a cross-section in a unit of time. From the concentration of pollutant $C(x, y, t)$ a mean value $\bar{C}$ is defined by

$$\bar{C} = \frac{1}{A} \int_0^W C(x, y, t)h(y)\,\mathrm{d}y \qquad (5.41)$$

With respect to the mean values one can write the precise values as

$$u(y) = \bar{u} + u'(y) \qquad (5.42)$$

from which it follows that

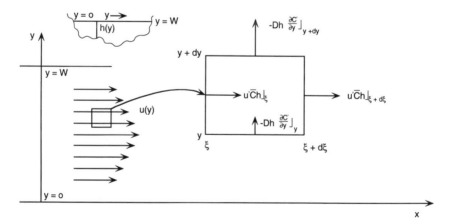

Figure 5.2 Interpretation of the one-dimensional dispersion equation. On the left-hand side a stream is sketched, of which a small part is enlarged on the right-hand side. The variables indicated there refer to a frame moving with the average flow

$$\frac{1}{A}\int u'(y)h(y)\,dy = 0 \tag{5.43}$$

Also we define

$$C(x, y, t) = \bar{C}(x, t) + C'(x, y, t) \tag{5.44}$$

The diffusion equation in three dimensions was already given in eq. (5.39) as

$$\frac{dC}{dt} = D\nabla^2 C \tag{5.45}$$

where on the left the total derivative is written. It is easiest to perform a Galilei transformation to a system in which the average velocity vanishes, which implies

$$\begin{aligned}\xi &= x - \bar{u}t \\ y &= y \\ t &= t\end{aligned} \tag{5.46}$$

The total derivative on the left of eq. (5.45) now may be written as

$$\frac{d}{dt} = \frac{\partial}{\partial t} + \frac{\partial\xi}{\partial t}\frac{\partial}{\partial\xi} = \frac{\partial}{\partial t} + (u - \bar{u})\frac{\partial}{\partial\xi} = \frac{\partial}{\partial t} + u'\frac{\partial}{\partial\xi} \tag{5.47}$$

where we assumed that the flow goes in the ξ direction only. This leads to

$$\frac{\partial}{\partial t}(\bar{C} + C') + u'\frac{\partial}{\partial\xi}(\bar{C} + C') = D\left[\frac{\partial^2}{\partial\xi^2}(\bar{C} + C') + \frac{\partial^2 C'}{\partial y^2}\right] \tag{5.48}$$

In this equation the diffusion in the ξ direction on the right-hand side can be neglected as the effect of change of velocity with diffusion in the y direction will dominate. With some arguments due to the British mathematician Taylor most of the remaining terms can be ignored, leading to

$$u'\frac{\partial\bar{C}}{\partial\xi} = D\frac{\partial^2 C'}{\partial y^2} \quad \text{with } \frac{\partial C'}{\partial y} = 0 \text{ at } y = 0, W \tag{5.49}$$

where the boundary conditions are indicated.

Instead of summarizing Taylor's arguments it is more illuminating to study the right-hand side of Fig. 5.2, which pictures the river in the moving frame (5.46). In that picture a rectangle is analysed as to its inflow and outflow. In the ξ direction only the advective flow $(u'\bar{C}h)$ is considered, which assumes that the advection is dominated by the mean concentration; the change of u' and h with ξ will be neglected. In the y direction only diffusion occurs, determined by the y derivative of C which equals the y derivative of C'. For any rectangle in the moving frame the total amount of pollutant remains the same, which by using the first terms of the series expansions in ξ and y immediately gives

$$u'h\frac{\partial \bar{C}}{\partial \xi} = \frac{\partial}{\partial y}\left(Dh\frac{\partial C'}{\partial y}\right)$$ (5.50)

For constant diffusion D and depth h this indeed gives eq. (5.49). We note that eq. (5.50) implies a stationary situation where ingoing and outgoing fluxes cancel, which will only be reached after sufficient time.

We further use eq. (5.50) which is more general than eq. (5.49) and integrate twice:

$$\frac{\partial \bar{C}}{\partial \xi}\int_0^y \frac{1}{Dh(y_2)}\int_0^{y_2} u'(y_1)h(y_1)\,dy_1\,dy_2 = C'(y) - C'(0)$$ (5.51)

where the boundary condition of eq. (5.49) was used. The flow of pollutants through a unit area perpendicular to the ξ direction becomes

$$J(\xi) = \frac{1}{A}\int_0^W Cu'h\,dy$$

$$= \frac{\partial \bar{C}}{\partial \xi}\frac{1}{A}\int_0^W u'(y)h(y)\int_0^y \frac{1}{Dh(y_2)}\int_0^{y_2} u'(y_1)h(y_1)\,dy_1\,dy_2\,dy$$ (5.52)

Here the concentration C was replaced by $C'(y) - C'(0)$. The difference with $C(x, y, t)$ is $\bar{C}(x, t) + C'(0)$ which is constant with respect to y. Its integral weighted by $u'h$ vanishes because of eq. (5.43). One may write eq. (5.52) as

$$J = -K\frac{\partial \bar{C}}{\partial \xi}$$ (5.53)

$$K = -\frac{1}{A}\int_0^W u'(y)h(y)\int_0^y \frac{1}{Dh(y_2)}\int_0^{y_2} u'(y_1)h(y_1)\,dy_1\,dy_2\,dy$$ (5.54)

where eq. (5.54) defines the *longitudinal dispersion coefficient* K. Finally, just as with diffusion in eq. (5.3), conservation of mass leads to

$$\frac{\partial \bar{C}}{\partial t} = -\frac{\partial J}{\partial \xi} = \frac{\partial}{\partial \xi}\left(K\frac{\partial \bar{C}}{\partial \xi}\right)$$ (5.55)

If K is independent of ξ this gives the one-dimensional equation

$$\frac{\partial \bar{C}}{\partial t} = K\frac{\partial^2 \bar{C}}{\partial \xi^2}$$ (5.56)

One may return to the fixed frame by realizing that on the left there is essentially the total derivative of the mean $\bar{C}$. On the right the partial derivative to ξ equals the one to x; this at last gives the *one-dimensional dispersion equation*

$$\frac{\partial \bar{C}}{\partial t} + \bar{u}\frac{\partial \bar{C}}{\partial x} = K\frac{\partial^2 \bar{C}}{\partial x^2}$$ (5.57)

which is analogous to eq. (5.5) with D replaced by K. This parameter is inversely proportional to the diffusion constant D and strongly dependent on the variation of velocity u' over the river cross-section as seen from eq. (5.54). If $u' = 0$, for example, this would imply an absence of transversal mixing, which is consistent with the algebraic result that $K = 0$ and $J = 0$ in the moving frame. In the absence of diffusion ($D = 0$) it follows that $K = \infty$; this would imply by eq. (5.56) that $\partial \bar{C}/\partial \xi$ would vanish, which agrees with eq. (5.49).

The molecular diffusion represented by D is very small. From eq. (5.21) and Table 5.1 it may be determined that after a release of salt in still water it would need 12 years before a root mean square dispersion of 1 metre is reached (cf. Exercise 5.7). In the case of an advective dispersion this number will depend on K, which in its turn depends on the velocity variation $u'(y)$. For a realistic case (cf. Exercise 5.8) K may be $0.0008 \, \text{m}^2 \, \text{s}^{-1}$ and the salt needs 10 minutes only to achieve a root mean square dispersion of 1 m. The latter number should not be taken literally as eq. (5.57) only holds after some time.

The Influence of Turbulence

When the water is in motion as in a river a dominant physical effect is turbulence. This is easily observed when some cans of dye are emptied at an interval of a minute. Each time they will disperse differently. This is due to instabilities of the flow pattern; they not only occur in rivers with a complicated geometry but also in long pipes, when the Reynolds number is above some critical value. In Section 5.5 a physical description of turbulence is given. Essentially turbulence is described as a statistical process just like molecular diffusion, which is the consequence of many collisions between molecules. In practice one just replaces the constant for molecular diffusion D by similar constants for turbulent diffusion ε.

As a preliminary it is necessary to look at the simple case of a uniform flow in a river of constant depth d and width W. Assume that the cross-section is rectangular and that a slope S causes a flow with constant velocities. The situation is sketched in Fig. 5.3. The two forces that act on a unit length and width W are the gravity and the tangential stress along the boundary.

Let τ_0 be the stress along the surface, which is the force acting per unit area as in Section 4.4.2. As there are no accelerations the stress τ_0 follows from

$$W\tau_0 = \rho W dg \sin \alpha \tag{5.58}$$

The slope S may be substituted for $\sin \alpha$, leading to

$$\tau_0 = \rho dg S \tag{5.59}$$

For the slope S any other force may be substituted which causes the flow to run such as a pressure difference. A characteristic velocity in describing the boundary layer at the bottom of the flow will be the friction velocity u_* already introduced in eq. (4.182) having the right dimension and being dependent on the relevant

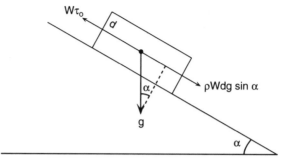

Figure 5.3   Stationary fluid movement under gravity and friction force in equilibrium

dynamical quantities. Therefore,

$$u_* = \sqrt{\frac{\tau_0}{\rho}} = \sqrt{gdS} \qquad (5.60)$$

In the hydraulics literature one usually gives this derivation for a cylindrical pipe with radius R. One finds eq. (5.59) with $R/2$ instead of d. Subsequently one defines a *hydraulic radius* r_h as the ratio of the cross-sectional area of the flow to its wetted perimeter. It follows that $r_h = R/2$, leading to eq. (5.59) with r_h instead of d. For a river the depth d is much smaller than its width. The hydraulic radius therefore becomes essentially the depth d, again leading to eq. (5.59).

For a real river one could go back to the definition of flux **F** in eq. (5.36) and represent turbulence by replacing the single diffusion coefficient D by three values $\varepsilon_x, \varepsilon_y$ and ε_z, representing the turbulent mixing in the three directions. One then writes

$$\mathbf{F} = \mathbf{u}C - \varepsilon_x \frac{\partial C}{\partial x}\mathbf{e}_x - \varepsilon_y \frac{\partial C}{\partial y}\mathbf{e}_y - \varepsilon_z \frac{\partial C}{\partial z}\mathbf{e}_z \qquad (5.61)$$

with $\mathbf{e}_x, \mathbf{e}_y$ and $\mathbf{e}_z$ unit vectors in the three dimensions. The ε values are called *eddy viscosities*. As turbulence will relate to the behaviour in the boundary layer, it seems acceptable that u_* appears in the expressions for ε; as the dimension of ε should be $m^2 s^{-1}$ the velocity should be multiplied by a length scale, for which the average depth d of the flow seems a good candidate. For the vertical ε_z one usually takes the value

$$\varepsilon_z = 0.067u_*d \qquad (5.62)$$

while for the transversal ε_y the value appears to be strongly dependent on the kind of flow:

$$\begin{array}{ll} \varepsilon_y = 0.15u_*d & \text{straight channel} \\ \varepsilon_y = 0.24u_*d & \text{irrigation canal} \\ \varepsilon_y = 0.6u_*d & \text{meandering streams} \end{array} \qquad (5.63)$$

All these estimates have large margins and the last decimal will not be significant. Although the vertical turbulence is smaller than the transversal one, the depth is usually so much smaller than the width that one assumes rapid mixing over the depth and ignores ε_z.

In the x direction, the longitudinal direction, one may ignore the direct turbulent diffusion, for by the argument depicted in Fig. 5.2 and leading to eqs. (5.53) it is again clear that the transversal turbulent diffusion will dominate dispersion in the direction of the stream. In fact, it is again eq. (5.57) that describes the one-dimensional dispersion. The factor K is again given by eq. (5.54) with the only adaptation that the molecular diffusion coefficient has to be replaced by ε_y. This gives

$$K = -\frac{1}{A} \int_0^W u'h(y) \int_0^y \frac{1}{\varepsilon_y h(y_2)} \int_0^{y_2} u'h(y_1)\,dy_1\,dy_2\,dy \qquad (5.64)$$

A few simple examples are discussed in order to illustrate the methods. It should be kept in mind that many approximations have been made and that in practice one uses empirical values for K. Some assumptions such as the constancy of the cross-sectional area A may be omitted, giving somewhat more complex equations.

Example: A Calamity Model for the Rhine River

Suppose that at a certain moment a factory ejects an amount M (kg) of poisonous substance in a river as the European Rhine River. As a first approximation one takes the one-dimensional equation (5.57) to describe the propagation of the concentration along the river. This equation reads

$$\frac{\partial \bar{C}}{\partial t} + \bar{u}\frac{\partial \bar{C}}{\partial x} = K\frac{\partial^2 \bar{C}}{\partial x^2} \qquad (5.65)$$

As this equation is the one-dimensional form of eq. (5.5) the solution is the one-dimensional form of eq. (5.26):

$$\bar{C}(x, t) = \frac{M}{2\sqrt{\pi K t}}e^{-(x-\bar{u}t)^2/(4Kt)} \qquad (5.66)$$

where of course the effective diffusion constant K from eq. (5.64) is used. At a certain point the amount of water passing will be S m^3 s^{-1} and the amount of pollution passing in a second will be $\bar{u}\bar{C}$ kg s^{-1}. Therefore the concentration φ passing per second will be

$$\varphi = \frac{\bar{u}\bar{C}}{S} = \frac{M/S}{\sqrt{4\pi Kt/\bar{u}^2}}e^{-(t-x/\bar{u})^2/(4Kt/\bar{u}^2)} \qquad (5.67)$$

In practice one divides a river in stretches in which the parameters K and S are

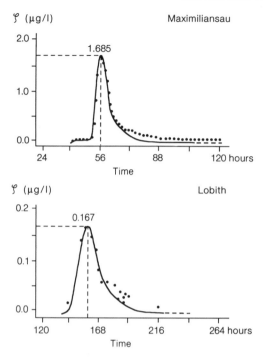

Figure 5.4 Calculations based on a Rhine calamity model, compared with calibration experiments. The upper figure refers to Maximiliansau, some 210 km from the 'calamity', and the lower figure to Lobith, some 710 km from the 'calamity'. (Reproduced by permission of Dr. A. van Mazijk, from Albert van Mazijk, Hardy Wiesner und Christian Leibundgut, *Das Alarmmodell für de Rhein—Theorie und Kalibrierung der Version 2.0-DGM 36*, 1992, Civiele Techniek, Delft, Netherlands, H2, p. 42–7)

essentially constant. Inflowing rivers may be taken into account by increasing S from stretch to stretch. The parameter K may be determined experimentally by injecting tracers into the river. The results of some model calculations are shown in Fig. 5.4.

One observes that the peak concentration lowers and the width of the distribution increases with time. As the width increases with time the concentration when measured at a fixed position will not be symmetrical in time. There will be a tail, though in practice more pronounced than in the simplest theory.

A Continuous Point Emission

Assume that an effluent is being emitted in a channel of constant width W and depth d at a rate of q kg m^{-3} s^{-1} at location $x = y = 0$. There will be rapid vertical mixing, so the source may be replaced by a vertical line source of strength q/d when d is the constant depth. If the boundaries of the river may be ignored the concentration after a long time is approximated by eq. (5.32) with the diffusion constant D replaced by the transverse ε_y and the velocity u by the average $\bar{u}$:

$$C = \frac{q}{d\sqrt{4\pi\varepsilon_y x\bar{u}}} e^{-y^2\bar{u}/(4\varepsilon_y x)} \tag{5.68}$$

The result would be exact when the velocity of the river over the cross-section is constant and when one could ignore the walls.

The walls of the river at $y = 0$ and $y = W$ (with constant W) will imply boundary conditions that the flow is zero, so $dC/dy = 0$. Let the source have the coordinates $x = 0$, $y = y_0$. Then the walls can be taken into account by introducing a double infinite series of mirror sources located at $y - y_0 = 0, \pm 2W, \pm 4W,\ldots$ and $y + y_0 = 0, \pm 2W, \pm 4W$, etc., as explained in eq. (5.35).

To work this out it is convenient to introduce dimensionless quantities:

$$C_0 = \frac{q}{\bar{u}dW}; \qquad x' = \frac{x\varepsilon_y}{\bar{u}W^2}; \qquad y' = \frac{y}{W}; \qquad y_0' = \frac{y_0}{W} \tag{5.69}$$

in which eq. (5.68) reads as

$$\frac{C}{C_0} = \frac{1}{\sqrt{4\pi x'}} e^{-y'^2/(4x')} \tag{5.70}$$

It is now easy to write down the series analogous to eq. (5.35) as

$$\frac{C}{C_0} = \frac{1}{\sqrt{4\pi x'}} \sum_{n=-\infty}^{\infty} [e^{-(y'-y_0'-2n)^2/(4x')} + e^{-(y'+y_0'-2n)^2/(4x')}] \tag{5.71}$$

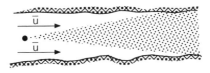

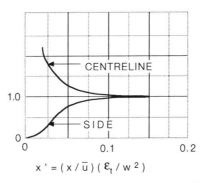

$$x' = (x/\bar{u})(\varepsilon_t/w^2)$$

Figure 5.5 Concentration profile for a continuous centreline injection. The curve 'centreline' shows the concentration as a function of distance along the centre of the river, the curve 'side' similarly at one of the sides

This result is plotted in Fig. 5.5 for a centreline injection ($y_0 = W/2$).

The profiles at the centreline and at the side are plotted. For $x' = 0.1$ the mixing is rather complete, within 5%. Taking $x' = 0.1$ as a standard the mixing length for centreline injection follows from eq. (5.69) as

$$L = \frac{0.1\bar{u}W^2}{\varepsilon_y} \tag{5.72}$$

Example: Dilution of Pollution*

When an industry discharges 10 million litres of effluent a day with a concentration of 200 ppm of pollutants, this would imply $q = 24\,\text{m}^3\,\text{s}^{-1}$ ppm. Assume a slowly meandering very wide river with a depth of 10 m, an average velocity of $1\,\text{m}\,\text{s}^{-1}$ and a friction velocity $u_* = 0.1\,\text{m}\,\text{s}^{-1}$. The problem is to calculate the width of the plume and the maximum concentration 300 m downstream. It follows from (5.63) that $\varepsilon_y = 0.6\,\text{m}^2\,\text{s}^{-1}$. Let the width of the plume be defined by two standard deviations on both sides. Equation (5.68) then gives a width b of

$$b = 4\sigma = 4\sqrt{2\varepsilon_y x/\bar{u}} = 76\,\text{m} \tag{5.73}$$

The maximum concentration is found where the exponential in eq (5.68) is one, leading to

$$C = \frac{q}{d\sqrt{4\pi\varepsilon_y x\bar{u}}} = 0.05\,\text{ppm} \tag{5.74}$$

Example: Mixing Length[†]

A plant discharges a pollutant in the middle of a straight rectangular channel with a depth $d = 2\,\text{m}$, a width $W = 70\,\text{m}$, a slope $S = 0.0002$ and an average velocity of $0.6\,\text{m}\,\text{s}^{-1}$. What is the length L for complete mixing? From (5.60) it follows that $u_* = 0.06\,\text{m}\,\text{s}^{-1}$ and from (5.63) that $\varepsilon_y = 0.019\,\text{m}^2\,\text{s}^{-1}$. According to eq. (5.72), $L = 15\,\text{km}$.

In view of the approximations made, the equations derived above can only give first-order estimates. Also, it should be realized that K is not easy to calculate from eq. (5.64) as the $u'(y)$ function in general will not be known. Thus, in practice

---

*Taken by permission of Academic Press from H. B. Fischer, E. J. List, R. C. Y. Koh, J. Imberger and N. H. Brooks, *Mixing in Inland and Coastal Waters*, Academic Press, New York, 1979, example 5.1, p. 117.
[†]Taken by permission of Academic Press from H. B. Fischer, E. J. List, R. C. Y. Koh, J. Imberger and N. H. Brooks, *Mixing in Inland and Coastal Waters*, Academic Press, New York, 1979, example 5.2, p. 118.

one makes some measurements in order to deduce a K value for a particular stream.

How can the equations be improved? The first possibility is to assume an x dependence for K, which would simulate an x dependence of the depth h or the cross-section A or the velocity field u'. If one repeats the derivation of eq. (5.55) one will see that when these are varying slowly with x, eq. (5.55) will be a good approximation, leading to

$$\frac{\partial \bar{C}}{\partial t} + \bar{u}\frac{\partial \bar{C}}{\partial x} = \frac{\partial}{\partial x}\left(K\frac{\partial \bar{C}}{\partial x}\right) \tag{5.75}$$

Another approach would be to return to eq. (5.36) and replace the single diffusion coefficient D by the three coefficients of turbulent diffusion ε_x, ε_y and ε_z. As vertical mixing happens quickly one may average over the depth by writing d times an averaged $\bar{C}$ instead of C, which amounts to looking at the total column of fluid. Similarly average fluxes are given by

$$\bar{F}_x = \left(u_x\bar{C} - \varepsilon_x\frac{\partial \bar{C}}{\partial x}\right)d$$

$$\bar{F}_y = \left(u_y\bar{C} - \varepsilon_y\frac{\partial \bar{C}}{\partial y}\right)d \tag{5.76}$$

where the fluxes now refer to the amount of mass passing through a unit width. In eq. (5.76) subtleties that one may not always put the average of a derivative equal to the derivative of the average are ignored. Conservation of mass in the xy plane gives

$$\frac{\partial(\bar{C}d)}{\partial t} + \frac{\partial}{\partial x}(u_x\bar{C}d) + \frac{\partial}{\partial y}(u_y\bar{C}d) = \frac{\partial}{\partial x}\left(\varepsilon_x d\frac{\partial \bar{C}}{\partial x}\right) + \frac{\partial}{\partial y}\left(\varepsilon_y d\frac{\partial \bar{C}}{\partial y}\right) \tag{5.77}$$

For ε_y the values given by eq. (5.63) may be inserted; for ε_x a value

$$\varepsilon_x = 5.93u_*d \tag{5.78}$$

is adopted as the best guess. When the y derivatives in eq. (5.77) are ignored, one would end up with eq. (5.75) where for K the value (5.78) is taken when there are no empirical data available. For a full two-dimensional calculation, using eq. (5.77), the precise value of ε_x does not matter much, as discussed before. Therefore one may as well use eq. (5.78).

5.3 GROUNDWATER FLOW

In contaminated soil pollutants may dissolve into the groundwater and disperse by groundwater flow. Downstream the groundwater then may become unsuitable for drinking or for agriculture and industry. Groundwater flow is described by

Table 5.2 Hydraulic conductivity k, particle size d and height of capillary rise h_c for some soils. (Data from various sources on hydraulics, going back to the 1950s)

	$k\,(\mathrm{m\,s^{-1}})$	$d\,(\mathrm{mm})$	$h_c\,(\mathrm{cm})$
Clay	$10^{-10}\!-\!10^{-8}$	<0.002	200–400
Silt	$10^{-8}\!-\!10^{-6}$	0.002–0.06	70–150
Sand	$10^{-5}\!-\!10^{-3}$	0.06–2	12–35
Gravel	$10^{-2}\!-\!10^{-1}$	>2	

Darcy's equations or *Darcy's law*, named after the French engineer Henry Darcy. He based these relations on experiments with groundwater percolation through filter beds in connection with the design of the Dijon water supply in 1856. These equations will be derived after a brief discussion of the structure of soil and unconsolidated sediments.

Natural soil consists of material in the form of solid grains with water and air in the pores in between. The porosity n is defined as the volume of the pores divided by the total (bulk) volume. For sandy soils one finds $0.35 < n < 0.45$; for clays and peat usually $0.40 < n < 0.60$. Not all pores necessarily contribute to groundwater flow (e.g. dead-end pores); if appropriate an effective porosity $\varepsilon \leqslant n$ is introduced. Some characteristic sizes of soil particles are displayed in Table 5.2.

Consider a steady flow of incompressible ground water and without change in storage in the soil. Take a volume element $d\tau = dx\,dy\,dz$ with the z direction of the Cartesian coordinate system pointing upwards. The pressure in the water is indicated by $p(x, y, z)$. The total force per unit of volume on the volume element is **F**. Then one has (cf. Fig. 5.6)

$$F_z\,d\tau = -\rho g\,d\tau + p(x, y, z)\,dx\,dy - p(x, y, z + dz)\,dx\,dy + f_z\,d\tau \qquad (5.79)$$

Here f_z is the z component of the friction force **f** per unit volume. In the x and y directions the gravity term on the right-hand side is absent. In eq. (5.79) ρ is the water density and g is the gravitational acceleration.

The friction force **f** is approximated by considering that the volume element contains many pores. Then Darcy's equations may be derived by assuming

$$\mathbf{f} = -\frac{\mu}{\kappa}\mathbf{q} \qquad (5.80)$$

Here **q** is the specific discharge vector, the discharge of fluid in $\mathrm{m^3\,s^{-1}}$ per unit area $\mathrm{m^2}$. Its dimension is $\mathrm{m\,s^{-1}}$ and it is proportional to the average velocity **u**

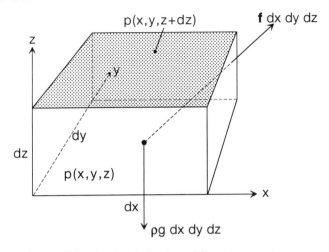

Figure 5.6 On the derivation of Darcy's equations

of the groundwater:

$$\mathbf{u} = \frac{\mathbf{q}}{n} \tag{5.81}$$

as the volume available to the flow is n times smaller than the total volume (ignoring $\varepsilon \leqslant n$). Of course the friction $\mathbf{f}$ is directed opposite to $\mathbf{q}$. The constant μ in eq. (5.80) is the dynamic viscosity of the fluid and κ the permeability of the medium, a soil parameter, usually increasing with the width of the pores.

Using eq. (5.80) one finds from eq. (5.79) that

$$F_z = -\rho g - \frac{\partial p}{\partial z} - \frac{\mu}{\kappa} q_z \tag{5.82}$$

and similar equations in the y and z directions but without the gravity term. The left-hand side of these equations is taken as zero, as the accelerations of ground water are negligible. This leads to

$$\frac{\partial p}{\partial x} + \frac{\mu}{\kappa} q_x = 0$$

$$\frac{\partial p}{\partial y} + \frac{\mu}{\kappa} q_y = 0 \tag{5.83}$$

$$\frac{\partial p}{\partial z} + \frac{\mu}{\kappa} q_z + \rho g = 0$$

These equations may be simplified further by introducing

$$\phi = z + \frac{p}{\rho g} \tag{5.84}$$

and assuming the density ρ constant. This gives

$$\mathbf{q} = -\frac{\kappa \rho g}{\mu} \operatorname{grad} \phi \tag{5.85}$$

The quantity ϕ is called the groundwater head and is the height of the water level in an open vertical test tube above the level $z = 0$, as illustrated in Fig. 5.7. Let the bottom end of the tube have the coordinate z and the water level be ϕ with respect to $z = 0$. The pressure p at the bottom then equals the hydrostatic pressure of the water column $(\phi - z)\rho g$ which agrees with eq. (5.84). Other names for ϕ are hydraulic head or hydraulic potential.

Darcy deduced eq. (5.85) experimentally and found values for the *hydraulic conductivity*

$$k = \frac{\kappa \rho g}{\mu} \tag{5.86}$$

which leads to Darcy's equations or Darcy's law

$$\mathbf{q} = -k \operatorname{grad} \phi \tag{5.87}$$

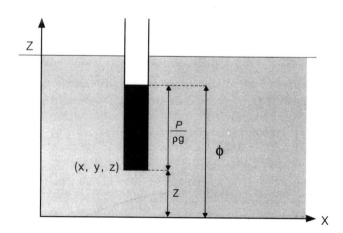

Figure 5.7 Illustration of the definition of the groundwater head ϕ. The arbitrary level $z = 0$ corresponds to the bottom of the picture

The discussion given above leads to the following comments:

(a) The assumption that the density ρ is constant is not trivial. A change of salinity as a function of position may well occur in practice. This effect is ignored.

(b) The hydraulic conductivity k is proportional to κ, the permeability of the medium. It is therefore strongly dependent on the size of the pores. Some empirical numbers are given in Table 5.2.

(c) Except near wells and sinks the horizontal components of the water discharge **q** (or velocity **u**) are much bigger than the vertical q_z (or u_z). Assume that at (x, y) the vertical discharge q_z vanishes: $q_z = 0$. Then at that position the groundwater head ϕ will be independent of the depth to which the test tube is lowered. By decreasing the depth one would reach the situation where $p = 0$, i.e. the local pressure is precisely the atmospheric pressure. These positions as a function of x, y define the so-called *phreatic surface*. This surface need not be horizontal for on a non-local scale q_z will not vanish completely.

It must be mentioned that in the absence of a vertical flow component, the phreatic surface is found by pushing the test tube to some depth as in Fig. 5.7 and deducing the height z for which $p = 0$ from eq. (5.84). In practice there may be capillary suction that pulls the water through narrow pores to above the phreatic surface (cf. Table 5.2). This gives an unsaturated zone in the soil which we will discuss below. The boundary between saturated and unsaturated soil is called the *groundwater table*. If one has shallow open tubes, wells or drains the water level will represent the phreatic surface which therefore is identical to the groundwater table.

(d) The derivation of eq. (5.87) assumed an isotropic medium. If the soil were anisotropic one usually generalizes the relation between **q** and grad ϕ into a tensor relation

$$\mathbf{q} = -\mathbf{k} \cdot \text{grad } \phi$$

with a tensor **k**. This may happen in a layered soil where the horizontal permeability κ may be larger than the vertical one. It can be shown that the tensor **k** is symmetrical. It can therefore be brought into diagonal form by a local rotation of the coordinate system. Equation (5.87) then reduces to three equations, each with their hydraulic conductivity. For a slowly varying layer structure one axis will be rather vertical with conductivity k_v; the two horizontal ones will be equal and are indicated by k_h.

(e) In deriving Darcy's equations (5.87) it was assumed that the inertia forces could be neglected with respect to the viscous forces. In terms of eq. (3.49) this implies a low Reynolds number Re. By definition (3.50) $Re = \rho U L / \mu = U L / v$. For water $v \approx 10^{-6} \, \text{m}^2 \, \text{s}^{-1}$. By taking a water velocity $U = 0.25 \times 10^{-2} \, \text{m s}^{-1}$ and particle size $L = 0.04 \times 10^{-2} \, \text{m}$ one finds $Re = 1$. By looking at Table 5.2 one may conclude that even for coarse sand Darcy's equations may hold provided that the groundwater velocity is not (much) greater than $0.25\,\mathrm{cm\,s^{-1}}$.

The comments given above should be kept in mind when Darcy's equation is applied to practical cases.

Vertical Flow in the Unsaturated Zone

The unsaturated soil is of essential importance to life. In a moist environment oxygen gives rise to many chemical reactions and many microorganisms influence the chemical and biological composition. Therefore we just mention a few aspects of the vertical water flow in the zone, although in the rest of the book we shall ignore it.

The water in the unsaturated zone is called soil moisture; it exerts a pressure which is negative with respect to the atmospheric pressure. Without flow the capillary suction in a point above the phreatic surface is equal to minus the height above that surface in metres of water pressure. More precisely, the negative suction head is expressed as $\psi = p/(\rho g)$.

Let us restrict ourselves to vertical flow only. Then Darcy's equations (5.87) give

$$q_c = -k\frac{d}{dz}(\psi + z) = -k\left(\frac{d\psi}{dz} + 1\right) \tag{5.88}$$

Especially in the unsaturated zone the hydraulic conductivity, $k = k(\psi)$, which is also called the capillary conductivity, is a function of position and time because of changing moisture content. The flow is upwards or downwards, depending on the derivative of the suction head. The limiting case (no flow) happens when

$$\frac{d\psi}{dz} = -1 \tag{5.89}$$

For more negative values the flow q_c will be positive, and so upwards, which will occur in the case of evaporation. For less negative values the flow will be downwards, which corresponds with the infiltration after rain. If the suction disappears the derivative will go to zero and eq. (5.88) reduces to $q_c = k_0$ where k_0 approaches the saturated hydraulic conductivity.

Conservation of Mass

It will be shown in Fig. 5.17 that conservation of mass within a volume leads to the equation of continuity:

$$\mathrm{div}(\rho\mathbf{q}) + \frac{\partial\rho}{\partial t} = 0 \tag{5.90}$$

For a steady flow the density is constant in time; in most applications variation of ρ with position may be ignored as well. This gives

$$\text{div } \mathbf{q} = 0 \tag{5.91}$$

When the hydraulic conductivity k is not dependent on position, eqs. (5.85) and (5.91) lead to the *Laplace equation*

$$\text{div grad } \phi = \frac{\partial^2 \phi}{\partial x^2} + \frac{\partial^2 \phi}{\partial y^2} + \frac{\partial^2 \phi}{\partial z^2} = 0 \tag{5.92}$$

This equation is discussed in any undergraduate course on electromagnetism. All properties of the equation discussed there hold again. One may even take the similarity further by introducing point wells and point sinks like positive and negative charges in a Coulomb field and use the method of images. As a matter of fact, the mathematical equation (5.92) together with its boundary conditions give unique solutions. It is the physicist who should give the interpretation and be aware of the assumptions underlying the equations.

Stationary Flow

A few relevant examples should illustrate the application of the equations. Assume that

(a) The ground is isotropic and saturated with fluid (water).
(b) The flow picture is independent of time, i.e. stationary.
(c) The flow essentially occurs in a vertical x, z plane, i.e. it is independent of the y coordinate: $\phi = \phi(x, z)$.

Darcy's equations (5.87) then read as

$$q_x = -k \frac{\partial \phi}{\partial x} \tag{5.93}$$

$$q_z = -k \frac{\partial \phi}{\partial z} \tag{5.94}$$

Vertical Flow

Consider the situation depicted in Fig. 5.8. There are two horizontal layers. The top layer is only a little permeable, e.g. clay, and the lower one is well permeable, e.g. sand. It forms a water-carrying layer, an *aquifer*. The sand may be under pressure from the top layer, resulting in a positive groundwater head $\phi = h$, as shown on the left in Fig. 5.8a. The top layer is saturated and its groundwater head precisely reaches the surface $z = 0$. There will be a vertical flow $q_z = q_0$ of unknown magnitude in the top layer. For the bottom layer with its much greater pores the flows are ignored.

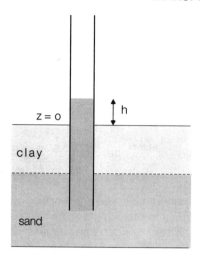

(a)

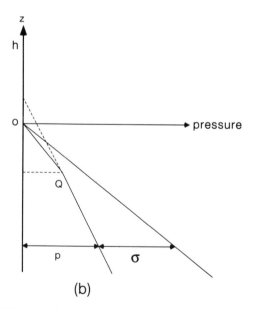

(b)

Figure 5.8 Vertical flow only. In (a) one observes two layers. The bottom layer is under pressure from the top layer or fed by the surroundings, resulting in a groundwater head above the surface $z = 0$. In (b) the groundwater pressure p and the internal grain pressure σ' are indicated. Note that $p_{tot} = p + \sigma'$

Let us first look at the top layer. From $q_x = 0$ and eq. (5.93) it follows that ϕ is independent of x. Substituting $q_z = q_0$ on the left of eq. (5.94) gives

$$\phi = -\frac{q_0 z}{k} + \text{constant}$$

As $\phi(0) = 0$ for the top layer, one finds

$$\phi = -\frac{q_0 z}{k} \tag{5.95}$$

For the pressure $p(z)$ one finds, with eq. (5.84),

$$p(z) = \rho g(\phi - z) = -\rho g z \left(\frac{q_0}{k} + 1\right) \tag{5.96}$$

Let us now look at the bottom layer. The pressure is just hydrostatic as flows are ignored:

$$p(z) = \rho g(h - z) \tag{5.97}$$

The curve $p(z)$ is drawn in Fig. 5.8b in such a way that the pressure joins continuously at the interface of both layers. As h can be measured, the discharge q_0 follows from the fit of both curves at point Q in Fig. 5.8. One should realize that $p(z)$ is the groundwater pressure in the pore system. One could also draw the total pressure p_{tot} as a function of depth: this pressure at certain z is the weight of the wet soil column on top of it. Assuming for simplicity that the density of wet soil is just a constant ρ_w one finds for the total pressure

$$p_{tot}(z) = -\rho_w g z \tag{5.98}$$

This is drawn in Fig. 5.8b as well. The total pressure p_{tot} may be interpreted as the sum of water pressure p and the intergranular soil pressure σ', as indicated in Fig. 5.8b. From eq. (5.97) it follows that the derivative of p in the sand is just ρg and therefore constant. When its groundwater head h increases the point Q in Fig. 5.8b moves to the right. Eventually the soil pressure in the clay vanishes and the clay will burst. In the limiting case that the intergranular soil pressure almost vanishes, the soil can no longer bear a weight: *quicksand*.

Flow Underneath a Wall

Consider a very thick permeable layer below the horizontal plane $z = 0$ as sketched in Fig. 5.9. At $x = 0$ there is a vertical wall separating a region with groundwater head $\phi(x < 0, z = 0) = 0$ on the left from a region with $\phi(x > 0, z = 0) = H$ on the right.

A practical example might be the effort to keep polluted water at the right of a wall. Figure 5.9 illustrates the extreme situation that no other boundaries are

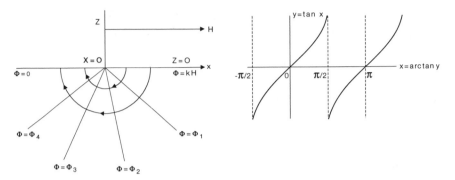

Figure 5.9   Flow underneath a wall. At the right of the wall the groundwater head equals H. Lines of constant ϕ and streamlines are indicated. The figure on the right gives the definition of $x = \arctan y$

present. Without derivation it will be shown that the solution is

$$\phi = \frac{H}{\pi} \arctan \frac{z}{x} \tag{5.99}$$

Here the $\arctan y = \tan^{-1} y$ is defined in an unconventional way. For y running from $-\infty$ to 0 the function runs from $\pi/2$ to π instead of from $-\pi/2$ to 0. At $y = 0$ there is a discontinuity and if y runs from 0 to ∞ the function goes from 0 to $\pi/2$ in the usual way. For the derivatives this does not make any difference. It is then easy to show that eq. (5.99) obeys Laplace's equation (5.92).

Next, study the boundary conditions of Fig. 5.9. For $x < 0$ and z approaching zero from below, ϕ goes to the limit $\arctan(+0) = 0$. For $x > 0$ and z approaching zero from below, ϕ goes to the limit H. Therefore the boundary conditions are satisfied with the unconventional definition of arctan. Substituting ϕ in Darcy's eqs. (5.93) and (5.94) gives

$$q_x = \frac{kH}{\pi} \frac{z}{x^2 + z^2} \tag{5.100}$$

$$q_z = -\frac{kH}{\pi} \frac{x}{x^2 + z^2} \tag{5.101}$$

For $z = 0$ we find

$$q_z = -\frac{kH}{\pi x} \tag{5.102}$$

Therefore, for $x > 0$ the water goes down and for $x < 0$ it comes up. The amount of water Q that leaves the reservoir at the right in the region $a < x < b$ per unit depth in the y direction is found by integration of eq. (5.102):

$$Q = \int_a^b q_z \, dx = \frac{kH}{\pi} \ln \frac{b}{a} \tag{5.103}$$

One notices that realistic values should be taken for a and b, as for $a = 0$ or $b = \infty$ the integral Q becomes infinite.

The Complex Variable Method

A general method of tackling complicated problems is found in the method of complex variables. We start by introducing a function

$$\Phi = k\phi \tag{5.104}$$

Darcy's equations (5.87) now become even simpler:

$$\mathbf{q} = -\operatorname{grad} \Phi \tag{5.105}$$

In order to apply complex functions one has to stick to two dimensions, either x and z as discussed above or x and y when we consider horizontal groundwater flow, e.g. between sinks and wells. By virtue of the definition (5.84) the mathematics for both cases is the same. As we want to keep the symbol z for a complex number, we take x and y as our coordinates.

Equation (5.105) suggests that Φ be interpreted as a two-dimensional *potential*. Lines $\Phi = $ constant would then be the aequipotential curves. The discharge vector $\mathbf{q}$ is everywhere perpendicular to these curves.

Another function of importance is the *stream function* Ψ. Conservation of mass led with appropriate assumptions to the continuity equation (5.91). In two dimensions this gives

$$\frac{\partial q_x}{\partial x} + \frac{\partial q_y}{\partial y} = 0 \tag{5.106}$$

As this equation holds for all (x, y), mathematics texts prove that there exists a function Ψ with the property that

$$q_x = -\frac{\partial \Psi}{\partial y} \tag{5.107}$$

$$q_y = +\frac{\partial \Psi}{\partial x} \tag{5.108}$$

It follows that the vector $\mathbf{q}$ is perpendicular to the vector $(\partial \Psi/\partial x, \partial \Psi/\partial y)$, as their scalar product vanishes. Therefore the curves $\Phi = $ constant and $\Psi = $ constant are perpendicular and consequently the lines $\Psi = $ constant are the streamlines having the direction of $\mathbf{q}$ everywhere. In Fig. 5.9 describing the flow underneath

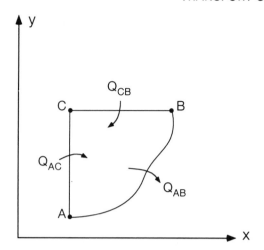

Figure 5.10 Interpretation of the stream function Ψ

a wall, the curves with $\Phi = \text{constant}$ and the perpendicular streamlines were indicated. The student should check that Ψ also obeys the Laplace equation.

The stream function Ψ relates to the discharge Q. Consider Fig. 5.10 and look at points A and B with an auxiliary point C such that $x_C = x_A$; $y_C = y_B$. In Fig. 5.10 the discharges Q_{AC}, Q_{CB} and Q_{AB} through the corresponding lines are indicated. Let H be the thickness of the layer through which the ground water flows, measured perpendicularly to the x, y plane. Then from eq. (5.108) the in-stream into ABC becomes

$$Q_{AC} = H \int_A^C q_x \, dy = -H \int_A^C \frac{\partial \Psi}{\partial y} \, dy = H(\Psi_A - \Psi_C)$$

$$Q_{CB} = -H \int_C^B q_y \, dx = H(\Psi_C - \Psi_B)$$

The minus sign in the lowest equation corresponds to the (arbitrary) drawing in Fig. 5.10. From continuity (no accumulation of mass) it follows that the out-flow equals the in-flow:

$$Q_{AB} = Q_{AC} + Q_{CB} = H(\Psi_A - \Psi_B) \tag{5.109}$$

Therefore the difference in stream function Ψ between two points A and B represents the amount of water flowing per unit thickness through *any* line connecting both points.

By combining eqs. (5.105), (5.107) and (5.108) we find

$$\frac{\partial \Phi}{\partial x} = \frac{\partial \Psi}{\partial y} \tag{5.110}$$

$$\frac{\partial \Phi}{\partial y} = -\frac{\partial \Psi}{\partial x} \qquad (5.111)$$

These relations may be recognized as the *Cauchy–Riemann equations*, which are 'necessary and sufficient' for the existence of an analytical function

$$\Omega = \Phi + i\Psi \qquad (5.112)$$

of a complex variable $z = x + iy$. The aim of the calculations then is to deduce the functions $\Phi(x, y)$ and $\Psi(x, y)$ in a water-carrying layer. One only needs their values on a boundary, which are specified from the physics of the problem. Streamlines, for example, can only start at a source or at a water surface and can only end at a sink or at another water surface.

Before the advent of the personal computer a lot of attention was paid to analytical methods to find $\Omega(z)$, comprising both $\Phi(x, y)$ and $\Psi(x, y)$. This was done by the method of *conformal mapping* by which complicated boundaries were transformed into simple, e.g. rectangular, ones. At present these methods are replaced by numerical solutions of the differential equations. Then it is easy to take into account variations of density ρ and conductivity k with position and solve a complete three-dimensional problem. Analytical methods, though, still keep their value in giving a first approximation and in getting a feeling about what the numerical solutions should look like. A few examples will be given below.

The Dupuit Approximation

In the third comment below eq. (5.87) the concept of groundwater head was explained by assuming that the main groundwater flow $\mathbf{q}$ is horizontal. Consequently, the groundwater head ϕ would be a function of the horizontal components only: $\phi = \phi(x, y)$. This simplification is called the *Dupuit approximation*.

Dupuit's approximation will not hold in the case of Fig. 5.8 where two layers were depicted with a different groundwater head. Neither will it apply near the wall in Fig. 5.9 where strong vertical components of the discharge are expected (cf. eq. (5.102)). The approximation holds good, however, for a broad class of problems connected with *unconfined aquifers*. These are water-carrying layers where the top surface of the ground water is, via pores, in open connection with the atmosphere. As remarked earlier, at this phreatic surface the groundwater pressure equals the atmospheric pressure (or vanishes when one only describes deviations from atmospheric pressure).

Unconfined aquifers are the topmost water-carrying layer. It is assumed that their bottom is formed by an impervious or semi-impervious layer. It is the unconfined aquifer where groundwater is in closest contact with the human environment. The equations that govern its flow may be derived by looking at Fig. 5.11.

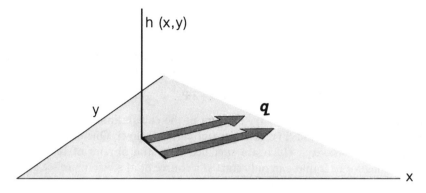

Figure 5.11   Flow through an unconfined aquifer with groundwater head $h(x, y)$.

The impervious layer is taken as $z = 0$ and the groundwater head is denoted as $h(x, y)$ instead of the usual $\phi(x, y)$. The total mass flow through the layer in a strip of unit width is given by using eq. (5.87) as

$$\rho h\mathbf{q} = - k\rho h \operatorname{grad} h \tag{5.113}$$

Conservation of mass leads to

$$\operatorname{div}(\rho h\mathbf{q}) = 0 \tag{5.114}$$

or, assuming that ρ and k are constant,

$$\frac{\partial}{\partial x}\left(h\frac{\partial h}{\partial x} \right) + \frac{\partial}{\partial y}\left(h\frac{\partial h}{\partial y} \right) = 0$$

$$\frac{\partial^2 h^2}{\partial x^2} + \frac{\partial^2 h^2}{\partial y^2} = 0 \tag{5.115}$$

This is a Laplace equation for h^2, whereas eq. (5.92) applies to $\phi = h$. The difference is related to the fact that Fig. 5.11 refers to the total movement through a vertical cross-section and takes the variation of h with position into account. Of course, eq. (5.92) is more general and should hold as well. However, then one should also take into account vertical motion explicitly.

One-Dimensional Flow

Consider two parallel canals with an aquifer between, all on the same impervious layer. The problem can be described in one dimension by $h = h(x)$ with $h(0) = h_1$ and $h(L) = h_2$. This is illustrated in Fig. 5.12.

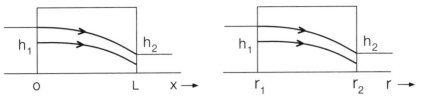

Figure 5.12 One-dimensional flow through a linear wall (left) or a circular wall (right). The parabola are solutions of the equations. In reality there is a seepage face at the right-hand side of both figures, just above height h_2.

The solution becomes

$$h^2 = h_1^2 - (h_1^2 - h_2^2)\frac{x}{L} \tag{5.116}$$

In Fig. 5.12 $h(x)$ will become a parabola connecting both water levels. The discharge per unit width over the total water column becomes, at position x,

$$hq = -kh\frac{dh}{dx} = -\frac{1}{2}k\frac{dh^2}{dx} = \frac{k}{2L}(h_1^2 - h_2^2) \tag{5.117}$$

which is independent of x because of conservation of mass. The Laplace equation (5.92) may now be used to estimate whether the vertical discharges q_z may indeed be ignored. One finds

$$\frac{\partial q_x}{\partial x} + \frac{\partial q_z}{\partial z} = 0 \tag{5.118}$$

$$q_z = kz\frac{\partial^2 h}{\partial x^2}$$

or, with eqs. (5.116) and (5.117) and some algebra

$$q_z = -\frac{zq^2}{kh^3} \tag{5.119}$$

These negative velocities are highest at the groundwater head where $z = h$. In reality the top streamlines finish on the right a little above height h_2, resulting in a *seepage face*.

Cylindrical Symmetry

For problems with a cylindrical symmetry one uses the coordinates (r, θ, z) and the Laplace operator from eq. (B.2) becomes

$$\nabla^2 = \text{div grad} = \frac{1}{r}\frac{\partial}{\partial r}\left(r\frac{\partial}{\partial r}\right) + \frac{1}{r^2}\frac{\partial^2}{\partial \theta^2} + \frac{\partial^2}{\partial z^2} \tag{5.120}$$

When the z and θ coordinates may be ignored, only the first term on the right-hand side of eq. (5.120) should be kept.

A Circular Pond

More relevant to the environment than the situation sketched on the left of Fig. 5.12 is perhaps the pond, depicted on the right of this figure. A pond with radius r_1 is surrounded by a soil wall with outer radius r_2. The groundwater level on the outside (h_2) is lower than on the inside (h_1). The discharge of possibly polluted water is calculated with Dupuit's equation (5.115) in cylindrical coordinates (5.120) and putting $h = h(r)$. This gives

$$h^2(r) = h_1^2 - (h_1^2 - h_2^2)\frac{\ln r - \ln r_1}{\ln r_2 - \ln r_1} \qquad (5.121)$$

The discharge vector $h\mathbf{q} = -kh\,\mathrm{grad}\,h$ is in the radial direction, which gives

$$hq_r = -kh\frac{\partial h}{\partial r} = \frac{k}{2r}\frac{h_1^2 - h_2^2}{\ln r_2 - \ln r_1} \qquad (5.122)$$

The total discharge is given by $Q = 2\pi r\,hq_r$ and is independent of r because of conservation of mass:

$$Q = \pi k\frac{h_1^2 - h_2^2}{\ln(r_2/r_1)} \qquad (5.123)$$

Simple Flow in a Confined Aquifer

A confined aquifer is a saturated water-carrying layer confined between two rather impervious layers. Assume that the layer is horizontal with uniform thickness H. It is not appropriate to apply Dupuit's equations as the layer is saturated. The groundwater head ϕ will normally be higher than its top and will be indicated by $\phi = h(x, y)$. Following the discussion earlier in this section one introduces a potential function $\Phi = kh(x, y)$ and a stream function $\Psi(x, y)$ to describe the flow. We shall discuss two examples of flows in confined aquifers.

Flow around a Source or Sink

The first example is a source or sink in the layer. This is assumed to be a cylindrical, vertical hole of radius $R \leqslant r_1$. In the layer itself, at $r = r_1$ the groundwater head will be $\phi = h_1$ and some distance away at $r = r_2$ it will be $\phi = h_2$. Depending on whether water is put into the cylinder or is taken out, one has $h_1 > h_2$ or $h_1 < h_2$. This is called a sink or a source respectively.

Cylindrical symmetry leads to the simplified Laplace equation within the layer $r_1 < r < r_2$:

$$\frac{1}{r}\frac{\partial}{\partial r}\left(r\frac{\partial \Phi}{\partial r}\right) = 0$$

Similar to the derivation of eq. (5.121) one now finds

$$\Phi(r) = kh_1 - k(h_1 - h_2)\frac{\ln r - \ln r_1}{\ln r_2 - \ln r_1} \qquad (5.124)$$

The radial discharge q_r becomes

$$q_r = -\frac{\partial \Phi}{\partial r} = \frac{k(h_1 - h_2)}{r(\ln r_2 - \ln r_1)} \qquad (5.125)$$

The total seepage Q *from outside to inside* at distance r is found by

$$Q = -2\pi r H q_r = -\frac{2\pi k H(h_1 - h_2)}{\ln r_2 - \ln r_1} \qquad (5.126)$$

We notice that eq. (5.126) reduces to eq. (5.123) when $H = (h_1 + h_2)/2$, which should not mislead us as the physical situation is different. Equation (5.126) is the basic equation for pump tests, applied to determine the so-called *transmissivity* kH from the seepage Q and the resulting drawdown $(h_1 - h_2)$.

The freedom of choice in the position of the $z = 0$ plane is used by taking $h_2 = 0$ at some big radius r_2. Substitution of eq. (5.126) into eq. (5.124) and replacing $r_1 = r$ then gives

$$\Phi = \frac{Q}{2\pi H}\ln\frac{r}{r_2} \qquad (5.127)$$

Looking at eqs. (5.125) and (5.126) one notices that for $Q < 0$ the water flows outwards (a sink) and for $Q > 0$ inwards (a source or a spring). The aequipotential lines $\Phi = $ constant are circles in the x, y plane around $x = y = 0$ and the streamlines $\Psi = $ constant must be radial lines. In polar coordinates they will read

$$\Psi = c_1 \theta \qquad (5.128)$$

The constant c_1 is found by realizing that the difference $-d\Psi$ between the points r, θ and $r, \theta + d\theta$ must be the mass flow for a unit thickness through the circle segment joining them (cf. eq. (5.109)). Hence

$$-d\Psi = r\, d\theta q_r = r\, d\theta \frac{\partial \Phi}{\partial r} \qquad (5.129)$$

$$\frac{1}{r}\frac{\partial \Psi}{\partial \theta} = \frac{\partial \Phi}{\partial r} \qquad (5.130)$$

Equation (5.130) will hold in general. With $\Psi = c_1\theta$ and eq. (5.127) it leads to

$$\Psi = \frac{Q}{2\pi H}\theta \qquad (5.131)$$

The complex function $\Omega = \Phi + i\Psi$ which was introduced in eq. (5.112) becomes

$$\Omega = \Phi + i\Psi = \frac{Q}{2\pi H}(\ln r + i\theta) - \frac{Q}{2\pi H}\ln r_2$$

$$\Omega = \frac{Q}{2\pi H}\ln z - \frac{Q}{2\pi H}\ln r_2 \qquad (5.132)$$

where it was used that $z = x + iy = re^{i\theta}$.

The advantage of working with the complex function Ω is that it helps to calculate more complicated problems. We notice that the Laplace equation is linear in Φ. This implies that for a complicated problem with, for example, many sources and sinks the sum of the corresponding functions Φ is again a solution of Laplace's equation. This is the *superposition principle*. Therefore, from the resulting Φ one may calculate the flow pattern using eq. (5.105).

Source or Sink in Uniform Flow

As an example of applying the superposition principle, consider a source or sink in a uniform flow with discharge U in the negative x direction. For the uniform flow one has

$$\mathbf{q} = -U\mathbf{e}_x = -\operatorname{grad}\Phi = -\frac{\partial\Phi}{\partial x}\mathbf{e}_x - \frac{\partial\Phi}{\partial y}\mathbf{e}_y \qquad (5.133)$$

where $\mathbf{e}_x$ and $\mathbf{e}_y$ are the unit vectors in the positive x and y directions respectively. It follows that $\Phi = Ux$ and therefore (cf. eq. (5.110)) $\Psi = Uy$ and, for a *uniform flow*,

$$\Omega = U(x + iy) = Uz \qquad (5.134)$$

For an extra source or sink one should add eq. (5.132) to this field, which leads to the complex function

$$\Omega = \frac{Q}{2\pi H}\ln z + Uz - \frac{Q}{2\pi H}\ln r_2 = \Phi + i\Psi \qquad (5.135)$$

From this equation it follows that

$$\Phi = Ux + \frac{Q}{4\pi H}\ln\frac{x^2 + y^2}{r_2^2} \qquad (5.136)$$

$$\Psi = Uy + \frac{Q}{2\pi H}\arctan\frac{y}{x} \qquad (5.137)$$

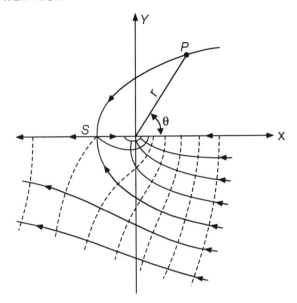

Figure 5.13 Flow picture for a well ($Q > 0$) in the origin and a uniform flow with discharge U in the negative x direction. Note the point of stagnation S where $q_x = 0$

From $z = x + iy = re^{i\theta}$ one may express the function Ψ as

$$\Psi = Ur\sin\theta + \frac{Q}{2\pi H}\theta \tag{5.138}$$

The lines $\Psi = $ constant are the streamlines, depicted in Fig. 5.13.

Simple Time Dependence in a Confined Aquifer

Let us consider a river or a lake in connection with a confined aquifer of constant thickness H. In equilibrium the groundwater head h_0 in the aquifer will be equal to the river level as indicated in Fig. 5.14. The situation is essentially one dimensional with variable x while $x = 0$ denotes the interface between aquifer and river.

Suppose that at time $t = 0$ the river level suddenly falls to $h_0 - s_0$ and remains at that level. The aquifer will empty into the river and a time-dependent head $h(x, t)$ will result. The total flow through 1 m width and total height H of the aquifer will be

$$\rho H\mathbf{q} = -k\rho H\nabla h \tag{5.139}$$

The so-called *storage coefficient* or *storativity* S of an aquifer is defined as the volume of water taken into or released from storage per unit horizontal area per

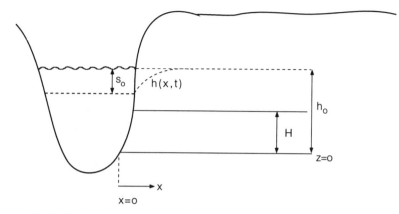

Figure 5.14   River in connection with an aquifer. At $t = 0$ the water level drops by s_0

unit rise or decline in groundwater head.* Change in storage in a confined aquifer is produced by the compressibility of water and of the particle skeleton. Therefore, the mass of water taken in for a rise dh of the groundwater head per unit area is

$$\rho S \, dh \tag{5.140}$$

Conservation of mass in the x, y plane requires

$$\text{div}(\rho H \mathbf{q}) + \frac{\partial(\rho S h)}{\partial t} = 0 \tag{5.141}$$

Assuming constant ρ and S and using eq. (5.139) leads to

$$-k\rho H \nabla^2 h + \rho S \frac{\partial h}{\partial t} = 0 \tag{5.142}$$

In the one-dimensional case considered here we find

$$\frac{\partial^2 h}{\partial x^2} = \frac{S}{kH} \frac{\partial h}{\partial t} \tag{5.143}$$

This is precisely the one-dimensional form (4.19) of the heat equation with $a = kH/S$. The sudden decrease s_0 in river level corresponds with the sudden change in temperature (4.23). Therefore the solution of eq. (5.143) corresponds with

---

*For a confined aquifer $S < 0.001$; the specific yield of sandy soils varies from 0.15 for fine sand to 0.30 for coarse grained soils. (From J. J. de Vries, lecture notes, Amsterdam.)

eq. (4.25) and we write

$$h(x, t) = h_0 - s_0 + s_0 \, \mathrm{erf}\left(\frac{x}{2} \sqrt{\frac{S}{kHt}} \right) \tag{5.144}$$

For $t = 0$ we indeed find $h = h_0$ and for $t \to \infty$ we find $h = h_0 - s_0$.

Transport of Pollutants

For a piece of contaminated ground water a first approximation to the question of how long it would take for the pollutants to travel a distance L in the ground water would be L/u. Here u will be the real water velocity $u = q/n$ according to eq. (5.81).

In practice one is rather interested in the more complicated question of how a concentration distribution $c(\mathbf{r}, t)$ travels in ground water of velocity $\mathbf{u}$. Here c describes the mass of the pollutant per unit volume ground water. One has to realize that the ground water is winding through pores, wide and narrow, which results in a so-called *hydrodynamic dispersion*. As the origin is similar to the origin of diffusion, collisions of particles with the surroundings, the resulting differential equations are similar. They are found from eq. (5.5) as

$$\frac{\partial c}{\partial t} = D \nabla^2 c - (\mathbf{u} \cdot \nabla) c \tag{5.145}$$

The first term of eq. (5.145) describes the dispersion, which is composed of a hydrodynamic dispersion and a proper diffusion as described in Section 5.1. When the ground water does not move ($\mathbf{u} = 0$) the dispersion coefficient D reduces to a diffusion coefficient. For many practical applications the hydrodynamic dispersion dominates and the molecular diffusion may be ignored.

The second term in eq. (5.145) describes the *advection* of the concentration c by the ground water, i.e. the motion of contaminants with the ground water. Equation (5.145) is, because of its two contributions, called the *dispersion–advection* equation. In many cases one may use a one-dimensional form of the equation with x a, possibly curvilinear, coordinate:

$$\frac{\partial c}{\partial t} = D \frac{\partial^2 c}{\partial x^2} - u \frac{\partial c}{\partial x} \tag{5.146}$$

Even a simple equation like (5.146) can only be solved analytically for simple cases. One often estimates D and u empirically, or calculates u by methods described in this section, and then solves eq. (5.146) numerically.

Besides dispersion and advection a third process often occurs: *adsorption* of contaminants to the solid soil of the ground. To describe the process one has to add a third term to eq. (5.146) which becomes

$$\frac{\partial c}{\partial t} = D\frac{\partial^2 c}{\partial x^2} - u\frac{\partial c}{\partial x} - \frac{\rho_b}{n}\frac{\partial S}{\partial t} \tag{5.147}$$

Here S is the mass of the pollutant adsorbed on the solid part of the medium per unit mass of solids; ρ_b is the bulk mass density of the porous medium. The ratio ρ_b/n then becomes the bulk mass per unit volume of pore space. Thus $S\rho_b/n$ describes the adsorbed mass of pollutant per unit volume of pore space, which is consistent with the definition of c. One may check eq. (5.147) by putting $D = u = 0$. The only process by which c can increase is the decrease in adsorption S.

In practice it appears that to a good approximation

$$S = \kappa c \tag{5.148}$$

where κ expresses the partitioning of pollutants between solids and water. Equation (5.147) may then be written as

$$\left(1 + \frac{\kappa\rho_b}{n}\right)\frac{\partial c}{\partial t} = D\frac{\partial^2 c}{\partial x^2} - u\frac{\partial c}{\partial x} \tag{5.149}$$

This equation applies to cleaning a contaminated piece of ground. Some holes are then usually put in the ground, one in the middle of the contamination where the water is pumped out and a few around where clean water is put in. The contaminated water which was pumped out is cleaned and put back into the ground water outside the contamination. In this way the ground will eventually be cleaned.

For certain chemicals the adsorption κ may be very strong, leading to a high factor $R = 1 + \kappa\rho_b/n$ on the left of eq. (5.149): for PCBs (polychlorinated biphenyls) it may be as large as $R \approx 1000$. Ignoring dispersion D in eq. (5.149) this would mean that the velocity u by which the ground is cleaned in a first approximation reduces from u to u/R. With a typical value $u = 1\,\text{m/day}$ this implies long periods of time, and high expense, to clean the ground.

It should be stressed that the discussion above applies to contaminants dissolved in ground water and possibly adsorbed in soil. Much more complicated is the motion of oil from a leaking tank in the ground. One has to describe several phases: oil, water and possibly air. This is beyond the scope of the present text.

5.4 THE EQUATIONS OF FLUID DYNAMICS

Fluid dynamics is the part of physics that deals with fluids, a term comprising the motion of gases like air (aerodynamics) and the mechanics of liquids (hydrodynamics). A fluid is characterized by the fact that its elements very easily change shape; to be more precise, if one considers a certain mass within a volume of a certain shape, forces will always be able to deform the volume. In the present section we discuss the basic properties of fluids and derive the fundamental Navier–Stokes differential equations. Finally, we come back to the concept of the Reynolds number which has already been introduced in Section 3.3 and we describe a few properties of moving fluids.

The objective of the present section is to appreciate the difficulty of solving the basic equations of motion and the frequent need for simplifications. It thus justifies the simplified treatments of the preceding two paragraphs and it introduces the next one.

Stress Tensor

Consider a fluid element with part of its surface area being a plane surface element $\delta\mathbf{A}$. Of course, the vector $\delta\mathbf{A} = \mathbf{n}\delta A$ points outwards, towards other fluid elements. The outer fluid exerts a force

$$\mathbf{\Sigma}(\mathbf{n}, \mathbf{r}, t)\delta A \qquad (5.150)$$

on the element under consideration. This force $\mathbf{\Sigma}$ per unit area (a vector!) is called *stress*. It is not necessarily perpendicular to the surface, as internal friction may cause tangential components. Because of Newton's law that action + reaction = zero, the stresses between two adjoining fluid elements are pointing in opposite directions. Therefore the stress $\mathbf{\Sigma}$ is an odd function of $\mathbf{n}$. It is a surface force as it acts through the surface and has a dimension of $N\,m^{-2}$.

Consider now a volume element in the shape of a tetrahedron with three orthogonal faces (Fig. 5.15) and a sloping face with surface area $\delta\mathbf{A} = \mathbf{n}\delta A$. From the figure it is clear that the orthogonal areas are essentially the projections of $\delta\mathbf{A}$ on the planes:

$$\delta A_1 = \mathbf{n}\cdot\mathbf{a}\,\delta A$$
$$\delta A_2 = \mathbf{n}\cdot\mathbf{b}\,\delta A \qquad (5.151)$$
$$\delta A_3 = \mathbf{n}\cdot\mathbf{c}\,\delta A$$

The sum of the surface forces acting on the fluid element can therefore be written as

$$\mathbf{\Sigma}(\mathbf{n})\delta A + \mathbf{\Sigma}(-\mathbf{a})\delta A_1 + \mathbf{\Sigma}(-\mathbf{b})\delta A_2 + \mathbf{\Sigma}(-\mathbf{c})\delta A_3$$
$$= (\mathbf{\Sigma}(\mathbf{n}) - \mathbf{\Sigma}(\mathbf{a})\mathbf{a}\cdot\mathbf{n} - \mathbf{\Sigma}(\mathbf{b})\mathbf{b}\cdot\mathbf{n} - \mathbf{\Sigma}(\mathbf{c})\mathbf{c}\cdot\mathbf{n})\delta A \qquad (5.152)$$

in which the property was used that $\mathbf{\Sigma}$ is an odd function of the argument vector. Now take the *i*th component of the vector (5.152) with $i = 1, 2, 3$ denoting x, y

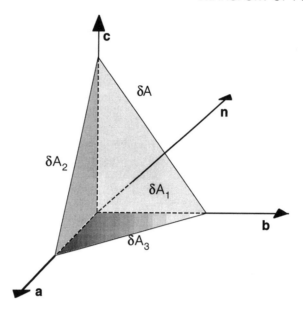

Figure 5.15   A volume element with three orthogonal faces and a fourth plane face
δA

and z components respectively and use the summation convention

$$\mathbf{a} \cdot \mathbf{n} = a_j n_j$$

which implies summing over repeated indices. Then the ith component of (5.152)
becomes

$$\{\Sigma_i(\mathbf{n}) - [a_j \Sigma_i(\mathbf{a}) + b_j \Sigma_i(\mathbf{b}) + c_j \Sigma_i(\mathbf{c})] n_j\} \delta A \qquad (5.153)$$

Consider now the equation of motion for the fluid element:

$$\text{mass} \times \text{acceleration} = \text{total body force} + \text{total surface force} \qquad (5.154)$$

The left-hand side of this equation is proportional to δV, as is the total body
force. They are proportional to L^3 if L is a linear dimension of the element. The
total surface force is proportional to the total surface area, i.e. proportional to
L^2 Let L become smaller and smaller, approaching zero. It follows that the total
surface force (5.153) dominates on the right-hand side of eq. (5.154) while the
others go to zero. For small elements the total surface force (5.153) should therefore vanish identically:

$$\Sigma_i(\mathbf{n}) = \sigma_{ij} n_j \qquad (5.155)$$

in which one should keep in mind the summation over j. The *stress tensor* σ_{ij} is a quantity with nine components, relating the vectors $\boldsymbol{\Sigma}$ and $\mathbf{n}$. It is called a tensor because eq. (5.155) will hold in any coordinate system (although of course the components will depend on the system). In mathematics courses it is proven that both indices behave like a vector under coordinate transformation.

Of course the nine components of σ_{ij} could be defined by means of eq. (5.153). It is more illuminating, however, to use (5.155) as the defining equation once we know how σ_{ij} behaves under transformation. Thus, once one knows the stress tensor σ_{ij} at position $\mathbf{r}$ one may deduce the surface force on any element by eq. (5.155). Consider a component σ_{ij} in a certain coordinate system. Take a plane surface element normal to the j direction and therefore with a normal $\mathbf{n} = \mathbf{e}_j$. According to eq. (5.155) $\Sigma_i = \sigma_{ij}$; therefore σ_{ij} is the i component of the force per unit area exerted across a plane normal to the j direction. The diagonal components σ_{11}, etc., are called the *normal* stresses and the non-diagonal elements the *tangential* stresses (some-times *shearing* stresses). For two dimensions this is illustrated in Fig. 5.16.

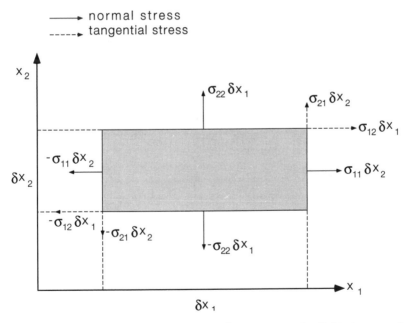

Figure 5.16 Normal and tangential forces for a rectangular fluid element of unit depth. The minus signs on the left/bottom vectors indicate that they point into the negative direction

For transformations between Cartesian coordinate systems two elementary tensors exist:

(a) *Kronecker delta tensor* with elements δ_{ij} such that $\delta_{ij} = 1$ if $i = j$ otherwise $\delta_{ij} = 0$.

(b) *Alternating tensor* with elements ε_{ijk} such that
   (i) $\varepsilon_{ijk} = + 1$ if the three numbers i, j and k are a positive permutation of the numbers 1, 2 and 3 (i.e. cyclic);
   (ii) $\varepsilon_{ijk} = - 1$ if i, j and k are a negative permutation of 1, 2, 3;
   (iii) $\varepsilon_{ijk} = 0$ otherwise, i.e. if two indices are equal.

Now take again a fluid element with a point O inside, and consider the torque τ of the surface forces about O. For a certain surface element δA one has a contribution

$$\tau = \mathbf{r} \times \Sigma \delta A \tag{5.156}$$

With the help of the alternating tensor one may write

$$\tau_i = \varepsilon_{ijk} x_j \Sigma_k \delta A = \varepsilon_{ijk} x_j \sigma_{kl} n_l \delta A \tag{5.157}$$

using again summation convention. The i component of the total torque becomes

$$\oint \varepsilon_{ijk} x_j \sigma_{kl} n_l \delta A = \oint \varepsilon_{ijk} x_j \sigma_{kl} (\delta A)_l \tag{5.158}$$

This is an integral around a closed surface. The integrand may be interpreted as the scalar product of a vector $\boldsymbol{\omega}$ and the surface element $\delta \mathbf{A}$ such as $\boldsymbol{\omega} \cdot \delta \mathbf{A}$. The scalar product is represented by the repeated index l. The fact that there is a loose index i does not affect this interpretation. From Gauss's divergence theorem the surface integral eq. (5.158) is therefore equal to a volume integral (in which the divergence summation is taken over l)

$$\int \frac{\partial}{\partial x_l} (\varepsilon_{ijk} x_j \sigma_{kl}) d\tau = \int \varepsilon_{ijk} \left(\delta_{ij} \sigma_{kl} + x_j \frac{\partial \sigma_{kl}}{\partial x_l} \right) d\tau$$

$$= \int \varepsilon_{ijk} \left(\sigma_{kj} + x_j \frac{\partial \sigma_{kl}}{\partial x_l} \right) d\tau \tag{5.159}$$

We now apply a line of argument similar to the one leading to eq. (5.155). We represent the linear dimensions of the fluid element by a length L and note that quantities such as σ_{kj}, its divergence, velocity and acceleration, will be independent of L. In the last part of eq. (5.159) the first term will behave just like L^3 and the

---

*The torque of the body forces, ignored in the main text, will also behave as L^4.

second one just like L^4.* Together they should give the time derivative τ_i of the angular momentum which behaves just like L^4, which is easily verified. For L approaching zero the L^3 term in eq. (5.159) will dominate, giving rise to contradictions unless it vanishes identically. Therefore

$$\varepsilon_{ijk}\sigma_{kj} = 0 \qquad (5.160)$$

If we write this out for $i = 1$ it follows that $\sigma_{23} - \sigma_{32} = 0$, and similarly for $i = 2$ and $i = 3$. Therefore $\sigma_{kj} = \sigma_{jk}$ and the stress tensor is symmetrical. This holds, of course, in any coordinate system.

Mathematics textbooks show that for a symmetrical tensor one can always find a coordinate system in which the non-diagonal elements vanish (by a so-called principal axis transformation). Then only normal stresses act, if one considers a rectangular fluid element with planes perpendicular to the local principal axis. Note, however, that the stress tensor σ_{ij} may be dependent on the position. In such a case the orientation of the principal axes would change with position as well.

It is easy to show that for a *fluid in rest* the stress tensor is isotropic everywhere with $\sigma_{11} = \sigma_{22} = \sigma_{33} = (\sigma_{ii})/3$ (cf. Exercise 5.22). This leads to the definition

$$\sigma_{ij} = -p\delta_{ij} \qquad (5.161)$$

in which p is called the static fluid pressure. This pressure p is usually positive, implying that the normal stresses are usually negative, corresponding to compression.

Equations of Motion

In deriving the equations of motion for a fluid element the acceleration of that element should be considered. The total derivative du/dt defined in eq. (5.9) should therefore be used. The mass of a volume element $d\tau$ is found from the product with its density ρ. The product of mass and acceleration becomes

$$\frac{du}{dt}\rho\,d\tau \qquad (5.162)$$

The *volume forces* are represented by $\mathbf{F}$, the resultant force per unit of mass. For the element considered this gives

$$\mathbf{F}\rho\,d\tau \qquad (5.163)$$

The *surface forces* are represented by the stress tensor σ_{ij}. Consider a part of the surface of the fluid element $\mathbf{n}\delta A$. The i component of the surface force is given by (5.155) as

$$\sigma_{ij}n_j\delta A \qquad (5.164)$$

The i component of the total surface force acting on the element will be the integral over the closed surface

$$\oint \sigma_{ij}(\delta A)_j \tag{5.165}$$

As was the case in reducing eq. (5.158) one may apply Gauss's divergence theorem, giving

$$\int \frac{\partial}{\partial x_j}(\sigma_{ij})\,\mathrm{d}\tau \tag{5.166}$$

For a small fluid element the integral sign may be omitted and one finds that the i component of eq. (5.162) should be equal to the i component of the total force

$$\frac{\mathrm{d}u_i}{\mathrm{d}t}\rho\,\mathrm{d}\tau = F_i\rho\,\mathrm{d}\tau + \frac{\partial}{\partial x_j}\sigma_{ij}\,\mathrm{d}\tau$$

This holds for elements of any shape. Dividing by $\mathrm{d}\tau$ gives

$$\rho \frac{\mathrm{d}u_i}{\mathrm{d}t} = \rho F_i + \frac{\partial \sigma_{ij}}{\partial x_j} \tag{5.167}$$

One notes that surface forces only contribute to change of momentum if the divergence of σ_{ij} on the second index is non-zero. Otherwise they may change the *shape* of the element but not its momentum.

Newtonian Fluids

In order to proceed with the equations of motion (5.167) one has to know more about the stress tensor σ_{ij}. For a fluid at rest it was found in eq. (5.161) that

$$\sigma_{ij} = -p\delta_{ij} \tag{5.168}$$

For a moving fluid one will usually have tangential stresses so eq. (5.168) will not apply. It is convenient to use an expression such as (5.168) as a reference. Therefore one introduces the scalar quantity

$$p = -\tfrac{1}{3}\sigma_{ii} \tag{5.169}$$

For a fluid at rest it reduces to (5.168). In general cases it is still a scalar, as the summation over i establishes its invariance under rotations. One again calls p the 'pressure' at a certain point in the fluid.

With the definition (5.169) the most general expression for the stress tensor becomes

$$\sigma_{ij} = -p\delta_{ij} + d_{ij} \tag{5.170}$$

The tensor d_{ij} is entirely due to the motion of the fluid and because of (5.169) it has the property that its trace vanishes: $d_{ii} = 0$.

In order to find d_{ij} one needs some approximations. The main one is to recognize that the non-diagonal elements of σ_{ij} are due to friction between adjacent fluid elements having different velocity—as with the same velocity there could be no friction. For a fluid with a simple shearing motion in the x direction only, and with an increase of x velocity into the y direction one would expect

$$d_{12} = d_{21} = \mu \frac{\partial u_1}{\partial x_2} \tag{5.171}$$

This assumption was already made by Newton, whence the name Newtonian fluids. In fact, without much discussion the same approximation was made in eq. (3.34). The constant μ is called the *viscosity* of the fluid; it is a measure for the local friction between fluid elements.

Let us assume that d_{ij} will be linear in the first-order derivatives of u_i. Then we expect

$$d_{ij} = 2\mu(e_{ij} - \tfrac{1}{3}\delta_{ij}\Delta) \tag{5.172}$$

$$e_{ij} = \frac{1}{2}\left[\frac{\partial u_i}{\partial x_j} + \frac{\partial u_j}{\partial x_i}\right] \tag{5.173}$$

$$\Delta = \operatorname{div} \mathbf{u} = e_{ii} \tag{5.174}$$

In the example of a simple shearing motion in the x direction eq. (5.172) indeed reduces to eq. (5.171). If one accepts (5.173) it is obvious that the divergence term should be subtracted in eq. (5.172) as we should stick to $d_{ii} = 0$.*

The Navier–Stokes Equation

With the expression (5.170) for σ_{ij} and (5.172) for d_{ij} it is now possible to tackle the equation of motion (5.167). We find

$$\rho \frac{du_i}{dt} = \rho F_i - \frac{\partial p}{\partial x_i} + \frac{\partial}{\partial x_j}[2\mu(e_{ij} - \tfrac{1}{3}\delta_{ij}\Delta)] \tag{5.175}$$

This is the famous *Navier–Stokes equation*. One should realize that most

---

*Batchelor [1] gives a general proof of eqs. (5.172), (5.173) and (5.174) within the assumption of linearity. They appear to hold only for isotropic fluids, i.e. with an isotropic intrinsic structure. For fluids consisting of long chain-like molecules the equation may not be valid.

mass outflow in x-direction=

$\rho(x+dx,y,z)u_x(x+dx,y,z)\ dydz$

$-\rho(x,y,z)u_x(x,y,z)\ dydz=$

$$\frac{\partial(\rho u_x)}{\partial x}\ dxdydz$$

total mass outflow=
div (ρ **u**) dxdydz

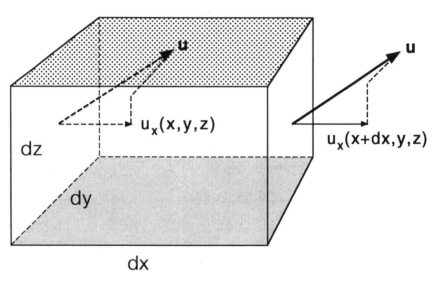

Figure 5.17 Mass outflow per unit of volume: div (ρu). Mass conservation requires this to be equal to $-\partial\rho/\partial t$

quantities between brackets on the right will be a function of position. Some approximations (which, however, are not always appropriate) may be useful:

(a) The viscosity μ does not depend on position. As μ is strongly dependent on the temperature this implies that temperature gradients should be small or zero.

(b) The fluid is incompressible, i.e. the density ρ is constant in position and in time. The equation of mass conservation

$$\frac{\partial\rho}{\partial t}+\text{div}(\rho\mathbf{u})=0 \tag{5.176}$$

then leads to div $\mathbf{u} = \Delta = 0$. For completeness the derivation of the equation of continuity (5.176) is given in Fig. 5.17.

With these two approximations one may write (5.175) as

$$\rho \frac{\partial \mathbf{u}}{dt} = \rho \mathbf{F} - \nabla p + \mu \nabla^2 \mathbf{u} \qquad (5.177)$$

We leave the derivation to Exercise 5.23.

Gravity Only

Let us assume that the volume forces $\mathbf{F}$ are due to gravity only. In eq. (5.163) $\mathbf{F}$ was defined as the force per unit of mass. Therefore

$$\mathbf{F} = \mathbf{g}$$

which by substitution into eq. (5.177) gives

$$\rho \frac{d\mathbf{u}}{dt} = \rho \mathbf{g} - \nabla p + \mu \nabla^2 \mathbf{u} \qquad (5.178)$$

We note in passing that the acceleration $d\mathbf{u}/dt$ is essentially determined by the ratio μ/ρ which is called the *kinematic viscosity* in distinction of the viscosity μ.

For a fluid at rest, it follows from eq. (5.178) that one may put, using a position vector $\mathbf{r}$,

$$p = p_0 + \rho \mathbf{g} \cdot \mathbf{r} \qquad (5.179)$$

Then $\rho \mathbf{g} - \nabla p = 0$ and the first two terms on the right-hand side of eq. (5.178) cancel. Apparently the pressure (5.179) is just hydrostatic. It is therefore useful to define the deviation of (5.179) as the *modified pressure* P by

$$p = p_0 + \rho \mathbf{g} \cdot \mathbf{r} + P \qquad (5.180)$$

leading to

$$\rho \frac{d\mathbf{u}}{dt} = - \nabla P + \mu \nabla^2 \mathbf{u} \qquad (5.181)$$

It is the convention to denote the modified pressure, the deviation from the hydrostatic pressure, by an undercase p. We will adopt this convention in the following.

Reynolds Number

Let us take the i component of eq. (5.181). This gives

$$\rho \frac{du_i}{dt} = - \frac{\partial p}{\partial x_i} + \mu \frac{\partial^2}{\partial x_j \partial x_j} u_i \qquad (5.182)$$

On the left we put $d/dt = \partial/\partial t + \mathbf{u} \cdot \mathbf{V}$ giving

$$\rho\left(\frac{\partial u_i}{\partial t} + u_j \frac{\partial}{\partial x_j} u_i\right) = -\frac{\partial p}{\partial x_i} + \mu \frac{\partial^2 u_i}{\partial x_j \partial x_j} \tag{5.183}$$

We note that in deriving eqs. (5.181) and (5.183) the following approximations were made:

(a) μ constant,
(b) incompressibility, so constant ρ,
(c) gravity as the only volume force.

We now write eq. (5.183) with its boundary conditions in terms of dimensionless parameters characterizing the physical problem. There will be an essential length L, e.g. a distance between enclosing boundaries, and some representative velocity U, e.g. the steady speed of a rigid boundary. We introduce

$$\mathbf{u}' = \frac{\mathbf{u}}{U}$$

$$\mathbf{r}' = \frac{\mathbf{r}}{L} \tag{5.184}$$

$$t' = \frac{tU}{L}$$

$$p' = \frac{p}{\rho U^2}$$

Substituting this in eq. (5.183) leads to

$$\frac{\partial u_i'}{\partial t'} + u_j' \frac{\partial u_j'}{\partial x_j'} = -\frac{\partial p'}{\partial x_i'} + \frac{1}{Re} \frac{\partial^2 u_i'}{\partial x_j' \partial x_j'} \tag{5.185}$$

$$Re = \frac{\rho L U}{\mu}$$

One may easily check that all variables appearing in eq. (5.185) are dimensionless, as is the Reynolds number Re, which we encountered earlier in eq. (3.50). This means that flows with the same boundary conditions in terms of L and U and with the same value of Re are governed by the same equation (5.185). They are *dynamically similar*. In practice, dynamic similarity is widely used to find models for complicated flows.

The Reynolds number was already introduced in eq. (3.49) as a measure for the ratio of the forces of inertia, on the left in (5.182), and the viscous forces on its extreme right. This ratio may be written in terms of dimensionless quantities as

$$\left| \frac{\rho du_i/dt}{\mu \partial^2 u_k/\partial x_j \partial x_j} \right| = Re \times \left| \frac{du'_i/dt'}{\partial^2 u'_k/\partial x'_j \partial x'_j} \right| \tag{5.186}$$

We presume that U and L in eq. (5.184) were chosen so as to make the nominator and denominator between the right-hand brackets in the order of one. Then Re is indeed the ratio of the inertia and the viscous forces. In practice, it appears that for large Reynolds numbers Re a flow easily creates eddies; even a small instability in a laminar flow then will cause turbulence.

A final remark to be made is that eq. (5.185) is highly non-linear. In fact, u'_j is multiplied by its derivative. The consequence is that it is impossible to find general analytic solutions. One cannot go further than qualitative discussions or turn to numerical approximations.

Example: a Flow through a Pipe

As a concluding example we consider the problem of a flow through a circular pipe. Suppose that the flow follows the x direction. In practice, one is interested in the loss of energy by friction. The origin of the friction is the fact that the velocity u_x along the boundary will be smaller than it is in the middle of the pipe.

For a straight pipe of circular cross-section the flow appears to be laminar up to $Re = 2000$. For spirals, eddies only start with $Re = 6000$ to $Re = 7600$, depending on the radius of curvature (which is 50 or 15 times the pipe diameter respectively). A curved pipe therefore stabilizes the flow, leading to smaller losses of energy.

5.5 TURBULENCE

Anybody watching a fast-flowing stream will enjoy the ever-changing picture of eddies and anybody looking at a smoking chimney will notice the change of the plume over time. One of the striking features is that eddies appear on all scales large and small. It seems clear therefore that these phenomena are too complicated to reproduce all at the same time by calculations and that one will have to make approximations for certain features to get reliable results for others. Even with the most sophisticated present-day computer programs one has to make assumptions that often go back to more traditional considerations.

Turbulence occurs at high Reynolds numbers Re, where the influence of viscous forces is small, as we discussed in eq. (5.186). Observations show that large eddies transfer their kinetic energy to smaller eddies, until eventually viscosity makes the eddies disappear while their energy is converted into heat.

This energy transport process is the essence of turbulence rather than its random appearance. This means that one may not ignore viscosity in the description of turbulence.

Another point to be kept in mind is that turbulence is a property of flow and not of the fluid itself. At low velocities there will be no turbulence, whereas viscosity, albeit relatively small, is always present.

In this section some of the traditional approaches will be followed first, in particular dimensional analysis and order of magnitude estimates. Next the Navier–Stokes equations will be written down and averaged over time. Similar equations for transport of heat or matter are added and it is indicated how one solves them numerically.

Dimensional Analysis and Scales

The fundamental dimensions in mechanics are kilogram, metre and second, or more generally mass M, length L and time T. A physicist is trained to check that his or her equations have the same dimensions on both sides of the equality sign. In fluid mechanics this requirement is used as a procedure to find physical entities adapted to the scale of the particular problem.

In this way the friction velocity was introduced in eqs. (4.181) and (4.182) to describe the velocity scale in a boundary layer. It again was used in eq. (4.183) to find a simple expression for the increase of velocity u with height:

$$\frac{\partial u}{\partial z} = \frac{u_*}{kz} \tag{5.187}$$

In this case a dimensionless constant k was introduced in the equation as dimensional analysis does not imply that such constants will be equal to one.

In order to compare scales of molecular and turbulent diffusion, we first take an imposed length scale to compare times and next an imposed time scale to compare lengths. In a system with an externally imposed length scale L one may compare the time scales of molecular diffusion and turbulent diffusion. Take as an example the diffusion of a gas in a room with dimension L. From the diffusion equation (5.10) one may write down an order of magnitude estimate

$$\frac{\delta C}{T_m} \approx D \frac{\delta C}{L^2} \tag{5.188}$$

where δC is a concentration difference which will be reached by molecular diffusion at distance L after time T_m. If the velocity of turbulent transport for the major eddies in the room is u, the corresponding time scale T_t becomes

$$T_t \approx \frac{L}{u} \tag{5.189}$$

The ratio of both times may be expressed as

$$\frac{T_t}{T_m} \approx \frac{L}{u}\frac{D}{L^2} = \frac{D}{uL} \approx \frac{v}{uL} = \frac{1}{Re} \tag{5.190}$$

where it was used that the diffusion coefficient D for air and the kinematic viscosity for air are v of the same order of magnitude (cf. Appendix C and Table 5.1). This kinematic viscosity determines the dynamics of the problem, as is seen from the simplified Navier–Stokes equation (5.181) and its definition $v = \mu/\rho$. The interesting point is that the Reynolds number appears, showing that turbulent transport usually is much quicker than molecular diffusion. One may check this by lighting a cigar in a lecture hall; after a few minutes it will be smelt in the farthest corner.

In a second example a time scale is imposed on the atmospheric boundary layer by the rotation of the earth. The Coriolis acceleration for a particle with velocity u is given by $2\Omega u \sin \beta$, where β is the geographical latitude. The time scale is given by $T = 1/f$, where $f = 2\Omega \sin \beta$. At middle latitudes $T \approx 10^4$ s. In the absence of turbulence the relation between length and time scales would again be given by eq. (5.188) leading to

$$L_m^2 \approx DT \tag{5.191}$$

With known values of D and T this gives $L_m \approx 20$ cm. The length scale for turbulence is determined by the velocity u of the biggest eddies in the boundary layer as in eq. (5.189):

$$L_t \approx uT \tag{5.192}$$

The Reynolds number will be defined by the turbulent motions. For the ratio of the length scales one may therefore write

$$\frac{L_t}{L_m} = \sqrt{\frac{L_t^2}{L_m^2}} \approx \sqrt{\frac{L_t u}{a}} \approx \sqrt{\frac{L_t u}{v}} = \sqrt{Re} \tag{5.193}$$

where again the Reynolds number shows up. For the atmospheric boundary layer it is in the order of $Re \approx 10^7$.

The discussion so far has focused on the largest eddies. Energy dissipation, however, happens in the smallest ones. The rate of energy dissipation ε per unit of mass has the dimension of $J\,s^{-1}\,kg^{-1}$; this works out as $m^2\,s^{-3}$. The dynamic factor governing this dissipation must be the kinematic viscosity v with a dimension of $m^2\,s^{-1}$. It is now possible to derive the so-called *Kolmogorov* scales for length η, time τ and velocity w as

$$\eta = \left(\frac{v^3}{\varepsilon}\right)^{1/4} \quad \text{(m)}$$

$$\tau = \left(\frac{v}{\varepsilon}\right)^{1/2} \quad (s)$$

$$w = (v\varepsilon)^{1/4} \quad (m\ s^{-1}) \tag{5.194}$$

where one may check that $w = \eta/\tau$. The Reynolds number on this scale becomes

$$Re = \frac{w\eta}{v} = 1 \tag{5.195}$$

which illustrates that viscosity dominates the processes on this scale.

Let us denote the length scale of the large eddies by l and the velocities of its parcels of fluid by u. The basic assumption, which agrees with observations, is that they lose their energy when a parcel has travelled a distance of the order of l. The associated time is l/u. The kinetic energy of a unit of mass is of the order u^2, so the energy a unit of mass is losing per second becomes u^3/l. On average this should be equal to the rate ε with which the smallest eddies are dissipating their kinetic energy into heat. Therefore,

$$\varepsilon = \frac{u^3}{l} \tag{5.196}$$

It is now possible to relate the size η of the smallest eddies to the size l of the largest. One finds

$$\frac{\eta}{l} = \left(\frac{v^3}{\varepsilon}\right)^{1/4}\left(\frac{1}{l^4}\right)^{1/4} = \left(\frac{v^3}{u^3 l^3}\right)^{1/4} = \left(\frac{ul}{v}\right)^{-3/4} = Re^{-3/4} \tag{5.197}$$

The length scales of the smallest eddies are therefore very small; the same holds for the time and velocity scales. With Reynolds numbers in the order of $Re \approx 10^4$, eq. (5.197) shows that length scales differ by an order of 10^3. Numerical grids to perform calculations should describe both eddies at the same time, which is impossible and illustrates the need for physically sensible approximations.

A similar relation holds for the time scales:

$$\frac{\tau}{t} = \frac{\tau}{l/u} = \left(\frac{vu^2}{\varepsilon l^2}\right)^{1/2} = Re^{-1/2} \tag{5.198}$$

The angular velocities in the eddies or, more precisely, the vorticities are inversely proportional to the time scales. Therefore, they will be largest for the smallest eddies, as one would imagine intuitively.

It must be mentioned that besides the Reynolds number there are quite a few other dimensionless numbers in use for describing flows. They are applied to design model experiments where the relevant dimensionless numbers are reproduced; from model experiments one draws conclusions as to the behaviour of the real flows. This remains outside the scope of this book.

Time-Averaged Equations of Motion

When describing the transport of pollutants one is usually interested in transport over a scale much bigger than the eddies, and certainly much bigger than the smallest eddies. One may therefore try to separate the flow in an average and a fluctuating part. Before this is done, the precise equations determining the motion of a parcel of fluid are recalled. Usually the fluid is assumed to be incompressible with a constant viscosity μ, so that the Navier–Stokes equations may be written in the form of eq. (5.177), which is given here in the index notation with the usual convention that repeated indices imply summation over that index:

$$\rho\frac{du_i}{dt} = \rho F_i - \frac{\partial p}{\partial x_i} + \mu\frac{\partial^2 u_i}{\partial x_j \partial x_j} \tag{5.199}$$

The condition of incompressibility (ρ = constant) leads with the help of eq. (5.176) to

$$\frac{\partial u_i}{\partial x_i} = 0 \tag{5.200}$$

Further, pollutants with concentration c will be transported, following eq. (5.145) or (5.57), and temperature differences will tend to smooth out following eq. (4.17); these equations may be summarized by

$$\frac{\partial \phi}{\partial t} + u_i\frac{\partial \phi}{\partial x_i} = \lambda\frac{\partial^2 \phi}{\partial x_i \partial x_i} + S_\phi \tag{5.201}$$

where ϕ may represent either concentration c or temperature T. It may be noted that eq. (5.201) generalizes heat transport in that it implies advection as well. The last term on the right-hand side describes a source of concentration (a sewage line) or of heat (cooling water).

Finally, there will be an equation of state

$$\rho = \rho(T, p) \tag{5.202}$$

For gases we will take the relation for ideal gases eq. (3.22); for liquids like water an empirical relation will do.

Incompressibility of a fluid does not necessarily mean that its density ρ is constant, for temperature differences may lead to density variations by eq. (5.202). The simplification one makes here is the so-called *Boussinesq approximation*. It makes the point that the main effect of density variations will be the vertical buoyancy when a parcel of fluid arrives in surroundings with a different temperature and therefore a different density. As the variations are small anyway, the inertia term on the left in eq. (5.199) is supposed to remain at the average level with an overall density ρ_0. The pressure term on the right of that equation will contain the hydrostatic pressure exerted by the surroundings; the first term on the right in eq. (5.199) will contain the actual density ρ. This gives buoyancy, for if the temperature of the parcel is higher than the surroundings, its density would (in general) be smaller and the gravity force ρF_i smaller than the upward pressure, resulting in an upward force on the parcel. Dividing eq. (5.199) by ρ_0 the Boussinesq approximation leads to

$$\frac{du_i}{dt} = \frac{\rho}{\rho_0} F_i - \frac{1}{\rho_0} \frac{\partial p}{\partial x_i} + v \frac{\partial^2 u_i}{\partial x_j^2} \tag{5.203}$$

where the kinematic viscosity v shows up again. The body force F_i was defined in eq. (5.163) as the force per unit of mass. In environmental flows it comprises both the gravitational force and the Coriolis force introduced in eq. (3.36):

$$\mathbf{F}_{\text{Coriolis}} = -2\mathbf{\Omega} \times \mathbf{u} \tag{5.204}$$

As the flows are mainly horizontal only the vertical component of the earth angular velocity matters, so $2\mathbf{\Omega}$ is replaced by the vector $(0, 0, f)$. In line with the Boussinesq approximation ρ is left in the gravity term only leading to

$$\frac{\rho}{\rho_0} F_i = \frac{\rho}{\rho_0} g_i - \varepsilon_{ijn} f_j u_n \tag{5.205}$$

in which ε_{ijn} is the alternating tensor defined around eq. (5.156) which takes care of the vector product (5.204).

The next step is to divide the flow in a mean flow which is slowly varying in time and a rapidly fluctuating part of the flow. One needs an unspecified time interval over which to perform the averaging—large compared with the time scales of the eddies but small enough to retain an overall time variation. All quantities are written as a sum of a mean part and a fluctuating part. The latter is indicated by a prime; this reads as

$$u_i = U_i + u_i'$$
$$p = P + p'$$
$$\phi = \Phi + \phi' \tag{5.206}$$
$$\rho = \bar{\rho} + \rho'$$

An average over the time interval meant above will be denoted by an overbar, so the definitions lead to

$$\bar{u}_i = U_i$$
$$\bar{u}_i' = 0 \tag{5.207}$$

When the time average of the continuity eq. (5.200) is taken, one finds that it also holds for the mean flow

$$\frac{\partial U_i}{\partial x_i} = 0 \tag{5.208}$$

and by subtracting eqs. (5.200) and (5.208) that it holds for the fluctuating part of the flow as well:

$$\frac{\partial u_i'}{\partial x_i} = 0 \qquad (5.209)$$

Before continuing, it is useful to write the left-hand side of eq. (5.203) more explicitly as

$$\frac{du_i}{dt} = \frac{\partial u_i}{\partial t} + u_j \frac{\partial u_i}{\partial x_j} = \frac{\partial u_i}{\partial t} + \frac{\partial}{\partial x_j}(u_j u_i) \qquad (5.210)$$

where eq. (5.200) was used. The last term on the right-hand side actually displays the origin of turbulence. Its non-linearity implies that cause and effect are not proportional. On the contrary, a 5% change in u will lead to a 10% change in the quadratic term, causing bigger disturbances, etc.

As the next step, one may substitute the decomposition (5.206) into eq. (5.203) and average over the time. We assume that differentiation and time averaging may be interchanged. On the right-hand side it then appears that all fluctuations average out; all quantities may therefore be replaced by their time average. On the left we consider the right-hand side of eq. (5.210) and find

$$\frac{\partial \overline{(U_i + u_i')}}{\partial t} + \frac{\partial}{\partial x_j}\overline{((U_j + u_j')(U_i + u_i'))} = \frac{\partial U_i}{\partial t} + \frac{\partial}{\partial x_j}(U_j U_i) + \frac{\partial}{\partial x_j}\overline{(u_j' u_i')} \qquad (5.211)$$

The equations of motion (5.203) can now be written as

$$\frac{\partial U_i}{\partial t} + \frac{\partial}{\partial x_j}(U_j U_i) = \frac{\bar{\rho}}{\rho_0}g_i - \varepsilon_{ijn}f_i U_n - \frac{1}{\rho_0}\frac{\partial P}{\partial x_i} + v\frac{\partial^2 U_i}{\partial x_j^2} - \frac{\partial}{\partial x_j}\overline{(u_i' u_j')} \qquad (5.212)$$

In order to interpret this equation we have to realize that the equation of motion (5.203) originates from eq. (5.167) which we rewrite as

$$\frac{du_i}{dt} = F_i + \frac{1}{\rho}\frac{\partial \sigma_{ij}}{\partial x_j} \qquad (5.213)$$

Here the stress tensor σ_{ij} represented the surface forces which gave a momentum transport in the x_i direction if the divergence of σ_{ij} on its second index is non-zero. Let us now look at the last two terms of eq. (5.212):

$$v\frac{\partial^2 U_i}{\partial x_j^2} - \frac{\partial}{\partial x_j}\overline{(u_i' u_j')} = \frac{1}{\rho}\frac{\partial}{\partial x_j}\left[\mu\frac{\partial U_i}{\partial x_j}\right] - \frac{\partial}{\partial x_j}\overline{(u_i' u_j')} \qquad (5.214)$$

It is clear that both terms may be interpreted as the divergence of a stress. The first term represents the effect of viscosity on the mean velocity field U_i, just as in eqs. (5.171) and (5.172). The second term $\overline{(u_i' u_j')}$ is called *Reynolds stress* after the scientist who wrote it down first. If the fluctuations in the flow were uncorrelated the time averages $(\overline{u_i' u_j'})$ would vanish for $i \neq j$. It appears that turbulent motion is highly correlated. Therefore the divergence of the Reynolds stress causes a transport of momentum of the mean flow in the direction x_i and couples turbulent motion with the average flow.

We should now get rid of the Reynolds stress term in eq. (5.212). In order to discuss this transport of momentum we define an *eddy viscosity* v_T by

$$-\overline{u_i' u_j'} = v_T \left(\frac{\partial U_i}{\partial x_j} + \frac{\partial U_j}{\partial x_i} \right) - \frac{2}{3} k \delta_{ij} \tag{5.215}$$

The term with the Kronecker delta is understood when one writes down the diagonal terms of eq. (5.215)

$$\overline{u_1'^2} = -2v_T \frac{\partial U_1}{\partial x_1} + \frac{2}{3} k$$

$$\overline{u_2'^2} = -2v_T \frac{\partial U_2}{\partial x_2} + \frac{2}{3} k \tag{5.216}$$

$$\overline{u_3'^2} = -2v_T \frac{\partial U_3}{\partial x_3} + \frac{2}{3} k$$

When one adds the three quantities, the sum of the first terms on the right will vanish because of the continuity eq. (5.208), leaving

$$\overline{u_1'^2} + \overline{u_2'^2} + \overline{u_3'^2} = 2k \tag{5.217}$$

which gives the interpretation of k as the *turbulent kinetic energy per unit of mass*.*

One may proceed by writing down a dimensional equation for the eddy viscosity v_T. It should have the dimension of $m^2 s^{-1}$. Looking at eqs. (5.215), (5.216) and (5.217) it seems appropriate to use $\sqrt{k}$ as a representative velocity and l as a length representing the large eddies. One obtains

$$v_T = c_\mu' l \sqrt{k} \tag{5.218}$$

where c_μ' is an empirical constant. One has to take into account the fact that the turbulent kinetic energy per unit of mass k will be dependent on position. Therefore, an equation is needed for k. As an illustration this equation is given here without derivations as[†]

$$\frac{\partial k}{\partial t} + \frac{\partial}{\partial x_j} (U_j k) = -(\overline{u_j' u_i'}) \frac{\partial U_i}{\partial x_j}$$

---

*The next step then could be to redefine P in eq. (5.212) as $P + (2k\rho_0)/3$ and ignore the k term; in the same vein one could absorb the gravity in the hydrostatic pressure, as was done in eq. (5.181).

[†]One may find the derivation of eq. (5.219) in Svensson ([2], p. 119, eq. (2.5)). In a somewhat more general way it is derived in Rodi ([3], p. 34, eq. (2.60)).

$$-\frac{\partial}{\partial x_j}\left[\frac{\overline{p'u_j'}}{\rho_0}+\overline{u_j'k'}-\nu\left(\frac{\partial k}{\partial x_j}+\frac{\partial}{\partial x_i}\overline{u_i'u_j'}\right)\right]$$

$$-\varepsilon_{ijn}f_j\overline{u_i'u_n'}+\frac{g_i}{\rho_0}\overline{(u_i'\rho')}-\varepsilon \qquad (5.219)$$

where ε is given by

$$\varepsilon=\nu\,\overline{\frac{\partial u_i'}{\partial x_j}\left(\frac{\partial u_i'}{\partial x_j}+\frac{\partial u_j'}{\partial x_i}\right)} \qquad (5.220)$$

and

$$k'=\frac{\overline{u_i'u_i'}}{2} \qquad (5.221)$$

Equation (5.219) may be summarized as follows:

$$A+B=C+D+E+F+G \qquad (5.222)$$

The interpretation of the terms is as follows:

$A =$ local rate of change
$B =$ advective transport (cf. eq. (5.201))
$C =$ stress production by interaction between Reynolds stresses and a mean velocity gradient; in a similar equation for the kinetic energy of the mean flow this terms appears with the other sign, representing a loss of energy
$D =$ diffusive transport (as it may be read as a gradient)
$E =$ Coriolis stress
$F =$ buoyancy production or destruction; note ρ', which indicates the deviation from ρ and will be coupled to temperature variations to be deduced from eq. (5.201).
$G =$ as its definition (5.220) contains the viscosity ν, it is seen as dissipation of turbulent kinetic energy.

There is some trade-off between the production of turbulent kinetic energy in term C, the buoyancy production/destruction of kinetic energy in term F and the disappearance of that energy in the form of heat represented by term G. The approach then is to solve the set of equations (5.212), (5.215), (5.218), (5.219) and (5.220) at the same time, while making approximations in eq. (5.219).

Another approach is to write down an equation for ε as well. This leads to the so-called $\varepsilon-k$ model, where one still needs many approximations, but which at least takes the behaviour of the dissipation ε into account.

Yet another approach is to put all computational effort into solving the mean flow equation (5.212) and to approximate the Reynolds stresses by simple models directly. This is useful when heat and mass fluxes may be ignored and consequently eq. (5.201) may be omitted.

5.6 GAUSSIAN PLUMES IN THE AIR

This section bears some resemblance to Section 5.2 where the transport of pollutants in rivers was discussed. Now we turn to transport in air and it is instructive to consider the time scales concerned. Table 5.3 shows the properties for horizontal transport and Table 5.4 for vertical transport of pollutants.

One observes that horizontal transport goes much quicker than vertical transport, because of the horizontal character of the wind velocities. However, even the vertical transport goes much more quickly than calculated by molecular diffusion alone (cf. Section 5.1); this is due to the convection in the atmosphere. For both types of transport the turbulence of the atmosphere will lead to rapid mixing and dispersion of the pollutants.

The discussion in the present section will focus on point sources like chimneys. In Fig. 5.18 the plume of a chimney is shown for different temperature gradients within the atmosphere. On the left-hand side in the figure the adiabatic temperature dependence on height is dashed. With a drawn line the real

Table 5.3 Horizontal transport of pollutants in the atmosphere. (Taken by permission of NV SDU Uitgeverij Semi-Officiële Publikaties from H. van Dop. *Luchtverontreiniging: Bronnen, verspreiding, Transformatie en Depositie*, KNMI, De Bilt, z.j., p. 11)

Distance	Time scale
1–10 m	Seconds
3–30 km	Hours
100–1000 km	Days
Hemisphere	Months
Globe	Years

Table 5.4 Vertical transport of pollutants in the atmosphere. Time scales for vertical transport of pollutants from the ground to the layers indicated. The boundaries of the layers vary with day and night. The troposphere determines the weather (cf. Section 3.2) and the stratosphere acts as a lid. (Taken by permission of NV SDU Uitgeverij Semi-Officiële Publikaties from H. van Dop, *Luchtverontreiniging: Bronnen, Verspreiding, Transformatie en Depositie*, KNMI, De Bilt, z.j., p. 11)

Layer	Extension	Time scale
Boundary layer	Ground to 100 m/3 km	Minutes to hours
Troposphere	100 m/3 km to 10 km/15 km	Days to weeks
Stratosphere	10 km/15 km to 50 km	Years

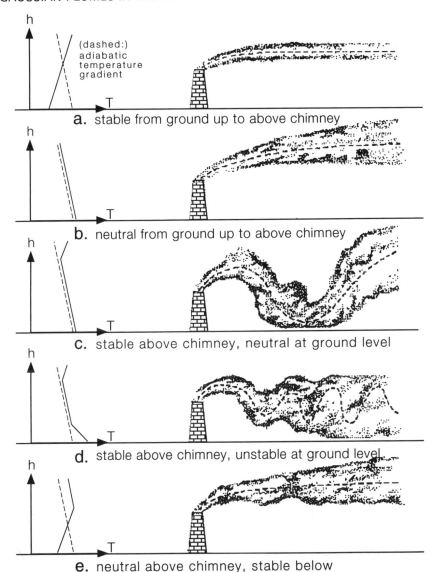

a. stable from ground up to above chimney

b. neutral from ground up to above chimney

c. stable above chimney, neutral at ground level

d. stable above chimney, unstable at ground level

e. neutral above chimney, stable below

Figure 5.18 Plume shapes from a high chimney (100 m) for several stability conditions. On the left the dashed lines are the adiabatic temperature profiles as a function of altitude and the drawn curves the 'real' profile for several cases (a little shifted to the right, for clarity). Following Fig. 3.9 one may understand the behaviour of a parcel of air, reflected in the plume shape on the right. (Taken by permission of KNMI from H. van Dop, *Luchtverontreiniging: Bronnen, Verspreiding, Transformatie on Depositie*, KNMI, De Bilt z.j., p. 18)

temperature curve is given, leading to the stability and instability conditions discussed in Fig. 3.9. One may assume that the air from the chimney is a little hotter than the surroundings, so it starts to rise. What happens next depends on the temperature profile, as we can see on the right-hand side of Fig. 5.18. The resulting plumes may be familiar from looking at a smoking chimney.

It is clear that a realistic calculation of the dispersion of pollution would require knowledge of the temperature profile over the region of interest and a realistic description of the geography. One should calculate the wind field from the meteorological variables, compute the solution of the Navier–Stokes equations, discussed in Section 5.4 for the emitted parcel of polluted air, while using some approximations to describe the turbulence.

Such a calculation (if done correctly) would be essential when, because of some industrial failure, there is a sudden emission of much dangerous material. Then one should be able to calculate quickly what measures to take, what warnings to issue, etc. When a series of these calculations and averages is performed, it appears that in an average way simple models are able to predict emissions of pollutants on the ground. For many purposes it is sufficient to know an average emission or, what amounts to the same thing, an accumulated emission over a year. For these purposes the simple models discussed below serve as a useful approximation [4].

It is assumed that there is a dominant horizontal wind velocity U, of which the direction is taken as the x direction of the coordinate frame. There will be advection to the air at this velocity and turbulent diffusion in the three x, y and z directions. Starting from eq. (5.61) and using conservation of polluting mass one will obtain a differential equation similar to eq. (5.77), where the eddy viscosity coefficients ε now refer to air:

$$\frac{\partial C}{\partial t} + U\frac{\partial C}{\partial x} = \varepsilon_x \frac{\partial^2 C}{\partial x^2} + \varepsilon_y \frac{\partial^2 C}{\partial y^2} + \varepsilon_z \frac{\partial^2 C}{\partial z^2} \qquad (5.223)$$

Here C is the concentration of pollutants in $kg\,m^{-3}$. Equation (5.223) is formally identical to eq. (5.5) which applied to molecular diffusion with the complication that the single coefficient D is replaced by three different coefficients for the three directions. The physical motivation for eq. (5.223) is that turbulent diffusion is a random process like molecular diffusion and to the extent that random or statistical processes are all one knows of the physics involved the differential equations will be similar.

It must be noted that it follows from detailed statistical analysis that the eddy viscosities ε in eq. (5.223) in general will be complicated functions of position, so the solution is not readily written down. If one ignores this complication and assumes constant eddy viscosities a solution to eq. (5.223) can readily be written down by looking at eq. (5.26) as

$$C = \frac{Q}{(2\pi)^{3/2}\sigma_x\sigma_y\sigma_z}\exp\left[-\frac{(x-Ut)^2}{2\sigma_x^2} - \frac{y^2}{2\sigma_y^2} - \frac{z^2}{2\sigma_z^2}\right] \qquad (5.224)$$

This equation is better than its derivation. When the eddy viscosity coefficients would indeed be constant, the expressions for σ_x, σ_y and σ_z would be found to be analogous to eq. (5.15), i.e. $\sigma_x^2 = 2\varepsilon_x t$, etc. In practice they are different and are found in several ways:

(a) from statistical analysis, combined with measurements, to be introduced directly below;
(b) from empirical relations, to be discussed around Fig. 5.20;
(c) by detailed calculations of the atmosphere, not to be discussed here;
(d) using models for the ε values usually called K-models, not to be discussed here.

Statistical Analysis of the Standard Deviations

Consider a point source from which particles are emitted in certain puffs. The actual velocities of the particles will vary around the mean flow $(U, 0, 0)$ and would be described by $(U + u, v, w)$. The time average of the velocities u, v and w of course will vanish, as will the average of these quantities over all particles. Still, in a real flow there will be a correlation between the velocities at a certain time t and the velocities at time $t + \tau$ a little later. What matters in the statistical models considered here is not the behaviour of a particle in a particular puff but the average behaviour of many particles in many puffs. The correlation should therefore also be described in an averaged way.

The concentration C discussed in this section is to be considered as an average over many emitted puffs from the source and is assumed to be the same as the probability of finding a particle at a certain position. Let us consider a coordinate system in which $U = 0$. The velocity field will then be described by (u, v, w) and overbars will denote the average over many puffs and particles. The time t is left as a parameter which indicates the time passed since the particles over which is averaged were emitted. For a stationary process, an average over many puffs will be independent of time:

$$\overline{u(t)} = 0$$
$$\overline{u^2(t)} = \overline{u^2} = \text{constant} \tag{5.225}$$

For the v and w components similar relations hold. Define the *autocorrelation function* $R(\tau)$ by

$$R(\tau) = \frac{\overline{u(t)u(t + \tau)}}{\overline{u^2}} \tag{5.226}$$

This function expresses how much the particle remembers of its previous velocity a time τ later. It is assumed that the statistical character of the motion is such that after averaging over all puffs the dependence on t disappears so that the autocorrelation function depends on τ only. One may then write down expression (5.226) for $t = 0$ and also for $t = -\tau$. Both should be equal to $R(\tau)$, but the latter expression may also be interpreted as $R(-\tau)$. The autocorrelation function $R(\tau)$ therefore should be an even function of τ. Other properties are that $R(0) = 1$ and $R(\infty) = 0$ for after a long period of time the particle will have forgotten its early velocities.

For a single particle with $x(t = 0) = 0$ one may write down

$$x(t) = \int_0^t u(t')\,dt' \tag{5.227}$$

and

$$\frac{d}{dt}x^2(t) = 2x\frac{dx}{dt} = 2\int_0^t u(t')u(t)\,dt' = 2\int_{-t}^0 u(t+\tau)u(t)\,d\tau \tag{5.228}$$

where in the last equality τ was chosen as an integration variable with $t' = t + \tau$. Equation (5.228) may now be averaged over many puffs as in eq. (5.225) which leads to

$$\frac{d}{dt}\overline{x^2} = 2\int_{-t}^0 \overline{u(t)u(t+\tau)}\,d\tau = 2\int_{-t}^0 \overline{u^2}R(\tau)\,d\tau = 2\overline{u^2}\int_0^t R(\tau)\,d\tau \tag{5.229}$$

where eq. (5.226) and the even character of $R(\tau)$ were used. Equation (5.229) is called *Taylor's theorem*.

Next, consider a cloud of particles around $x = 0$. Its size would be described by the average of x^2 over all particles and can therefore be found by integrating eq. (5.229) as

$$\overline{x^2} = 2\overline{u^2}\int_0^t dt' \int_0^t R(\tau)\,d\tau \tag{5.230}$$

Integration by parts leads to

$$\overline{x^2} = 2\overline{u^2}\int_0^t (t-\tau)R(\tau)\,d\tau \tag{5.231}$$

This shows that the autocorrelation function $R(\tau)$ determines the size of the cloud. The next step is to approximate the autocorrelation function taking into account the even character of the function and the properties at $\tau = 0$ and $\tau = \infty$, discussed above. For times that are not too small one may approximate the autocorrelation function by

$$R(\tau) = e^{-\tau/t_L} \qquad (\tau \to \infty) \tag{5.232}$$

This relation has the right behaviour for $\tau \to \infty$ but it is not even at $\tau = 0$. Substitution of eq. (5.232) into eq. (5.231) gives, for the y direction after time integration,

$$\sigma_y^2 = 2\overline{v^2}t_L^2\left(\frac{t}{t_L} + e^{-t/t_L} - 1\right) \tag{5.233}$$

For the x and z directions one may write down similar equations. For times that are small compared with the time scale of turbulence t_L, eq. (5.233) may be approximated by

$$\sigma_x = u_m t = \sqrt{\overline{u^2}}\, t$$
$$\sigma_y = v_m t = \sqrt{\overline{v^2}}\, t \qquad (5.234)$$
$$\sigma_z = w_m t = \sqrt{\overline{w^2}}\, t$$

Thus for small times the standard deviations vary linearly with the time.

Before discussing experimental and empirical ways to determine the standard deviations more accurately, we return to the Gaussian plume (5.224). The normalization is such that at any time t the space integral $\int C\,dV = Q$. This implies as before that solution (5.224) describes a point source at $x = y = z = 0$ emitting at $t = 0$ a mass of particles Q.

The result for a continuous point source emitting q particles per second may be found in a similar way as that following eq. (5.25). One assumes the equations (5.234) to hold for all times and substitutes them in an equation like (5.26) and integrates over all times:

$$C(x, y, z) = \frac{q}{(2\pi)^{3/2}} \int_0^\infty \exp\left[-\frac{(x - Ut')^2}{2\sigma_x^2} - \frac{y^2}{2\sigma_y^2} - \frac{z^2}{2\sigma_z^2} \right] \frac{dt'}{\sigma_x \sigma_y \sigma_z} \qquad (5.235)$$

A somewhat cumbersome time integration gives

$$C = \frac{q u_m}{(2\pi)^{3/2} v_m w_m r^2} \exp\left(-\frac{U^2}{2u_m^2} \right)\left[1 + \sqrt{\frac{\pi}{2}} \frac{Ux}{u_m r} \exp\left(\frac{U^2 x^2}{2u_m^2 r^2} \right) \operatorname{erfc}\left(-\frac{Ux}{u_m r \sqrt{2}} \right) \right]$$
$$(5.236)$$

with

$$r^2 = x^2 + \frac{u_m^2}{v_m^2} y^2 + \frac{u_m^2}{w_m^2} z^2 \qquad (5.237)$$

where one should note the negative argument of the erfc function. For some distance downstream one may write $r \to x$ and (ignoring diffusion in the direction of the flow) $U/u_m \to \infty$. One observes that the second term in eq. (5.236) dominates. Using eq. (5.234) then leads to

$$C = \frac{q}{2\pi \sigma_y \sigma_z U} \exp\left(-\frac{y^2}{2\sigma_y^2} - \frac{z^2}{2\sigma_z^2} \right) \qquad (5.238)$$

which is very similar if not identical to eq. (5.31) which was derived with σ values behaving as the square root of the time t. The reason may be that the interpretation is similar, for in both cases the x dependence is lost as the diffusion in the direction of the main flow is neglected. Further, in both cases the integral over a cross-section perpendicular to the flow may be calculated as

$$\int C \, dy \, dz = q/U \tag{5.239}$$

The essential point both derivations have in common is that the (turbulent) diffusion takes place over a slice perpendicular to the main flow. Consider a time dt in which $q \, dt$ particles are emitted as a slice. After a time t the slice has travelled a distance $x = Ut$ and the width of the slice will be $dx = U \, dt$. If one multiplies the integral (5.239) by the width of the slice dx one indeed finds the number of particles $q \, dt$ which were emitted in the time dt.

Gaussian Plume from a High Chimney

Consider a chimney with height h, which emits q particles per second continuously. Take the vertical as the z direction; the coordinates of the source are therefore $(0, 0, h)$. At the ground level $z = 0$ we assume complete reflection: as many particles are going down as are bouncing back. This means that like in eq. (5.34) we have to add a fictitious source with coordinates $(0, 0, -h)$. The plume of eq. (5.238) is now generalized as

$$C = \frac{q}{2\pi U \sigma_y \sigma_z} \left[\exp\left(-\frac{y^2}{2\sigma_y^2} - \frac{(z-h)^2}{2\sigma_z^2} \right) + \exp\left(-\frac{y^2}{2\sigma_y^2} - \frac{(z+h)^2}{2\sigma_z^2} \right) \right] \tag{5.240}$$

From a health point of view one is interested in the concentration C of pollutants

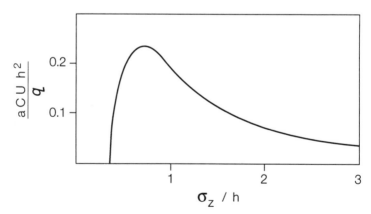

Figure 5.19 Ground level concentration underneath the axis of a horizontal plume. Here $a = \sigma_y/\sigma_z$. (Reproduced by permission of Kluwer Academic Publishers from G. T. Csanady, *Turbulent Diffusion in the Environment*, Reidel Publishers, Dordrecht, Netherlands 1973, Fig. 3.6, p. 68)

in the air at ground level, which becomes*

$$C = \frac{q}{\pi U \sigma_y \sigma_z} \exp\left(-\frac{y^2}{2\sigma_y^2} - \frac{h^2}{2\sigma_z^2} \right) \tag{5.241}$$

The largest concentrations will be found just below the axis of the plume at $y = 0$:

$$C = \frac{q}{\pi U \sigma_y \sigma_z} \exp\left(-\frac{h^2}{2\sigma_z^2} \right) \tag{5.242}$$

The distance to the chimney is hidden in the parameters σ_y and σ_z. The concentration C of eq. (5.242) is sketched in Fig. 5.19 as a function of σ_z/h, which should be a measure of the distance to the chimney. The ordinate is written as $CUah^2/q$, in which $a = \sigma_y/\sigma_z$.

In Fig. 5.19 one observes a maximum at $\sigma_z/h = 1/\sqrt{2}$. The value of that maximum is $0.23 \approx 2/\pi e = aCUh^2/q$. This implies that the maximum ground level concentration C is inversely proportional to the square of the height of the chimney—a good reason to build high chimneys. It must be noted that Fig. 5.19 applies to the position underneath the plume; when the distance to this position increases eq. (5.241) shows an extra Gaussian exponential which implies that the concentration will go down rapidly.

The parameters σ_y and σ_z in eqs. (5.241) and (5.242) will strongly depend on:

(a) The height. The turbulence will decrease farther from its cause—the ground.
(b) The roughness of the ground. It will increase with the roughness represented by some parameter z_0 defined in eq. (4.184).
(c) The stability of the atmosphere. More stability implies lower turbulence.

We now discuss two of the ways to determine the parameters σ_y and σ_z. The first is empirical; it assigns stability categories A (very unstable) to F (moderately stable) to atmospheric conditions. They are defined in Table 5.5. Empirical relations between σ_y and σ_z on the one hand and the distance x from the source on the other hand have been found, using the atmospheric conditions A to F. For ground level release and flat terrain one finds the curves reproduced in Fig. 5.20. In order to correct for the height h and roughness z_0 one has to correct the numbers deduced from Fig. 5.20 in a semi-empirical way (this complication is usually ignored).

One will be aware of the many approximations made above. We just draw attention to the fact that the wind direction **U** is not constant, as was assumed. Observing any smoking chimney or any child's wind vane will confirm this. Another point is that plumes resemble the bent shapes in Fig. 5.18 rather than the horizontal plumes assumed here.

---

*If the plume is completely absorbed by the ground one should not add a mirror source. The air concentration at ground level becomes half of eq. (5.241).

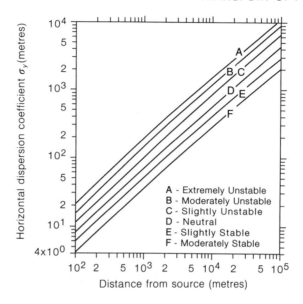

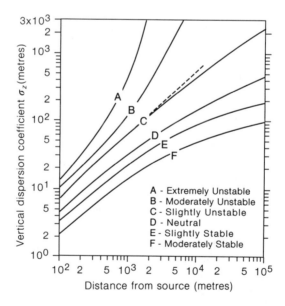

Figure 5.20 Dispersion coefficients σ_y and σ_z as a function of the distance from the source for the atmospheric stability conditions defined in Table 5.5. (From F. A. Gifford, *Nuclear Safety*, **2** (1961) 47, Fig. 3.9a/b)

Table 5.5 Pasquill stability categories. (Reproduced by permission of Kluwer Academic Publishers from G. T. Csanady, *Turbulent Diffusion in the Environment*, Reidel, Dordrecht, Holland, 1973)

Surface wind speed $(m\,s^{-1})$	Day with insolation			Night	
	Strong	Moderate	Slight	Overcast or $\geqslant 4/8$ low cloud	$\leqslant 3/8$ low cloud
2	A	A–B	B	—	—
2–3	A–B	B	C	E	F
3–5	B	B–C	C	D	E
5–6	C	C–D	D	D	D
6	C	D	D	D	D

A-extremely unstable, B-moderately unstable, C-slightly unstable, D-neutral, E-slightly stable, F-moderately stable.

Direct Determination of σ_y and σ_z

A more physical way to determine the standard deviations σ_y and σ_z is to divide the turbulence in a quickly varying part and a more slowly varying contribution [5, 6]. Let us first look at the quickly varying part. It originates from convection and changes in stability. One may apply eq. (5.233) and either measure or calculate from meteorological models $\overline{v^2}$ and $\overline{w^2}$. The parameter t_L in eqs. (5.232) and (5.233) refers to the autocorrelation of particles when one follows the turbulence in the course of time. Thus when wind velocity measurements $v(t)$ and $w(t)$ and performed at a fixed position they will be measured in a so-called Eulerian time scale. One uses the empirical relation

$$t_L = 3T_e \qquad (5.243)$$

to connect both time scales. It appears that to a good approximation one may write

$$\sigma_{zq} \approx \sigma_{yq} \qquad (5.244)$$

for altitudes $0.1z_i < z < 0.9z_i$ where z_i is the mixing height, the thickness of the earth's boundary layer (cf. Table 5.4). The index q in eq. (5.244) indicates that it refers to the quickly varying part of turbulence. The slowly varying part of the standard deviations originates from changes in the horizontal wind field. It is assumed that only σ_y is influenced so

$$\sigma_{zs} \approx 0 \qquad (5.245)$$

and, as the wind field is slowly varying, eq. (5.234) is appropriate with

$$\sigma_{ys} = v_m t \qquad (5.246)$$

Here and in eq. (5.233) the time t indicates the time since the particles were emitted. As the mean square velocity v_m can be measured or calculated, one finds σ_{ys}.

The standard deviations σ_y and σ_z are found from

$$\sigma_y^2 = \sigma_{yq}^2 + \sigma_{ys}^2 \tag{5.247}$$

$$\sigma_z^2 = \sigma_{zq}^2 \approx \sigma_{yq}^2 \tag{5.248}$$

When one has found σ_y and σ_z in this way, Fig. 5.20 may be used to deduce the Pasquill category with which they correspond. One can also deduce this category directly from Table 5.5. It then appears that simple application of Table 5.5 gives erroneous results. In a flat country one finds category D (neutral) in 70% of the cases whereas from eqs. (5.247) and (5.248) one finds only 35%.

Building a Chimney

If one is planning to build a factory and use a chimney to get rid of pollutants, the important question is what ground level concentrations are to be expected. Taking it the other way around, given some legal norms for the ground level concentration one could deduce the minimum height of the chimney.

One could measure the velocity field $u'(t)$, $v'(t)$ and $w'(t)$ at the location to be built, using the eqs. (5.243) to (5.248). Then one should take into account the fact that the wind velocity increases from ground level $z = 0$ to the chimney height h. Here one usually assumes a logarithmic velocity increase derived in eq. (4.184) and applicable for neutral atmospheric situations:

$$U(z) = \frac{u_*}{k} \ln \frac{z}{z_0} \tag{5.249}$$

where z_0 is the ground surface roughness introduced before. When one takes the average of $U(z)$ between $z = z_0$ and $z = h$ one finds the velocity U at a height $z = h/e$. This is the height at which the preparatory measurements should take place.

5.7 TURBULENT JETS AND PLUMES

In problems of sewage disposal one may discuss a pipe, which discharges a volume Q of water per second with a certain concentration C_0 of pollutant. We will restrict ourselves to round pipes with radius R. The pipe enters a lake or an ocean at a depth H. The question then is how the discharge dilutes and disperses in the ambient waters and at what distance an acceptably low concentration of pollutant is reached. One may assume that the ejected flow is turbulent for in this way the mixing with the surroundings will be optimal.

At the point of discharge the ejected flow will have a momentum defined as ρM and an average velocity W. This leads to the equations

$$Q = \pi R^2 W \tag{5.250}$$

$$M = \pi R^2 W^2 \tag{5.251}$$

Note that the density ρ could have been put at both sides of eq. (5.251) and is divided out. Besides volume and momentum the ejected flow may have a buoyancy B as well. This would be caused by a difference in density from a difference in temperature and/or salinity with the receiving waters. An expression for the buoyancy at the end of the pipe may be found by considering a volume Q of effluent with density ρ entering ambient water with a density $(\rho + \Delta \rho_0)$. Thus, $\Delta \rho_0$ is the density deficiency of the effluent at the end of pipe, whence the subscript 0. The downward gravity force will be $\rho Q g$ and the upward Archimedes force will be $(\rho + \Delta \rho_0) Q g$, resulting in an upward buoyancy force ρB defined by

$$B = \frac{\Delta \rho_0}{\rho} g Q \tag{5.252}$$

where again the density ρ has been brought to the right-hand side.

Outside the pipe one could still follow the flow; we will for simplicity assume that the density ρ is constant over a cross-section. Thus we define a mass flux $\rho \mu$ by

$$\rho \mu = \int \rho w \, dA \qquad \text{or} \qquad \mu = \int w \, dA \tag{5.253}$$

where μ is the volume flux of the flow, the integral runs over the cross-section of the flow and w is the local velocity perpendicular on the cross-section. Turbulent velocities are ignored as they amount to not more than 10% of the average velocities. Similarly the momentum flux ρm is given by

$$\rho m = \int \rho w^2 \, dA \qquad \text{or} \qquad m = \int w^2 \, dA \tag{5.254}$$

and the buoyancy flux $\rho \beta$ by

$$\rho \beta = \int g \Delta \rho w \, dA \qquad \text{or} \qquad \beta = \int \frac{\Delta \rho}{\rho} g w \, dA \tag{5.255}$$

where here $\Delta \rho$ is the difference between the density of the flow and of the surrounding fluid.

As a matter of definition one talks about a *pure plume* if at the end of pipe $Q = M = 0$, so that buoyancy B is the only initial value of interest. A smoke plume above a fire would be an example of such a case. Similarly one talks about a simple jet if a flow has initial momentum flux M and volume flux Q, but without buoyancy, the density being the same of that of the surrounding fluid. Finally one speaks about a buoyant jet if there is a buoyancy B as well.

When the equations of motion need to be solved for a parcel of ejected fluid one should not only take into account the initial conditions (5.250) to (5.252) but the physical properties of the surrounding fluid as well, such as a possible stratification (a varying density with height or depth, whence the subscript 0 in eq. (5.252)) and cross flows such as described in the previous section for a plume of smoke. One would have to follow the methods indicated in Section 5.5, while many empirical parameters would have to be introduced.

In the present section we will present a zero-order approximation, in use by engineers, where empirical parameters are readily measured and widely applicable. A slightly better way would be to solve the equations of motions in some regions where the flow exhibits a simple physical behaviour and next in some way relate the solutions to each other.*

Below we illustrate the engineering method, as it should be used as a zero-order check by physicists as well. Only a few simple cases are discussed and the receiving fluid will be described as simply as possible, i.e. without stratification or cross flows.

Dimensional Analysis

A physical engineer will identify the dominant physical quantities which should enter into the required equations and identify their dimensions: length L, time T and mass m. If there are n variables the dimensions of the equations should check and one could in this case construct $n - 3$ dimensionless groups of variables which should relate to each other. In the simple cases shown below one finds essentially $n - 3$, or sometimes $n - 2$ relations, and these examples may serve as a 'proof' of the general theorem.

The dimensions of the dominant variables are easily found from eqs. (5.250) to (5.255) as

$$[\mu] = [Q] = \frac{L^3}{T}$$

$$[m] = [M] = \frac{L^4}{T^2} \tag{5.256}$$

$$[\beta] = [B] = \frac{L^4}{T^3}$$

where the mass does not occur as it has been divided out in the definitions.

---

*This is worked out in the book by Fischer et al. ([7], sec. 9.3.2). A more complete treatment of buoyant jets will be found in the paper of Hussain and Rodi [8].

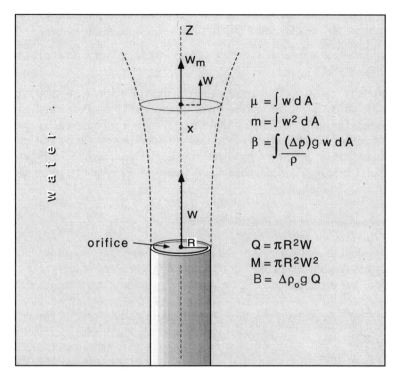

$$\mu = \int w\, dA$$

$$m = \int w^2\, dA$$

$$\beta = \int \frac{(\Delta \rho)g\, w\, dA}{\rho}$$

$$Q = \pi R^2 W$$

$$M = \pi R^2 W^2$$

$$B = \Delta \rho_0 g\, Q$$

Figure 5.21   Defining the parameters of a simple jet or (vertical) plume

Figure 5.21 sketches the relevant parameters: z as the distance from the orifice and x as the distance from the axis of the jet or plume, which for simplicity has been drawn vertically. Also shown is the local velocity w, which should be taken as a time-averaged value. Both for the velocity distribution $w(x)$ and the concentration of pollutant $C(x)$ we will assume a Gaussian shape with its maximum on the axis. This is in line with the arguments of preceding sections and certainly is the simplest symmetrical shape. Thus,

$$w = w_m e^{-(x/b_w)^2} \tag{5.257}$$

and

$$C = C_m e^{-(x/b_T)^2} \tag{5.258}$$

where w_m and C_m are the values on the axis of the jet or plume and b_w and b_T indicate the width of the Gauss distribution.

The Simple Jet

For simplicity we discuss the simple jet as a horizontal discharge with parameters Q and M, while $B = 0$. From the three quantities Q, M and z and the two dimensions L and T it follows that here $3 - 2 = 1$ dimensionless quantity determines the physics. As we want to find the dependence of the parameters on the distance z to the source we choose to find a relevant length scale. This should depend on the only two initial parameters Q and M. They can be combined into a length scale by defining

$$l_Q = \frac{Q}{\sqrt{M}} \qquad (5.259)$$

which is equal to $\sqrt{A}$ where A is the initial cross-sectional area of the jet.

As w_m has the dimension of a velocity and M/Q is the only velocity in the initial conditions the dependence of w_m on the relevant parameters can only be of the form

$$w_m \frac{Q}{M} = f\left(\frac{z}{l_Q}\right) \qquad (5.260)$$

where f is an unknown function of the dimensionless distance to the source. We may, however, deduce the asymptotic properties of f. For small z the function f should approach the value 1 because w_m should approach W and eqs. (5.250) and (5.251) will hold. More interesting is the limit where z/l_Q approaches infinity. This could happen in three ways:

(a) $z \to \infty$ with fixed l_Q and so with fixed Q and M;
(b) $l_Q \to 0$, as $Q \to 0$ with z and M fixed;
(c) $l_Q \to 0$, as $M \to \infty$ with z and Q fixed.

The point is that each of these three limits should give the same physical flow. This implies that a large z is in a sense equivalent to a small Q, so that for large z the magnitude of the momentum M determines the flow. Therefore Q should leave the relation (5.260) but M should be kept, which means that

$$w_m \frac{Q}{M} = a_1 \frac{l_Q}{z} \qquad (z \gg l_Q) \qquad (5.261)$$

Substitution of definition (5.259) shows the properties just mentioned. The empirical constant a_1 is found as 7.0 ± 0.1, which implies that data points very nicely fit eq. (5.261).

In a similar vein relations for the width parameters b_w and b_T are found. Indicating both by b one again finds that in general

$$\frac{b}{l_Q} = f\left(\frac{z}{l_Q}\right) \tag{5.262}$$

Again, for large z the value of Q should disappear from the equations, which implies that b/l_Q and z/l_Q should be proportional. The experiments indicate

$$\frac{b_w}{z} = 0.107 \pm 0.003 \qquad (z \gg l_Q) \tag{5.263}$$

and

$$\frac{b_T}{z} = 0.127 \pm 0.004 \qquad (z \gg l_Q) \tag{5.264}$$

The volume flux μ at some distance should be proportional to the initial volume flux Q, which leads to

$$\frac{\mu}{Q} = f\left(\frac{z}{l_Q}\right) \tag{5.265}$$

where f again is an unknown function. For large z the value of Q should be insignificant, as before, which leads to

$$\frac{\mu}{Q} = c_j \frac{z}{l_Q} \qquad (z \gg l_Q) \tag{5.266}$$

Of course this quantity is to be interpreted as the *mean dilution* of the jet. The value of c_j can be calculated from the earlier estimates. Substituting eq. (5.257) into eq. (5.253) and integrating gives

$$\mu = \pi w_m b_w^2 \qquad (z \gg l_Q) \tag{5.267}$$

Using eqs. (5.261), (5.259) and (5.263) leads to a value of 0.25 for c_j in eq. (5.266).

Consider an initial concentration C_0 of pollutant in the jet. With the volume output per second of Q the initial pollutant mass per second Y simply is

$$Y = QC_0 \tag{5.268}$$

The concentration C_m on the axis divided by Y has the dimension of Q^{-1}, which is T/L^3. For large z the variable Q may not occur in the equations but M should. Thus one should construct from M and z a variable of the same dimension. That leads to

$$\frac{C_m}{Y} = \frac{a_2}{z\sqrt{M}} \qquad (z \gg l_Q) \tag{5.269}$$

or

$$\frac{C_m}{C_0} = a_2 \frac{Q}{z\sqrt{M}} = a_2 \frac{l_Q}{z} \qquad (z \gg l_Q) \tag{5.270}$$

The value of a_2 has been estimated experimentally as 5.64.

Example of a Simple Jet*

Consider a jet with $Q = 1\,\text{m}^3/\text{s}$ and a velocity $W = 3\,\text{m/s}$ discharging into a fluid with the same density. This implies $M = QW = 3\,\text{m}^4/\text{s}^2$. The length scale of this problem would be $l_Q = Q/\sqrt{M} = 0.58\,\text{m}$. The question is to calculate the characteristics of the jet at a horizontal distance of 60 m.

The first thing is to calculate $z/l_Q = 104$. From eq. (5.261) and the value of a_1 it follows that $w_m = 0.20\,\text{m/s}$. The decrease in concentration of a pollutant on the axis will be found from eq. (5.270) and the value of a_2. This gives $C_m/C_0 = 0.054$. The dilution (5.266) becomes $\mu/Q = 26$, where $c_j = 0.25$ was used.

The Simple Plume

The simple plume is defined by the properties that $Q = M = 0$, so it only has a buoyancy B and the plume is assumed to be rising vertically. The physical variables concerned are B and the distance to the source z. Again one wants to calculate the mass and momentum fluxes some distance away. The viscosity of the fluid is ignored again on the basis of the argument that turbulence will be dominant.

Looking at the dimensions of B and z it appears that the only velocity scale is found as $(B/z)^{1/3}$. For the longitudinal velocity w_m on the axis one finds

$$w_m = b_1 \left(\frac{B}{z}\right)^{1/3} \tag{5.271}$$

where experimentally it is found that $b_1 = 4.7$. The integrated momentum flux m can be constructed from z and B in one way only, which leads to the expression

$$m = b_2 B^{2/3} z^{4/3} \tag{5.272}$$

where it is found that $b_2 = 0.35$. Finally the volume flux μ is written as

$$\mu = b_3 B^{1/3} z^{5/3} \tag{5.273}$$

with experimentally $b_3 = 0.15$. For later use these formulas are rearranged a little bit by expressing μ in terms of m as

$$\mu = \frac{b_3}{\sqrt{b_2}} \sqrt{mz} = c_p \sqrt{mz} \qquad \text{for a } simple\ plume \tag{5.274}$$

where the last equality defines $c_p = 0.254$. For a simple jet we found a similar

---

*Taken by permission of Academic Press from H. B. Fischer, E. J. List, R. C. Y. Koh, J. Imberger and N. H. Brooks, *Mixing in Inland and Coastal Waters*, Academic Press, New York, 1979, example 9.1, p. 328.

formula in eq. (5.266), which may be written as

$$\mu = c_j \sqrt{M} z \qquad \text{for a } \textit{simple jet} \qquad (5.275)$$

The difference between the two equations is the appearance of the initial momentum flux M in the equation of the jet and the local flux in the equation for the plume. Indeed, for a plume the flux increases as the 5/3 power of z, as can be seen from eq. (5.273). This originates in the buoyancy which gives extra velocities. Therefore, c_p may be regarded as the growth coefficient for a plume and c_j as the growth coefficient for a jet.

Equations (5.272) and (5.273) may be combined in a dimensionless number R_p as

$$\frac{\mu B^{1/2}}{m^{5/4}} = \frac{b_3}{b_2^{5/4}} = R_p \qquad (5.276)$$

where $R_p = 0.557$ by virtue of the values quoted above.

Finally, for environmental reasons it is important to find the decrease in the tracer concentration on the plume axis C_m by comparing it with the initial flux Y. This should be a function of the buoyancy B and the distance z from the source. Looking at the dimension C_m/Y of the only possible relation is

$$\frac{C_m}{Y} = \frac{b_4}{B^{1/3} z^{5/3}} \qquad (5.277)$$

with the empirical number $b_4 = 9.1$.

Example of a Simple Plume*

A freshwater discharge with $Q = 1 \text{ m}^3/\text{s}$ is located at a depth of 70 m in an ocean. The concentration of a pollutant is $C_0 = 1 \text{ kg/m}^3$, which leads to an output of pollutant $Y = 1 \text{ kg/s}$. The buoyancy B has to be calculated with eq. (5.252). It is given that the discharge has a temperature of 17.8 °C and the sea has a temperature of 11.1 °C and a salinity of 3.25%. Using tabulated values of seawater density and freshwater density one finds a buoyancy $B = 0.257 \text{ m}^4/\text{s}^3$.

The question is to calculate the tracer concentration C_m of eq. (5.277), the volume flux μ of (5.273) and the dilution μ/Q at 60 m from the source and thus 10 m below sea level. With the equations indicated and the parameters given one finds $C_m = 0.016$ and $\mu = 88 \text{ m}^3/\text{s}$ with a dilution $\mu/Q = 88$. The fact that the volume flux is so much higher than the initial flux of $1 \text{ m}^3/\text{s}$ originates from the increase of velocity w from the buoyancy.

---

*Taken by permission of Academic Press from H. B. Fischer, E. J. List, R. C. Y. Koh, J. Imberger and N. H. Brooks, *Mixing in Inland and Coastal Waters*, Academic Press, New York, 1979, example 9.2, p. 332.

It should be noted that real ocean diffusers usually consist of a long pipe with small holes releasing fluid. This should be described as a line source and the plume subsequently may be described as a two-dimensional plume, resulting in different equations. Indeed, even the dimensions will be different with Q in $L^2 T^{-1}$, M in $L^3 T^{-2}$ and B in $L^3 T^{-3}$.

A Vertical Buoyant Jet

In this case all three initial quantities B, Q and M will be different from zero. To simplify matters the jet will be taken vertically and with discharge into stagnant and homogeneous surroundings, which are slightly denser, resulting in the buoyancy of the jet. Therefore, B, Q, M and the distance z from the source will enter into the equations. As they all contain only the dimensions length and time there are $4 - 2 = 2$ dimensionless quantities which will determine the physics of the process. For reasons of comparison it is easiest to use length scales as frames of reference, as the distance from the source has to be measured in terms of physical scales. It then is obvious to use again

$$l_Q = \frac{Q}{\sqrt{M}} \tag{5.278}$$

The other scale must contain B and we will use

$$l_M = \frac{M^{3/4}}{B^{1/2}} \tag{5.279}$$

although other choices would be possible. Again the velocity w_m on the jet axis will be the relevant quantity to find. It will depend on B, Q and M and a function

$$f\left(\frac{z}{l_Q}, \frac{z}{l_M}\right) \tag{5.280}$$

of the dimensionless lengths z/l_Q and z/l_M.

First consider a flow with $Q = 0$. In this case the only quantity with the dimension of a velocity is $B^{1/2} M^{-1/4}$ and the only length scale l_M; therefore w_m should be of the form

$$w_m \frac{M^{1/4}}{B^{1/2}} = f\left(\frac{z B^{1/2}}{M^{3/4}}\right) \tag{5.281}$$

This equation should hold generally. For small B we have something like the simple jet and the velocity should not depend on B. Small B can be represented as the limit $B \to 0$. The function at the right-hand side of eq. (5.281) does not distinguish between $B \to 0$, $z \to 0$ or $M \to \infty$, so for small z the B dependence on both sides of eq. (5.281) should cancel. This leads to

$$w_m \frac{M^{1/4}}{B^{1/2}} \rightarrow c_1 \frac{M^{3/4}}{zB^{1/2}} = c_1 \frac{l_M}{z} \qquad (z \ll l_M) \qquad (5.282)$$

where 'small' z has to be interpreted in terms of the only length scale present, i.e. l_M. In the limit for very large z the right-hand side of eq. (5.281) does not distinguish between large z and small M; therefore in this limiting case the M dependence should cancel. This leads to

$$w_m \frac{M^{1/4}}{B^{1/2}} \rightarrow c_2 \left(\frac{M^{3/4}}{zB^{1/2}} \right)^{1/3} = c_2 \left(\frac{l_M}{z} \right)^{1/3} \qquad (z \gg l_M) \qquad (5.283)$$

When one only looks at the z dependence of w_m one finds from comparison of eq. (5.283) with eq. (5.271) that for large z the buoyant jet behaves like a plume, so its initial momentum M is forgotten. Now consider the more general case that $Q \neq 0$, where the length scale l_Q shows up. We expect that for $z \gg l_Q$ again the initial momentum will be forgotten, so we return to the simple jet. Large z again would be equivalent to small Q and therefore the expressions for w_m should become independent of Q. Thus, if z is big enough eq. (5.283) will give its asymptotic behaviour.

Similarly, from comparison of eq. (5.282) with eq. (5.261) it is apparent that for small z the buoyant jet behaves like a jet. Apparently the buoyancy has not had time to influence the flow. This will be true for the more general case that $Q \neq 0$ as well. Therefore we expect eq. (5.266) to be a good approximation, ignoring the fact that eq. (5.266) will hold only when z is not very small.

The ratio of l_Q and l_M will be an important determinant of the jet. This quantity is defined as

$$R_0 = \frac{l_Q}{l_M} = \frac{QB^{1/2}}{M^{5/4}} \qquad (5.284)$$

Two other quantities may be defined as well in order to discuss the limiting cases of the simple jet and the simple plume. First, define a dimensionless volume flux $\bar{\mu}$ by

$$\bar{\mu} = \frac{\mu B^{1/2}}{R_p M^{5/4}} \qquad (5.285)$$

where μ is the local volume flux from eq. (5.253) and R_p is defined by eq. (5.276). With the help of eq. (5.284) it is found that

$$\bar{\mu} = \frac{\mu}{Q} \frac{R_0}{R_p} \qquad (5.286)$$

A second dimensionless quantity will be

$$\zeta = c_p \frac{z}{R_p l_M} = c_p \frac{z}{l_Q} \frac{R_0}{R_p} \qquad (5.287)$$

Consider eq. (5.266) for the volume flux of a well-developed jet and rewrite it with the definitions of $\bar{\mu}$ and z. The ratio c_j/c_p which occurs is taken as 1, because of the close empirical values given previously. This gives

$$\bar{\mu} = \zeta \qquad (\zeta \ll 1) \tag{5.288}$$

The range $\zeta \ll 1$ is added because as discussed above the jet formula (5.266) would hold for small z. The corresponding equation for the volume flux in a simple plume was given in eq. (5.273) and can be rewritten again with $\bar{\mu}$ and z and should represent large ζ. One finds that

$$\bar{\mu} = \frac{b_3 R_p^{2/3}}{c_p^{5/3}} \zeta^{5/3} = \zeta^{5/3} \qquad (\zeta \gg 1) \tag{5.289}$$

For a real buoyant jet one can measure the dependence of $\bar{\mu}$ on ζ and finds indeed eqs. (5.288) and (5.289) as limiting cases. This is shown in Fig. 5.22. The dotted lines represent situations just outside the source where the flow still has to be established.

Example of a Simple Buoyant Jet*

Consider the example discussed before in connection with the simple plume, but with momentum added. A fresh water discharge with $Q = 1$ m³/s is located at a depth of 70 m in an ocean. The concentration of a pollutant is $C_0 = 1$ kg/m³, which leads to an output of pollutant $Y = 1$ kg/s. The buoyancy B again is found as $B = 0.257$ m⁴/s³. The momentum M is given as $M = QW = 3$ m⁴/s².

The question is to calculate the volume flux μ and the dilution μ/Q 60 m from the source, and thus 10 m below sea level. First one calculates the determining parameters as $l_Q = Q/M^{1/2} = 0.577$ m and $l_M = M^{3/4}/B^{1/2} = 4.5$ m which shows that very quickly one is in the situation where $z \gg l_M$ and has a plumelike behaviour of the jet. Further, $R_0 = l_Q/l_M = 0.128$. At a distance of 60 m from the source eq. (5.287) and the numerical values given above lead to $\zeta = 6.0$. From Fig. 5.22 one finds that $\bar{\mu} \approx 20$ and from eq. (5.286) with the known values of R_0 and R_p one finds for the increase in volume flux $\mu/Q \approx 87$.

For the simple plume with the same parameters but with $M = 0$ we found essentially the same value as 87.7. One would expect the former value (87) to be a little bit larger than the second one, but the values were found by reading a double logarithmic graph with the inaccuracies inherent there. For a jet without buoyancy the value was only 26, so it is buoyancy that leads to an increase in flux and correspondingly a dilution of pollutants.

---

*Taken by permission of Academic Press from H. B. Fischer, E. J. List, R. C. Y. Koh, J. Imberger and N. H. Brooks, *Mixing in Inland and Coastal Waters*, Academic Press, New York, 1979, example 9.3, p. 335.

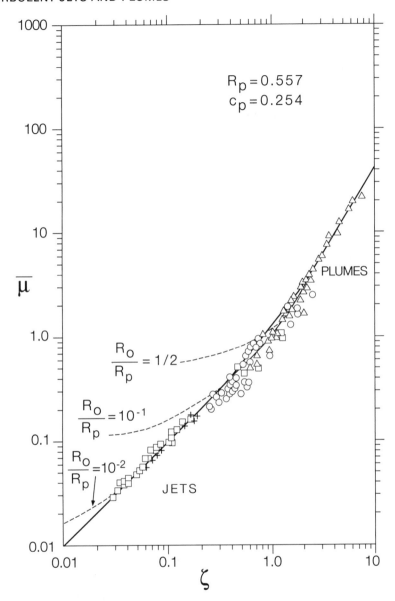

Figure 5.22 The relation between $\bar{\mu}$ and ζ for a real buoyant jet. The dotted lines indicate positions close to the source where the flow is not yet established. (Data from Ricou and Spalding, *J. Fluid Mech.*, **11** (1961) 21–32, p. 31 reproduced by permission of Cambridge University Press, adapted by Fischer *et al.* [7] p. 336)

5.8. PARTICLE PHYSICS

The particles in the atmosphere have diverse consequences: they scatter back radiation into space, they may become the nucleus of raindrops in which a variety of chemical reactions can take place and they may cause irritation to human lungs and throats. They also reduce the visibility and in high concentrations may be observed as a haze in the atmosphere.

The origin of particles may be natural, originating from volcanoes or wind-blown dusts, or they may be man-made such as all types of combustion products. In this section we give some data and discuss two aspects: time scales for the formation of an acid drop and the drag on a single particle.

Data

In Fig. 5.23 we give the volume fraction of aerosol particles as a function of the logarithm of their diameter D_p. The horizontal scale therefore is logarithmic and the vertical gives the volume fraction.

One notices that motor vehicles produce rather small particles whereas natural causes (deserts and oceans) produce larger particles. It must be mentioned that

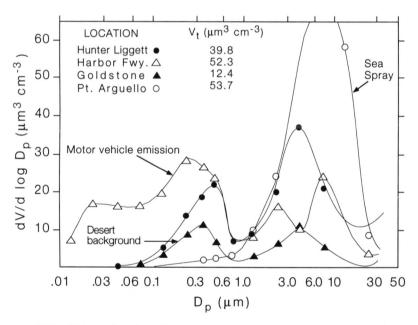

Figure 5.23 Volume fraction of aerosol particles in four quite different locations. (Reproduced by permission of the Journal of the air pollution control association from G. M. Hidy, Summary of the California aerosol characterization expperiment, *Journal of the Air Pollution Control Association*, **25** (1975) 1106–14)

the effect of smaller particles may be even larger than the curve suggests, as their relative surface area increases rapidly with decreasing sizes and the surface determines the growth of the particles.

Uptake of SO_2 in a Raindrop

Let us consider SO_2 gas in the atmosphere (perhaps originating from a coal-fired power plant) and look at the time scales with which this will be absorbed by a raindrop with radius R (in this section we draw extensively on Seinfeld [9], ch. 6). There are five characteristic times in play, indicated in Fig. 5.24:

(a) diffusion in the gas phase to the surface of the drop,
(b) passing the air–water interface,
(c) dissociation in water, formation of ions, HSO_3^- and SO_3^{2-},
(d) diffusion into the drop,
(e) oxidation of HSO_3^- and SO_3^{2-} by intermediate chemical reactions to SO_4^{2-}.

Process (c) happens very quickly, so we may assume that the chemical equilibrium exists everywhere where the SO_2 is coming. Processes (a) and (d) are governed by the same diffusion equation (5.10), which in spherical symmetry may be

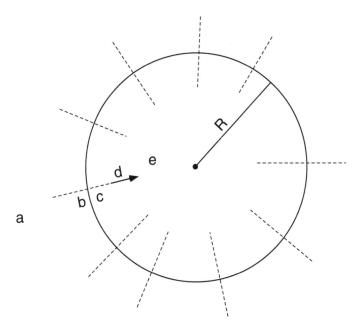

Figure 5.24   Stages in the absorption of gaseous SO_2 in a raindrop

written as

$$\frac{\partial c}{\partial t} = D_g\left(\frac{\partial^2 c}{\partial r^2} + \frac{2}{r}\frac{\partial c}{\partial r}\right) \tag{5.290}$$

where we used the expression for the Laplacian in spherical coordinates from eq. (B.3). We start by considering process (a), diffusion from the gas phase with a concentration c_∞ to the surface of the drop with a smaller concentration $c_s(t)$, where the subscript s refers to "surface". We put an extra index g to the diffusion constant as we are operating in the gas phase. As we are interested in time scales only we assume that the drop suddenly appears at $t = 0$ and is a perfect absorber with $c_s(t) = 0$ for all t. Thus the boundary conditions may be written as

$$c(r > R, t = 0) = c_\infty$$
$$c(r \to \infty, t) = c_\infty \tag{5.291}$$
$$c(R, t) = c_s(t) = 0$$

The physical problem resembles the problem of a continuous point source of Section 5.1, but it is rather its time reverse, a continuous sink with finite dimensions. From eq. (5.24) we would guess the solution

$$c = A + \frac{B}{r}\,\mathrm{erfc}\left(\frac{r - R}{2\sqrt{D_g t}}\right) \tag{5.292}$$

which happens to be correct, as one may check by substitution. The boundary conditions with the properties of the complementary error function erfc (x) lead to values for A and B and to the solution

$$c(r, t) = c_\infty - \frac{c_\infty R}{r}\,\mathrm{erfc}\left(\frac{r - R}{2\sqrt{D_g t}}\right) \tag{5.293}$$

A characteristic time is found by putting the argument of the erfc function equal to one:

$$\tau = \frac{(r - R)^2}{4D_g} \tag{5.294}$$

which means that the time scale on which changes take place increases with lárger distances. The obvious distance to choose for an estimate of the gaseous diffusion as a whole is $r = 2R$, which gives

$$\tau_{air} = \frac{R^2}{4D_g} \tag{5.295}$$

Let us now consider process (d), the diffusion of ions into the drop. The differential equation again is (5.290) but with a diffusion constant D_a referring to the aqueous phase. The boundary conditions are rather different, however.

Before $t = 0$ the concentration in the drop is zero, but then suddenly it obtains a surface concentration c_s which is constant and different from zero. When one puts $c_s = 0$ nothing happens inside the drop. This gives the boundary conditions:

(a) $c(r, t = 0) = 0$ $(r \leqslant R)$

(b) $\dfrac{\partial c}{\partial r} = 0$ $(r = 0, t > 0)$

$$(5.296)$$

(c) $c(R, t) = c_s$ $(t > 0)$

(d) $c(r, t \to \infty) = c_s$ $(r \leqslant R)$

The physics concerned is like an implosion with the concentration going to the centre from all points of the sphere. One could use a formula similar to (5.295) as an estimate for the time scale involved as it has the right dimension anyway. It is not cumbersome to be more accurate, however. We may adopt a structure for the solution similar to (5.292) by writing

$$c(r, t) = A + \frac{B}{r} f(r, t) \tag{5.297}$$

which leads to a simple equation for $f(r, t)$ as

$$\frac{\partial f}{\partial t} = D_a \frac{\partial^2 f}{\partial r^2} \tag{5.298}$$

One next uses the method of separation of variables by writing

$$f(r, t) = g(r)h(t) \tag{5.299}$$

and separating the r and t dependence. This leads to

$$\frac{1}{D_a h} \frac{\partial h}{\partial t} = \frac{1}{g} \frac{\partial^2 g}{\partial r^2} = -q \tag{5.300}$$

where q is a constant to be determined from the boundary conditions. Equation (5.300) leads to

$$f = \sin(\sqrt{q} r) e^{-q D_a t} \tag{5.301}$$

which should be substituted in eq. (5.297). Boundary condition (d) in eq. (5.296) gives $A = c_s$, where we note in passing that $q > 0$ in order to get finite results for large times t. Boundary condition (c) in eq. (5.296) implies that the sine function should vanish at the surface:

$$c_s + \frac{B}{R} \sin(\sqrt{q} R) e^{-q D_a t} = c_s \qquad \text{all } t > 0 \tag{5.302}$$

and

$$\sqrt{q} = \frac{n\pi}{R} \tag{5.303}$$

with integer n. The time constant for the problem follows from the exponential with $n = 1$ as

$$\tau_{drop} = \frac{R^2}{\pi^2 D_a} \tag{5.304}$$

which seems to be a little smaller than eq. (5.295). In practice, however, the diffusion constants are quite different. Table 5.1 gives values like $D_g = 10^{-5}\,m^2\,s^{-1}$ and $D_a = 10^{-9}\,m^2\,s^{-1}$. Table 5.6 shows the typical results for a few drop sizes; diffusion within the drop is slow.

In order to describe process (b), the passage of the interface between air and water, we have to use some results from the kinetic gas theory. Let us start by quoting Henry's law, which states that in equilibrium there is a simple relation between the concentration of a substance A (in our case SO_2) in the drop $[A\,(aq)]$ and in the air $[A\,(g)]$. Therefore, between both sides of the interface

$$\frac{[A\,(aq)]}{[A\,(g)]} = H_A RT \tag{5.305}$$

Here H_A is Henry's constant for substance A and it may be remarked that the constant varies widely, from 1.3×10^{-3} for O_2 to 2.1×10^5 for HNO_3 with SO_2 in the middle with 1.24, all in units of mole litre^{-1} atm^{-1}. These numbers hold at $T = 298$ K. In eq. (5.305) R is the universal gas constant and T the temperature in K.

Another way of expressing the constant H_A is by using the ideal gas law in the form of eq. (3.22) for a pure gas to relate the concentration of A in $kg\,m^{-3}$ to the vapour pressure p_A:

$$H_A = \frac{[A\,(aq)]}{RT[A\,(g)]} = \frac{[A\,(aq)]}{p_A} = \frac{c_s}{p_s} = \frac{c^*}{p_\infty} \tag{5.306}$$

Table 5.6   Characteristic diffusion times in air and water for several drop sizes. Calculated using eqs. (5.295) and (5.304), $D_g = 10^{-5}\,m^2\,s^{-1}$ and $D_a = 10^{-9}\,m^2\,s^{-1}$

$R\,(m)$	$10^{-3}\,(1\,mm)$	10^{-4}	10^{-5} $(10\,\mu m)$
τ_{air}	2.5×10^{-2}	2.5×10^{-4}	2.5×10^{-6}
τ_{drop}	10^2	1	10^{-2}

where the concentration in the drop has been rewritten as the surface concentration $c_s(t)$ and p_A as the surface vapour pressure $p_s(t)$; as Henry's coefficient is time independent one may also put in the equilibrium concentration c^* and pressure p_∞, which both will be realized after a long time.

In order to calculate the characteristic time for passage of the interface we no longer assume a constant surface concentration c_s but rather consider the fluxes passing the surface. The situation is sketched in Fig. 5.25, where the flux R_{-g} leaves the gas, the flux R_{+g} leaves the drop and enters the gas and R_{+1} is the resulting net flux inside.

For a molecule the surface of the drop is so enormous that it may be considered flat, so we may use the fact that the number of molecules striking a unit area per

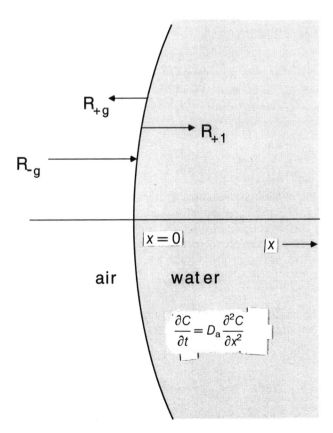

Figure 5.25 Fluxes at the interface air–water drop. The interface may be considered flat at $x = 0$

unit of time is

$$\tfrac{1}{4}[A(g)]\bar{u} \tag{5.307}$$

and $\bar{u}$ is the average velocity of the molecules given by

$$\bar{u} = \sqrt{\frac{8kT}{\pi m}} = \sqrt{\frac{8kN_A T}{\pi m N_A}} = \sqrt{\frac{8RT}{M\pi}} \tag{5.308}$$

where m is the mass of the molecules, N_A is Avogadro's number and M the mass of a mole of A. These results may be derived from the Maxwell–Boltzmann distribution for the velocities

$$f(u_x, u_y, u_z) = f(\mathbf{u}) = \left(\frac{m}{2\pi kT}\right)^{3/2} e^{-mu^2/(2kT)} \tag{5.309}$$

The concentration $[A(g)]$ is given by the ideal gas law (3.22) with p_A the partial vapour pressure of A in the gas phase. Therefore the number of molecules striking the area may be written as

$$\frac{1}{4}\frac{p_A}{RT}\sqrt{\frac{8RT}{M\pi}} = \frac{p_A}{\sqrt{2\pi MRT}} \tag{5.310}$$

The net flux R_{-g} able to pass the interface requires an extra factor of α, the accommodation coefficient:

$$R_{-g} = \alpha\frac{p_A}{\sqrt{2\pi MRT}} = \alpha\frac{p_\infty}{\sqrt{2\pi MRT}} \tag{5.311}$$

where p_A is the overall pressure of A in the air and equals p_∞. The return flux R_{+g} will obey the same equation with the vapour pressure $p_s(t)$ corresponding to the actual surface concentration $c_s(t)$, for if p_s happened to be p_∞ both fluxes would be the same. Therefore the net flux into the drop becomes

$$R_{+1} = R_{-g} - R_{+g} = \frac{[p_\infty - p_s(t)]\alpha}{\sqrt{2\pi MRT}} \tag{5.312}$$

By means of eq. (5.306) vapour pressures may be replaced by concentrations and the net flux may be rewritten in terms of a concentration gradient following eq. (5.22), giving

$$-D_a\frac{\partial c}{\partial x}\bigg|_{x=0} = R_{+1} = \frac{\alpha(c^* - c_s)}{H_A\sqrt{2\pi MRT}} \tag{5.313}$$

The characteristic time should follow from the diffusion equation with eq. (5.313) as one of the boundary conditions. We may deduce the characteristic time from a simple dimensional argument. We know the dimension of D_a, being $m^2\,s^{-1}$;

from eq. (5.313) we find

$$\left(\frac{\alpha}{H\sqrt{2\pi MRT}}\right) = \left(\frac{m}{s}\right) \tag{5.314}$$

In order to get the dimension of a time we write

$$\tau_p = D_a \left(\frac{H\sqrt{2\pi MRT}}{\alpha}\right)^2 = \frac{2\pi MRTD_aH^2}{\alpha^2} \tag{5.315}$$

This result indeed appears to be correct. We notice that Henry's factor appears squared, which implies a strong dependence on the substance concerned. Remembering that M is expressed in $g\,mol^{-1}$ we find $\tau_p = 1.51 \times 10^{-12}MH^2/\alpha^2$ s, which implies $\tau_p = 1.5 \times 10^{-10}$ s for SO_2. Therefore the crossing of the interface goes much more quickly than the diffusion (Table 5.6).

We do not go into the fifth process, the oxidation of HSO_3^- and SO_3^{2-} in the raindrops. It will be clear that for oxidation with ozone one has to consider the entry of ozone into the water drop as well and next consider the reaction rates. Rather we finish this section with a derivation of Stokes' law for the drag on a falling raindrop.

Drag on a Single Particle

Assume a spherical shape for the particle with radius R. We now derive *Stokes' law* for the drag of such a particle in a fluid (air, water or any other fluid) at low velocities i.e. at low Reynolds number *Re*. In that case we may use only the pressure term and the viscous term in the equations of motion. Combining eqs. (3.33) and (3.34) or consulting eq. (5.177) leads to

$$\nabla\left(\frac{p - p_0}{\mu}\right) = \nabla^2\mathbf{u} \tag{5.316}$$

and

$$\nabla\cdot\mathbf{u} = 0 \tag{5.317}$$

Here $\mathbf{u}$ is the velocity of the fluid $\mathbf{u} = \mathbf{u}(x, y, z, t)$ and p_0 denotes the pressure at a point far away, where the influence of the sphere on the velocity field is negligible. Equations (5.316) and (5.317) follow from the Navier–Stokes equations which imply a Newtonian viscosity μ independent of location and time and we assumed an incompressible fluid.

The situation has been sketched in the left of Fig. 5.26. It is clear that there will be axial symmetry around a vertical axis. The obvious coordinates are therefore the spherical polar coordinates (r, θ, ϕ) where $\mathbf{u}$ should be independent of ϕ, because of the axial symmetry, and should not have a ϕ component either.

The fact that the problem is essentially two dimensional makes it easy to introduce a stream function Ψ in a similar way as was done in Section 5.3 when

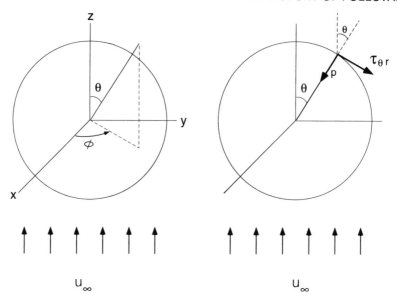

Figure 5.26 Derivation of Stokes' law. Fluid is flowing from bottom to top with a non-slip condition at the surface. On the left one finds the general coordinate system, on the right the forces at the surface of the sphere. The drag is determined by the components of the pressure force and tangential force in the direction of the flow

we were discussing groundwater flow. In eq. (5.106) we wrote down the equation of continuity, which was essentially eq. (5.317) of the present section. In Section 5.3 we then introduced a stream function $\Psi(x, y)$ and expressed the flow in terms of derivatives of Ψ such that the equation of continuity was satisfied.

This procedure can be followed in spherical polar coordinates as well. Expressing eq. (5.317) in these coordinates gives

$$\frac{1}{r^2} \frac{\partial}{\partial r} (r^2 u_r) + \frac{1}{r \sin \theta} \frac{\partial}{\partial \theta} (u_\theta \sin \theta) = 0 \qquad (5.318)$$

where the vector relations (B.6) of Appendix B were used. The continuity equation in the form of (5.318) will be satisfied for all (r, θ, ϕ) if one introduces a stream function Ψ with

$$u_r = \frac{-1}{r^2 \sin \theta} \frac{\partial \Psi}{\partial \theta} \qquad (5.319)$$

$$u_\theta = \frac{1}{r \sin \theta} \frac{\partial \Psi}{\partial r} \qquad (5.320)$$

The procedure is that one takes the r component of eq. (5.316) as well as the θ

component and differentiates both equations with respect to the other coordinate, thus eliminating p. After a lot of algebra this leads to an equation for Ψ, in which up to fourth derivatives occur. We do not give that derivation here, but find a short-cut by looking at the boundary conditions for the problem.

The first boundary condition is that at the surface of the sphere the fluid does not compress nor slip over the surface, which means that both components of $\mathbf{u}$ vanish for $r = R$:

$$u_r = -\frac{1}{r^2 \sin \theta} \frac{\partial \Psi}{\partial \theta} = 0 \qquad \text{at } r = R \tag{5.321}$$

$$u_\theta = \frac{1}{r \sin \theta} \frac{\partial \Psi}{\partial r} = 0 \qquad \text{at } r = R$$

The other condition is the asymptotic behaviour at $r \to \infty$ where $\mathbf{u}$ should approach the unperturbed velocity $u_\infty \mathbf{e}_z$. This gives

$$u_\infty \cos \theta = u_r = -\frac{1}{r^2 \sin \theta} \frac{\partial \Psi}{\partial \theta} \tag{5.322}$$

$$-u_\infty \sin \theta = u_\infty = \frac{1}{r \sin \theta} \frac{\partial \Psi}{\partial r}$$

Both equations will be satisfied for

$$\Psi = -\tfrac{1}{2} u_\infty r^2 \sin^2 \theta \qquad (r \to \infty) \tag{5.323}$$

This suggests that for smaller values of r one may try a similar separation:

$$\Psi = f(r) \sin^2 \theta \tag{5.324}$$

and indeed substitution of the corresponding velocity components in eq. (5.316) and elimination of p leads to

$$f'''' - \frac{4f'''}{r^2} + \frac{8f'}{r^3} - \frac{8f}{r^4} = 0 \tag{5.325}$$

Trying $f = r^4$ as a solution gives four roots $n = -1, +1, +2, +4$. Therefore the general solution becomes

$$f(r) = \frac{A}{r} + Br + Cr^2 + Dr^4 \tag{5.326}$$

For $r \to \infty$, eq. (5.323) shows that the dominant term should be quadratic in r, which gives

$$\Psi = \left(\frac{A}{r} + Br - \frac{1}{2} u_\infty r^2 \right) \sin^2 \theta \tag{5.327}$$

The no-slip conditions (5.321) can be used with the definitions (5.319) and (5.320) to find the remaining constants A and B, leading to

$$u_r = u_\infty \left[1 - \frac{3}{2}\frac{R}{r} + \frac{1}{2}\left(\frac{R}{r}\right)^3 \right] \cos\theta$$ (5.328)

$$u_\theta = -u_\infty \left[1 - \frac{3}{4}\frac{R}{r} - \frac{1}{4}\frac{R^3}{r} \right] \sin\theta$$

The pressure distribution $p(r,\theta)$ can be found from the equations of motion (5.316). This gives

$$p = p_0 - \frac{3}{2}\frac{\mu u_\infty}{R}\left(\frac{R}{r}\right) \cos\theta$$ (5.329)

This can be checked by substitution, but eqs. (B.8) and (B.9) should be consulted as the Laplace operator does not simply commute with the derivatives with respect to r and θ.

In order to calculate the drag we should consider both the pressure, which acts normal to the surface of the sphere, and the shear, which acts tangentially. The situation is sketched on the right-hand side of Fig. 5.26. The pressure operates to the inside of the sphere and the tangential force along the surface, as indicated. Only the component of the pressure in the direction of the flow gives a contribution as the rest cancels by virtue of the axial symmetry. From eq. (5.329) with $r = R$ we find

$$-p\cos\theta = -p_0\cos\theta + \frac{3}{2}\frac{\mu u_\infty}{R}\cos^2\theta$$ (5.330)

The first term averages out when integrating over the surface; the remaining integral becomes

$$\frac{3}{2}\frac{\mu u_\infty}{R}\int_0^\pi \cos^2\theta\, 2\pi R^2 \sin\theta\, d\theta = 2\pi\mu R u_\infty$$ (5.331)

The shear is determined by eq. (5.171) or, equivalently, in eq. (3.34) where the sign convention in the definition is such that $\tau_{\theta r}$ is the force a fluid element at small r would exert in the θ direction on an element at larger r because of an increase of θ velocity in the r direction. We are interested in the force of the fluid on the sphere and therefore add a minus sign:

$$-\tau_{\theta r} = -\mu \frac{\partial u_\theta}{\partial r}\bigg|_{r=R} = \frac{3}{2}\mu u_\infty \frac{1}{R}\sin\theta$$ (5.332)

As follows from the right of Fig. 5.26 we need to take the $+z$ component of this force, giving an extra factor of $\sin\theta$, which results in the integral

$$\int_0^\pi \frac{3}{2}\mu u_\infty \frac{1}{R}\sin^2\theta\, 2\pi R^2 \sin\theta\, d\theta = 4\pi\mu R u_\infty$$ (5.333)

The total drag force is the sum of both

$$F_{drag} = 6\pi\mu R u_\infty \qquad (5.334)$$

which is Stokes' law. For large particles one has to add the gravity force

$$F_{gravity} = \tfrac{4}{3}\pi R^3 \rho g \qquad (5.335)$$

Stokes' law ignored inertia terms in its derivation. It appears empirically that for small Reynolds numbers $Re < 0.1$ the law holds, while at $Re = 1$ the real value of the drag is 13% higher than would follow from Stokes' law. We should add that for very small particle diameters the fluid will no longer behave as a continuum. In such a case one has to add another empirical factor in order to describe the particle behaviour.

As a final remark to this chapter it may be stated that it is indeed remarkable how empiricism enters into transport physics.

Exercises

5.1 The diffusion of CO_2 in N_2 at 15 °C is determined by $D = 0.158 \times 10^{-4}\,m^2\,s^{-1}$. Calculate the times after which the mean square distance σ is 0.01, 1 and 100 m. Find respectively 3.16 s, 8.8 h and 10 yr. From these numbers it is once more obvious that other mechanisms account for most of the atmospheric distribution of gases.

5.2 For mathematically interested students, show that for small times t the centre concentration C of a finite size cloud (5.20) at $x = 0$ remains $C = C_0$, whereas for large times t at $x = 0$ the instantaneous plane source gives a good approximation. Give estimates for the terms 'small times' and 'large times'.

5.3 Verify solutions (5.21) by direct substitution in the diffusion equation (5.10). Show that in all cases Q is the amount of material released from the source at $t = 0$. Realize that the last statement is incomplete; define Q more precisely.

5.4 Look at eq. (5.24) at a fixed position r and let $t \to \infty$. Calculate the concentration $C(r)$. It does not go to infinity. Perform the same calculation for continuous line and plane sources. Find there that $C \to \infty$ for $t \to \infty$. Explain these facts physically.

5.5 Verify that the concentration C from eq. (5.26) satisfies the differential equation (5.5).

5.6 The interpretation of (5.31) given there should also follow from the differential equation obeyed by the approximate C. Study the diffusion equation (5.5) or (5.10) and adjust it so that (5.31) is a solution.

5.7 Consider an instantaneous point source of NaCl in three dimensions. Use eq. (5.21) to calculate the time in which the drop has a root mean square distance of 1 m from the origin. Calculate the time in which at a distance of 1 m the concentration is a fraction $f = 10^{-3}$ of the concentration at the origin at that moment.

5.8   Consider two parallel flat plates with a distance h. The upper one moves with a velocity of $U/2$ to the right, the lower one with the same velocity to the left. Assume a fluid between the plates which follows the plates, with $\bar{u} = 0$. Calculate $K = U^2 h^2/(120D)$ from eq. (5.54). Substitute $U = 10^{-2}\,\mathrm{m\,s^{-1}}$, $h = 10^{-3}\,\mathrm{m}$ and $D = 10^{-9}\,\mathrm{m^2\,s^{-1}}$. Find K. Use eqs. (5.57) and (5.21) with $n = 1$ to calculate the time in which a drop of NaCl released at $t = 0$ has a root mean square extension of 1 m in the direction of a local flow.

5.9   Apply eq. (5.65) to the river Rhine (Fig. 5.4). Apply the same equation to Maximiliansau and to Lobith with $K = 1760\,\mathrm{m^2\,s^{-1}}$. Deduce a value for the average velocity $\bar{u}$ and discuss the behaviour of the curves from the more precise Rhine model as compared with your simplest calculation.

5.10 Make a sketch of the source and mirror images leading to eq. (5.71). Compare centreline injection $y_0 = W/2$ with side injection where $y_0 = 0$. Argue why substitution $W \rightarrow 2W$ corresponds to a change from centre injection to side injection. Show that in Fig. 5.5 one should then substitute 'source side' for centreline and 'opposite side' for 'side'. What will be the mixing length (5.72) in this case?

5.11 Consider two channels each having $2\,\mathrm{m^3\,s^{-1}}$ of water. They join in a single rectangular channel of 7 m wide, a depth of 0.70 m and a slope of 0.001. In one stream there is a tracer of concentration C_0; in the other one its concentration is zero. At the beginning of the joint channel one may assume a tracer concentration of C_0 over the half-width. The question is how far downstream there is a mixing of at least 95%. Make a first estimate with the parameter L of eq. (5.72), corrected according to Exercise 5.10. A precise calculation can be made by generalizing eq. (5.71) to a series of line sources $0 < y_0' < 0.5$ and integrating over y_0'. One ends up with a series of erf functions.

5.12 Assume that sand consists of small spheres with radius R. Consider a volume in which the spheres form a cubic array. Prove that the porosity becomes $n = 0.48$. Note that one may show that for the most compact assemblage, a rhombohedral shape, one finds $n = 0.26$.

5.13 Explain the name hydraulic conductivity for k in eq. (5.86).

5.14 Discuss the similarities and if they exist—the differences in applying Laplace's equations to electrostatics and to groundwater flow. In particular, consider two two-dimensional examples by the method of images:
   (a) a well near a straight canal; sketch the stream lines from the well to the canal;
   (b) a well near a straight impermeable boundary.

5.15 According to the discussion below eq. (5.98) quicksand will arise for a certain groundwater head h. Let the depth of the top layer in Fig. 5.8 be d. Derive a condition for h/d in terms of ρ and ρ_w for quicksand to occur.

5.16 Consider a confined aquifer of thickness T and take a source of strength Q at $x = -d$ and a sink of strength $-Q$ at $x = +d$. Write down the complex function Ω for each, add and indicate how to find the streamlines.

5.17 Calculate Q of eq. (5.103) with $H = 10\,\text{m}$, $b = 30\,\text{m}$, $a = 5\,\text{m}$ and $k = 10^{-9}$ m s^{-1} (clay). Repeat the calculation for $k = 10^{-4}\,\text{m s}^{-1}$ (sand).

5.18 Calculate hq of eq. (5.117) with $L = 10\,\text{m}$, $h_1 = 5\,\text{m}$, $h_2 = 3\,\text{m}$ and the same values of k as Exercise 5.17.

5.19 In the example of a source ($Q > 0$) in a uniform flow with discharge U in the negative x direction there will be a point on the negative x axis where $q_x = 0$. Find this point of stagnation. Determine the value of the stream function Ψ for the streamline through S. Determine its intersection with the y axis.

5.20 The dispersion–advection equation (5.146) should describe the point source in a uniform wind, given by eqs. (5.25) and (5.26). Prove this.

5.21 A river is flowing through a field; the level of the river coincides with the groundwater head of a confined aquifer in connection with the river. At $t = 0$ the level of the river suddenly decreases with δh. Use conservation of mass to derive a differential equation and show that the solution may be written in terms of the erfc function.

5.22 Check that for a fluid at rest the stress tensor is isotropic. To prove this go to a local coordinate system where the non-diagonal elements of the stress tensor are zero. Consider a small sphere and discuss the forces on a surface element with normal (n_1, n_2, n_3). Show that a non-isotropic stress tensor would lead to deformation of the sphere: a contradiction.

5.23 Check the derivation of eq. (5.177). Note that one needs the relation $\operatorname{div} \mathbf{u} = 0$ twice.

5.24 Show that eq. (5.169) may be interpreted as the average of the normal component of the stress on a surface element, if one averages over all directions in space.

5.25 Consider a vertical plane; a thin film of vertical fluid with thickness d is streaming downwards under the influence of gravity. There are no accelerations but the viscosity is $\mu \neq 0$. Write down the equations of motion for a volume element with thickness $(d - x)$ at the outside of the flow using gravity and viscosity only and calculate the downward velocity u_y as a function of the distance x to the wall. The flow is laminar, the velocity u_y being a function of x only and $u_y(0) = 0$.

5.26 Use the eqs. (5.188), (5.189) and (5.190) to calculate the times for molecular diffusion in a heated room and the time for turbulent diffusion. To calculate the latter one may assume that air heated above a radiator for some $10\,°\text{C}$ is accelerated up to some $10\,\text{cm}$ above. The resulting velocity u should be reduced by a reasonable factor as it has to work its way against the hotter air on top of the room.

5.27 Derive expressions for the ratio of time scales and of velocity scales of the smallest Kolmogorov eddies with respect to the largest ones defined by u, l and $t = l/u$. Find expressions similar, but not identical, to eq. (5.197).

5.28 Make an estimate of the errors induced by the Boussinesq approximation that density variations only show up in buoyancy effects. Compare two fluid particles of the same volume V at a distance Z submerged in a lake of depth $B = 200$ m. The gradient in density is given by $\varepsilon' = -(d\rho/dz)/\rho_0$ where ρ_0 is the density at the bottom of the lake: ε' is about 10^{-4}. Write down the mass of both fluid particles and submit them to the same pressure gradient β. Estimate the relative difference in accelerations for one particle at the bottom and the other at the top of the lake due to the pressure gradient β.

5.29 Make a graph of eq. (5.241) as a function of x/h and y/h. Introduce a quantity $K = (CUh^2)/q$ and draw a contour with $K = 0.10$. Use $i_y = 0.2$ and $i_z = 0.1$. Notice the very 'slender' plume.

5.30 Reproduce Fig. 5.19 from eq. (5.242).

5.31 Find eqs. (5.234) directly from eq. (5.231) by approximating for small times $R(\tau) = 1 - (\tau/t_m)^2$. Notice that this function is even in τ.

5.32 The total output of a pollutant from a simple jet is given by $Y = QC_0$. At a certain position z far away one may calculate the throughput of pollutant per second as the integral over a cross-section of C_w, where for C and w the asymptotic expressions may be used. Do this and put in the values of the experimentally determined parameters given in Section 5.7. You will find $0.83QC_0$, which is smaller than the expected QC_0. Where does the difference originate from?

5.33 In the example of the simple plume it was noted that real ocean diffusers usually consist of a long pipe with small holes that release fluid. This should then be described as a line source and the plume subsequently should be described as a two-dimensional plume, resulting in different equations. Check the basic dimensions with Q in $L^2 T^{-1}$, M in $L^3 T^{-2}$ and B in $L^3 T^{-3}$.

5.34 Argue from the boundary conditions why in eq. (5.301) the cos solution has been ignored. Write down the complete series expansion solution for the problem of a spherical drop with 'imploding' concentration solute.

5.35 Derive eq. (5.307).

5.36 Derive eq. (5.308) from the Maxwell–Boltzmann equation (5.309).

References

1. G. K. Batchelor, *An Introduction to Fluid Dynamics*, Cambridge University Press, 1970. A fine text for all mathematical aspects of flow discussed in Section 5.4 and the Exercises.

2. Urban Svensson, A mathematical model of the seasonal thermocline, Department of Water Resources and Engineering, Report 1002, Lund, Sweden, 1978. To use for Section 5.5. Its appendix contains detailed derivations.

3. Wolfgang Rodi, *Turbulence Models and Their Application in Hydraulics, A State of the Art Review*, International Association for Hydraulic Research, Delft, The Netherlands, June 1980.

4. Gerhard Zuba, Air pollution modelling in complex terrain (in German), in *Computer Science for Environmental Protection* (eds. M. H. Hälker and A. Jaeschke), Springer, Berlin, 1991, pp. 375–84.
5. J. J. Erbrink, Simple determination of the atmospheric stability class for application in dispersion modelling, using wind fluctuations, *Kema Scientific & Technical Reports*, 7(6) (1989), 361–99.
6. J. J. Erbrink, A practical model for the calculation of σ_y and σ_z for use in an on-line Gaussian dispersion model for tall stacks, based on wind fluctuations, *Atmospheric Environment*, 25A (2) (1991), 277–83.
7. Hugo B. Fischer, E. John List, Robert C. Y. Koh, Joerg Imberger and Norman H. Brooks, *Mixing in Inland and Coastal Waters*, Academic Press, New York, 1979. For Sections 5.1, 5.2, 5.5 and 5.7. We draw extensively on this book for examples and illustrations.
8. M. S. Hussain and W. Rodi, A turbulence model for buoyant flows and its application to vertical buoyant jets, in *Turbulent Buoyant Jets and Plumes* (ed. Wolfgang Rodi), Pergamon, Oxford, 1982. This comprised the state of the art at the time of publication and gives extensive derivations elaborating on Section 5.5.
9. John H. Seinfeld, *Atmospheric Chemistry and Physics of Air Pollution*, John Wiley, New York, 1986

Bibliography

Csanady, G. T., *Turbulent Diffusion in the Environment*, Reidel, Dordrecht, Holland, 1973. For Sections 5.1 and 5.6 with much attention to the fundamentals of random motion.

Cunge, J. A., F. M. Holly Jr and A. Verwey, *Practical Aspects of Computational River Hydraulics*, Pitman, London, 1980. For use in the study of Section 5.2.

de Marsily, G., *Quantitative Hydrogeology, Groundwater Hydrology for Engineers*, Academic Press, 1986. A fine text in the best French tradition. Helpful for Section 5.3.

Harr, Milton E., *Groundwater and Seepage*, Dover, New York, 1991. A reprint of a 1962 text. Much attention is given to complex functions. Useful for Section 5.3, but note the different signs for Φ.

Tennekes, H. and J. L. Lumley, A first course in turbulence, MIT Press, Cambridge, Mass., 1972. A standard and lucid text to use for Section 5.5.

Tritton, D. J., *Physical Fluid Dynamics*, Clarendon Press, Oxford, 1988. For Sections 5.4 and 5.5.

Verruyt, A., *Theory of Groundwater Flow*, Macmillan, London, 1982. A practical text for Section 5.3 with numerical solution methods explained.

Vreugdenhil, Cornelis B., *Computational Hydraulics*, Springer, Berlin, 1989. Useful for computational applications of Sections 5.2 and 5.3.

6

Noise

Noise nuisance is a phenomenon to which some people are more sensitive than others. The owner of a tool shop may sleep well when his machines are operating (he would in fact awake when they stop) whereas the neighbours may experience it as a tremendous nuisance. Nevertheless, governments are using measurable criteria in order to control noise.

This chapter deals with noise and noise abatement. In Section 6.1 we present the basics of acoustics, the properties that determine sound and sound waves. In Section 6.2 we briefly discuss the human sound perception and the corresponding scales and criteria to define noise nuisance.

Noise abatement can be done by insulation of buildings against sound penetration, by scattering away sound and by absorption. These topics will be discussed in Section 6.3. The best action, however, is to reduce the sound level at its source. For problems with neighbours one is well advised to invite them to a cup of coffee and discuss social solutions to the experienced nuisance. For machine nuisance one has to apply technical fixes. In Section 6.4 we discuss the principle of active control of sound, where sound with opposite phase is added to the incident sound in order to reduce sound levels.

In order to simplify our discussions, we will most of the time use plane waves as examples, wherever appropriate. The formalism is a little more general than required for plane waves only, in order to make it easier for the reader to consult the literature.

6.1 BASIC ACOUSTICS

Consider a homogeneous medium at rest with pressure p_0 and density ρ_0. A local pressure fluctuation will have a pressure $p_0 + p(r, t)$ and will propagate through the medium with a velocity c_0. This fluctuation will give rise to a density $\rho_0 + \rho(r, t)$ and particle velocities $\mathbf{u}(\mathbf{r}, t)$. This produces an acoustic disturbance with, as we shall see, $p \ll p_0$ and $\rho \ll \rho_0$.

The simplest example of the propagation of a disturbance is shown on the left of Fig. 6.1, where one will notice a fluid (liquid or gas) with a piston on the left. The piston moves to the right with velocity u, compresses air a little and forces all little parcels of air immediately to the right to move with velocity u as well (a higher velocity would lead to vacuum and a lower velocity to penetration of the piston). As density fluctuations propagate with a velocity $c_0 \gg u$, as we shall see, one observes in Fig. 6.1 (b and c, left) that the right-hand edge of the compressed air runs faster than the piston. When the piston stops (Fig. 6.1c, left) the disturbance continues to the right. In the absence of friction the small parcels of air all have velocity u to the right. When the disturbance has passed, all individual parcels return to their initial conditions with pressure p_0, density ρ_0 and velocity $u = 0$. They have just been displaced a little bit to the right.

In a somewhat more complicated set-up the piston would perform one cycle of a harmonic motion, shown on the right-hand side of Fig. 6.1. The piston moves as

$$x(t) = A \sin(\omega t) \tag{6.1}$$

$$u(t) = \omega A \cos(\omega t) \tag{6.2}$$

In this case the disturbance shown on the bottom right of Fig. 6.1 would consist of compression, rarefaction and again compression. At time $t = T = (2\pi)/\omega$, the period of the motion, the piston is again at its initial position. The front edge of the disturbance then has just moved over a wavelength λ, which leads to the well-known relation

$$\lambda = c_0 T \tag{6.3}$$

Decibel Scale for Sound Pressure Level

In general the local pressure fluctuations $p(t)$ will be irregular not harmonic. It therefore has no well-defined amplitude, so the pressure level is represented by its root mean square value p_{rms}, defined by

$$p_{rms}^2 = \overline{p^2(t)} = \lim_{T \to \infty} \frac{1}{T} \int_{-T/2}^{T/2} p^2(t)\, dt \tag{6.4}$$

The values of p_{rms} encountered in practice vary between 10^{-5} and 10^3 Pa. One therefore uses a $\log_{10}$ scale to define a *sound pressure level* L_p as

$$L_p = 10 \log_{10} \frac{p_{rms}^2}{p_{ref}^2} = 20 \log_{10} \frac{p_{rms}}{p_{ref}} \tag{6.5}$$

where we will suppress the subscript 10 as that will be implied in the notation log. As a reference pressure one takes $p_{ref} = 2 \times 10^{-5}$ Pa, which is the pressure

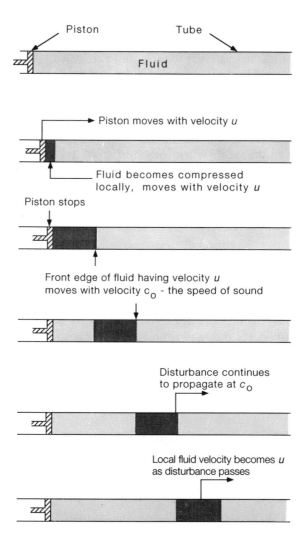

Figure 6.1 Acoustic propagation. The simplest case is shown on the left: a piston moves with velocity *u*, then stops. The compressed parcel of liquid moves with propagation velocity c_0. On the right the piston performs one harmonic motion. One notices the compression and rarefaction of the liquid. (Reproduced, with permission, from *Active Control of Sound*, P. A. Nelson and S. J. Elliott, Academic Press, 1992, Figs. 1.1 and 1.2, p. 2 and p. 4)

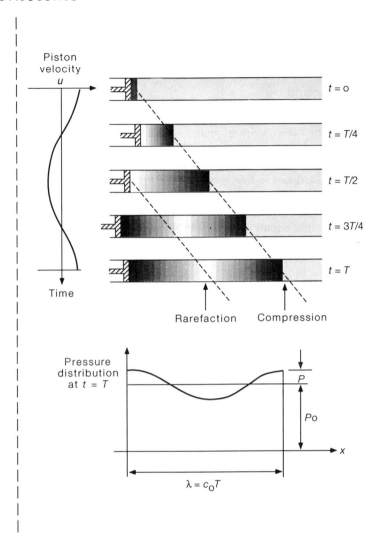

Figure 6.1 (*continued*)

fluctuation just audible to the human ear at a frequency of 1000 Hz.* The unit of sound pressure level is the *decibel* (dB). Therefore the threshold of hearing at 1 kHz corresponds with $p_{rms} = p_{ref}$, giving $L_p = 0$ dB. Another consequence of the

---

*The audibility threshold depends on the individual person (cf. Fig. 6.5). The choice at 1000 Hz was made before accurate standards were defined.

definition is that doubling of p^2_{rms} implies an increase of L_p with 3 dB. Examples of sound pressure levels are given in Table 6.1 where for completeness the value of p_{ref} is added.

Velocity of Sound

Let us once more study the one-dimensional problem of Fig. 6.1 in order to find an expression for the velocity of sound c_0. The essential points are shown in Fig. 6.2. The piston on the left moves with velocity u. In a time t it has moved to the right over a distance ut. The right edge of the compression is at position c_0t. The mass indicated in the lower half of the figure may be deduced from the unperturbed situation as $\rho_0 c_0 t S$, where S is the cross-sectional area of the piston. Newton's law may be applied to the situation shown in the bottom half of Fig. 6.2; the force to the right equals $(p_0 + p)S$ and to the left it reads $p_0 S$. Therefore

$$pS = \frac{d}{dt}(\rho_0 c_0 t S u) = \rho_0 c_0 S u \qquad (6.6)$$

giving

$$p = \rho_0 c_0 u \qquad (6.7)$$

Thus there is a simple proportional relationship between the essential acoustic

Table 6.1 Typical pressure fluctuations p_{rms} and their sound pressure levels L_p (dB re 2×10^{-5} Pa). Numbers only give an order of magnitude. (Reproduced, with permission, from *Active Control of Sound*, P. A. Nelson and S. J. Elliott, Academic Press, 1992, table 1.1, p. 5)

	p_{rms} (Pa)	L_p (dB)
3 m from jet engine	200	140
Pneumatic hammer	60	130
Car horn at 1 m	20	120
Rock/beat band	6	110
Heavy machine shop; heavy truck	2	100
Train of 120 km h^{-1} at 25 m; orchestra	0.6	90
Vacuum cleaner; near highway	0.2	80
TV	0.06	70
Conversation	0.02	60
Office space	0.006	50
Library	0.002	40
Hospital	0.0006	30
Broadcasting studio	0.0002	20
Falling of leaves	0.00006	10
Threshold of hearing at 1 kHz	0.00002	0

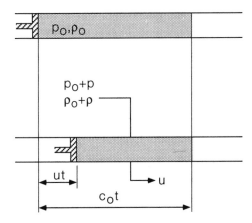

Figure 6.2 The piston on the left compresses a volume in front of the piston. Note that the right edge moves with velocity c_0. (Reproduced with permission from *Active Control of Sound*, P. A. Nelson and S. J. Elliott, Academic Press, 1992, Fig. 1.4, p. 7)

properties of the pressure fluctuation p and the parcel velocities u. In general one defines an *acoustic impedance z* by

$$z(x) = \frac{p(x)}{u(x)} \tag{6.8}$$

In the present case this is a real number.

We continue our derivation of an expression for the velocity of sound c_0 by looking at conservation of mass. In Fig. 6.2 the mass in the two cases shown on top and at the bottom should be equal. This gives

$$\rho_0 c_0 t S = (\rho_0 + \rho)(c_0 t - ut)S \tag{6.9}$$

or

$$\rho_0 c_0 = (\rho_0 + \rho)(c_0 - u) \tag{6.10}$$

As $\rho \ll \rho_0$ and $u \ll c_0$ the cross term ρu may be neglected, leading to

$$u = \frac{\rho c_0}{\rho_0} \tag{6.11}$$

With eq. (6.7) we find

$$c_0^2 = \frac{p}{\rho} \tag{6.12}$$

We remember that the total pressure is indicated by $(p_0 + p)$ and the total density by $(\rho + \rho_0)$. The pressure fluctuations will pass so quickly that (except for extremely high intensities) a parcel of air changes its state adiabatically. Thus, for ideal gases Poisson's law applies, giving in the usual notation $pV^\kappa = $ constant with $\kappa = c_p/c_V$, the ratio of the specific heats at constant pressure and at constant volume. For a given mass Poisson's law implies that $p\rho^{-\kappa}$ is constant or rather in the present notation that

$$p_0\rho_0^{-\kappa} = (p_0 + p)(\rho_0 + \rho)^{-\kappa} \tag{6.13}$$

This may be simplified by dividing by $\rho_0^{-\kappa}$ and using the first term of the Taylor expansion in $\rho/(\rho_0)$,

$$p_0 = (p_0 + p)\left(1 + \frac{\rho}{\rho_0}\right)^{-\kappa} \approx (p_0 + p)\left(1 - \kappa\frac{\rho}{\rho_0}\right) \tag{6.14}$$

giving

$$p = \kappa\frac{\rho p_0}{\rho_0} \tag{6.15}$$

and, with eq. (6.12),

$$c_0^2 = \frac{\kappa p_0}{\rho_0} \tag{6.16}$$

It should be noted that this equation holds for ideal gases only.

Data for Standard Air

For air at atmospheric pressure with $p_0 = 1.013 \times 10^5$ Pa and temperature of $20\,°C$ we have a density $\rho_0 = 1.205\,\mathrm{kg\,m^{-3}}$. With $\kappa = 1.4$ we find a velocity $c_0 = 343\,\mathrm{m\,s^{-1}}$. In water one would find with a different equation of state $c_0 = 1500\,\mathrm{m\,s^{-1}}$.

As we shall discuss in Section 6.2, the human ear is sensitive to frequencies $f = 1/T$ in the range between 20 and 20 000 Hz. With eq. (6.3) the corresponding wavelengths are between 17 m and 17 mm, which is the range of the size of objects in human society. In water the wavelengths would vary between 75 m and 75 mm.

The characteristic acoustic impedance $z = \rho_0 c_0$ is easily found as $413\,\mathrm{kg\,m^{-2}\,s^{-1}}$ and for fresh water as around $1.5 \times 10^6\,\mathrm{kg\,m^{-2}\,s^{-1}}$. From Table 6.1 it appears that the highest pressure fluctuations met in practice will have $p_{rms} = 0.6$ Pa. With the acoustic impedance $z = 413\,\mathrm{kg\,m^{-2}\,s^{-1}}$, eq. (6.7) then gives a corresponding root mean square value for the velocity $u_{rms} = 1.5\,\mathrm{mm\,s^{-1}}$, which is low enough for the approximation that products $u\rho$ may be neglected.

The Wave Equation

A parcel of air experiencing pressure fluctuations will obey the equations of motion derived earlier in this book. We may either look at eq. (3.32), leaving only the pressure gradient forces, or at eq. (5.177), putting viscosity $\mu = 0$. We find in our notation with $\rho + \rho_0$ as density and $\mathbf{u} = \mathbf{u}(x, y, z, t) = \mathbf{u}(x(t), y(t), z(t), t)$ as velocity that

$$(\rho + \rho_0)\frac{d\mathbf{u}}{dt} = -\nabla(p + p_0) \tag{6.17}$$

The left-hand side shows the acceleration of the parcel of air given by (cf. eqs. (5.6) and (5.9))

$$\frac{d\mathbf{u}}{dt} = \frac{\partial\mathbf{u}}{\partial t} + (\mathbf{u}\cdot\nabla)\mathbf{u} \tag{6.18}$$

Earlier we noticed that for most practical purposes the velocities u are very small. Therefore we may *linearize* the equations by omitting the last term on the right in eq. (6.18); we moreover ignore ρ in the term $(\rho + \rho_0)(\partial\mathbf{u}/\partial t)$ which leads to

$$\rho_0\frac{\partial\mathbf{u}}{\partial t} + \nabla p = 0 \tag{6.19}$$

The next step is to use conservation of mass in any element of volume. The basic equation has been derived in Fig. 5.17. This gives

$$\nabla\cdot((\rho + \rho_0)\mathbf{u}) = -\frac{\partial}{\partial t}(\rho + \rho_0) \tag{6.20}$$

where the right-hand side describes the decrease in mass of a unit of volume and the left-hand side the net outflow of mass. Ignoring again ρ in $(\rho + \rho_0)\mathbf{u}$ gives

$$\rho_0\nabla\cdot\mathbf{u} + \frac{\partial\rho}{\partial t} = 0 \tag{6.21}$$

We now take the divergence of eq. (6.19) and use eq. (6.21) to find

$$-\frac{\partial^2\rho}{\partial t^2} + \nabla^2 p = 0 \tag{6.22}$$

For propagation of density fluctuations we found eq. (6.12) as the simple relationship $p = c_0^2\rho$ which after substitution into eq. (6.22) leads to the *wave equation*

$$\nabla^2 p - \frac{1}{c_0^2}\frac{\partial^2 p}{\partial t^2} = 0 \tag{6.23}$$

This is different from the heat equation (4.15) or (4.19), where the first derivative to the time appears and only damped solutions occur.

Harmonic Waves in Complex Notation

A solution of the wave eq. (6.23) is the propagating plane wave in the positive x direction:

$$p(\mathbf{r}, t) = f\left(t - \frac{x}{c_0}\right) \tag{6.24}$$

For $t \to t + 1$ the substitution $x \to x + c_0$ leads to the same argument of the arbitrary function f, so the propagation velocity indeed equals c_0. An example of the plane wave is an *harmonic function* with frequency ω:

$$p(x, t) = |A| \cos(\omega t - kx + \phi_A) \tag{6.25}$$

where the superfluous y and z dependences have been omitted. Comparing eqs. (6.24) and (6.25) we immediately find $k/\omega = 1/c_0$ or

$$c_0 = \frac{\omega}{k} = 2\pi \frac{f}{k} \tag{6.26}$$

The harmonic function (6.25) repeats itself with wavelength λ and we see that

$$k\lambda = 2\pi \tag{6.27}$$

where k is the wave number. Both equations (6.26) and (6.27) together with $\omega T = 2\pi$ again lead to eq. (6.3).

The phase ϕ_A is important when one considers the interference between two or more waves. It turns out to be convenient to represent the harmonic wave (6.25) by a complex function

$$p(x, t) = A \, e^{j(\omega t - kx)} \tag{6.28}$$

$$A = |A| e^{j\phi_A} \tag{6.29}$$

with $j = \sqrt{-1}$. It is understood that the physical pressure $p(x, t)$ is found by taking the real part of eq. (6.28). Note that the sign convention in the exponential of eq. (6.28) is opposite to the one used in quantum mechanics.

In a more general presentation one would consider just one frequency ω but would not insist on a harmonic behaviour in space. Then we simplify eq. (6.24) as

$$p(\mathbf{r}, t) = p(\mathbf{r}) e^{j\omega t} = |p(\mathbf{r})| e^{j(\omega t + \phi)} \tag{6.30}$$

Here $p(\mathbf{r})$ may be a complex function and in the final analysis again the real part of the resulting $p(\mathbf{r}, t)$ should be taken. Substitution of eq. (6.30) into the wave eq. (6.23) gives the *Helmholtz equation*

$$\nabla^2 p + \left(\frac{\omega}{c_0}\right)^2 p = 0 \tag{6.31}$$

or

$$\nabla^2 p + k^2 p = 0 \qquad (6.32)$$

Superposition of Waves

Let us consider two solutions of the wave eq. (6.23), say $p_1(\mathbf{r}, t)$ and $p_2(\mathbf{r}, t)$. As the equation is linear, or rather has been linearized, the sum

$$p(\mathbf{r}, t) = p_1(\mathbf{r}, t) + p_2(\mathbf{r}, t) \qquad (6.33)$$

will also be a solution. When the two waves have the same frequency ω both can be written in the form (6.30) and from the linear character of the Helmholtz equation (6.32) their sum

$$p(\mathbf{r}) = p_1(\mathbf{r}) + p_2(\mathbf{r}) \qquad (6.34)$$

will also be a solution.

In Section 6.4 we will discuss noise abatement by adding a signal $p_2(\mathbf{r})$ to the unwanted sound $p_1(\mathbf{r})$. Writing

$$p_1(\mathbf{r}) = |p_1| e^{j\phi_1} \qquad (6.35)$$

$$p_2(\mathbf{r}) = |p_2| e^{j\phi_2} \qquad (6.36)$$

we find

$$L_{p,\text{sum}} = L_{p,1} + 10 \log \left[1 - 2 \frac{|p_2|}{|p_1|} \cos(\phi_2 - \phi_1) + \frac{|p_2|^2}{|p_1|^2} \right] \qquad (6.37)$$

Here we used the definition (6.5) of the sound pressure level and the fact that time averages of $\sin^2$ and $\cos^2$ functions have the simple value of 1/2.

In Fig. 6.3 we show the increase in sound pressure level $L_{p,\text{sum}} - L_{p,1}$ as a function of the ratio of the moduli or amplitudes $|p_1|/|p_2|$ and phase differences $\phi_2 - \phi_1$. It appears that amplitudes and phases should match accurately in order to obtain an appreciable reduction of some 15 dB, corresponding to a reduction factor of roughly $2^5 = 32$.

Acoustic Impedance

Let us stick to a single frequency ω. In the same way as the pressure $p(\mathbf{r}, t)$, the local velocities $\mathbf{u}$ may also be represented in complex notation by

$$\mathbf{u}(\mathbf{r}, t) = \mathbf{u}(\mathbf{r}) e^{j\omega t} = |\mathbf{u}(\mathbf{r})| e^{j(\omega t + \phi)} \qquad (6.38)$$

where again at the end the real part of $\mathbf{u}(\mathbf{r}, t)$ should be taken. Substitution into eq. (6.19) gives

$$\rho_0 j \omega \mathbf{u}(\mathbf{r}) + \nabla p = 0 \qquad (6.39)$$

where the time dependence $e^{j\omega t}$ has been omitted.

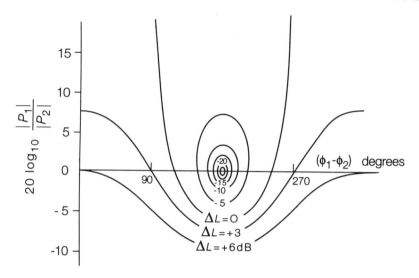

Figure 6.3 Increase of sound pressure level $L_{p,\text{sum}} - L_{p,1}$ on adding a secondary sound with modulus $|p_2|$ and relative phase angle $\phi_2 - \phi_1$ to sound p_1

Let us apply this to the *one-dimensional harmonic* case, where $p(x) = A\,e^{-jkx}$ (cf. eq. 6.28)). This gives

$$\rho_0 j\omega u + \frac{\partial p}{\partial x} = \rho_0 j\omega u - jkp = 0 \tag{6.40}$$

or

$$u(x) = \frac{k}{\rho_0 \omega} p(x) = \frac{1}{\rho_0 c_0} p(x) \tag{6.41}$$

The acoustic impedance

$$z(x) = \frac{p(x)}{u(x)} = \rho_0 c_0 \tag{6.42}$$

has the value found in eq. (6.7).

A different expression will be found for outgoing spherically symmetric waves where

$$p(r) = \frac{A\,e^{-jkr}}{r} \tag{6.43}$$

By using eq. (B.4) to find the ∇^2 operator in spherical coordinates one readily checks that $p(r)$ satisfies the Helmholtz eq. (6.32). We now substitute eq. (6.43) into the radial component of eq. (6.39), which gives

$$j\omega\rho_0 u_r(r) - jkp(r) - \frac{1}{r}p(r) = 0 \tag{6.44}$$

The acoustic impedance in this case reads

$$z(r) = \frac{p(r)}{u_r(r)} = \rho_0 c_0 \frac{jkr}{1+jkr} \tag{6.45}$$

For $kr \gg 1$ or $r \gg \lambda/(2\pi)$ this reduces to the plane wave value (6.42); at large distances from the source we therefore apparently have the plane wave impedance. For $kr \ll 1$ the impedance becomes purely imaginary and pressure $p(r)$ and parcel velocity $u_r(r)$ exhibit a 90° phase difference.

Acoustic Intensity and Acoustic Power

The energy of a parcel of air will consist of kinetic and potential energy parts. Let us consider a parcel with unperturbed volume V_0. Its kinetic energy may be written as

$$E_k = \tfrac{1}{2}\rho_0 V_0 u^2 \tag{6.46}$$

where products of small terms again have been omitted. Its potential energy may be found by the work done on the parcel of air as

$$E_p = -\int_{V_0}^{V} p\,dV \tag{6.47}$$

Conservation of mass implies that $(\rho + \rho_0)V = \rho_0 V_0$ and $V\,d\rho + (\rho + \rho_0)dV = 0$. In first order it follows that

$$dV = -\frac{V}{\rho + \rho_0}d\rho \approx -\frac{V_0}{\rho_0}d\rho \tag{6.48}$$

In order to calculate the integral (6.47) we have to change the variable to p, which is done by using eq. (6.12). We find $d\rho = dp/c_0^2$ which together with eq. (6.48) gives, for the potential energy (6.47),

$$E_p = \int_0^p \frac{pV_0\,dp}{\rho_0 c_0^2} = \frac{1}{2}\frac{p^2}{\rho_0 c_0^2}V_0 \tag{6.49}$$

The energy density ε is found as the kinetic energy (6.46) added to the potential energy (6.49) and divided by the volume V_0. This gives

$$\varepsilon = \frac{1}{2}\rho_0 u^2 + \frac{1}{2}\frac{p^2}{\rho_0 c_0^2} \tag{6.50}$$

We now want to derive an expression for the *acoustic intensity* **I**, which is the energy leaving a volume element in a unit of time through a unit of area. We derive this expression for the one-dimensional case. The energy leaving a unit of length should be equal to the decrease of the energy density. This one-dimensional form of conservation of energy implies

$$\frac{\partial I_x}{\partial x} = -\frac{\partial \varepsilon}{\partial t} = -\rho_0 u \frac{\partial u}{\partial t} - \frac{p}{\rho_0 c_0^2} \frac{\partial p}{\partial t} \tag{6.51}$$

We now write the terms on the right as derivatives with respect to the position x. For the first term we use eq. (6.19) and find that it equals $u \, \partial p / \partial x$. For the second term we use eq. (6.12) to switch to the variable p and the one-dimensional form of eq. (6.21) to go to $\partial u / \partial x$, giving

$$\frac{\partial I_x}{\partial x} = u \frac{\partial p}{\partial x} + p \frac{\partial u}{\partial x} = \frac{\partial(pu)}{\partial x} \tag{6.52}$$

Integration with respect to x gives

$$I_x = pu \tag{6.53}$$

For the general three-dimensional case it seems appropriate to define

$$\mathbf{I} = p(\mathbf{r}, t)\mathbf{u}(\mathbf{r}, t) \tag{6.54}$$

as the local power flow. The dimension of this quantity appears to be $\mathrm{W\,m^{-2}}$, which is indeed an energy flux per unit area.

For a wave with a single frequency ω the time-averaged acoustic intensity $\mathbf{I}$ is defined as

$$\mathbf{I} = \frac{1}{T} \int_{-T/2}^{T/2} p(\mathbf{r}, t)\mathbf{u}(\mathbf{r}, t) \mathrm{d}t \tag{6.55}$$

Here, for $p(\mathbf{r}, t)$ and $\mathbf{u}(\mathbf{r}, t)$ the real parts have to be inserted. By using eqs. (6.30) and (6.38) and separating the real and imaginary parts of $p(\mathbf{r})$ and $\mathbf{u}(\mathbf{r})$ one finds

$$\mathbf{I} = \frac{1}{T} \int_{-T/2}^{T/2} (p_R \cos \omega t - p_I \sin \omega t)(\mathbf{u}_R \cos \omega t - \mathbf{u}_I \sin \omega t)\mathrm{d}t$$

$$= \tfrac{1}{2}(p_R \mathbf{u}_R + p_I \mathbf{u}_I) = \tfrac{1}{2} \mathrm{Re}[p^*(\mathbf{r})\mathbf{u}(\mathbf{r})] \tag{6.56}$$

For *plane harmonic waves* in the x direction we use eq. (6.42) and find

$$I = \frac{|p(x)|^2}{2\rho_0 c_0} \tag{6.57}$$

From eq. (6.28) it is clear that for a harmonic plane wave $|p(x)|$ equals the amplitude of the wave. From the definition (6.4) it follows that $p_{\text{rms}}^2 = |p(x)|^2/2$, giving

$$I = \frac{p_{\text{rms}}^2}{\rho_0 c_0} \tag{6.58}$$

with dimension $W\,m^{-2}$. As the values of p_{rms}^2 cover many powers of 10 (cf. Table 6.1), we again use a logarithmic intensity level

$$L_I = 10\log\frac{I}{I_{ref}} \tag{6.59}$$

where $I_{ref} = 10^{-12}\,W\,m^{-2}$. For a single frequency we find from eq. (6.5) that

$$L_p = 10\log\frac{p_{rms}^2}{p_{ref}^2} = 10\log\left(\frac{I\rho_0 c_0 I_{ref}}{I_{ref}p_{ref}^2}\right) \tag{6.60}$$

$$L_p = L_I + 10\log\frac{\rho_0 c_0 I_{ref}}{p_{ref}^2} = L_I + 0.14 \tag{6.61}$$

The small number 0.14 holds for air of 20 °C and would vanish for an air temperature for which $\rho_0 c_0 = 400\,kg\,m^{-2}\,s^{-1}$. It may be added that measuring devices usually react to the pressure field and give L_p. Locally, this field often behaves like a plane wave; therefore eqs. (6.57) and (6.61) hold, whereas for most practical purposes the small number 0.14 may be ignored.

For outgoing *harmonic spherical waves* we use eq. (6.45) and find

$$I_r = \frac{1}{2}\mathrm{Re}[p^*(r)u_r(r)] = \frac{1}{2}\mathrm{Re}\left[p^*(r)\frac{p(r)}{jkr\rho_0 c_0}(1+jkr)\right]$$

$$= \frac{1}{2\rho_0 c_0}|p(r)|^2\,\mathrm{Re}\left(\frac{1+jkr}{jkr}\right) = \frac{|p(r)|^2}{2\rho_0 c_0} \tag{6.62}$$

The total sound power output in watts is found by integrating over a sphere with radius r, which simply gives

$$W = \frac{4\pi r^2|p(r)|^2}{2\rho_0 c_0} = \frac{2\pi|A|^2}{\rho_0 c_0} \tag{6.63}$$

where eq. (6.43) for $p(r)$ was used. The sound power W varies between 10^{-9} and $10^4\,W$, a scale similar to the one for p_{rms}^2 and twice as wide as the one for p_{rms}. One therefore again uses a logarithmic scale to describe the *sound power level* L_W as

$$L_W = 10\log\frac{W}{W_{ref}} \tag{6.64}$$

Here the reference level is $W_{ref} = 10^{-12}\,W$. Some typical values are given in Table 6.2. From the example of the chain-saw it is obvious that only a fraction of the power of a device is converted into acoustic power.

The sound power output W is a property of a given device, in eq. (6.63) taken as a point source. From this equation it follows that

$$|p(r)|^2 = \frac{\rho_0 c_0 W}{2\pi r^2} = 2p_{rms}^2 \tag{6.65}$$

Table 6.2 Typical sound power level output. (Reproduced, with permission,. from *Active Control of Sound*, P. A. Nelson and S. J. Elliott, Academic Press, 1992, Table 1.2, p. 29, and L. E. Kinsler, A. R. Frey, A. B. Coppers and J. S. Sanders, *Fundamentals of Acoustics*, 3rd edn, Wiley, New York, 1982, p. 275)

	Output W	L_W (dB)
Jet engine	10 000	160
Chain-saw	1	120
Shouting voice	10^{-3}	90
Loudly talking voice	2×10^{-4}	83
Normal voice	10^{-5}	70
Whisper	10^{-9}	30

The sound pressure level $L_p(R)$ at a distance R from a point source in air becomes

$$L_p(R) = 10 \log \frac{p_{rms}^2}{p_{ref}^2}.$$

$$= 10 \log \left(\frac{\rho_0 c_0 W_{ref}}{p_{ref}^2} \right) - 10 \log(4\pi R^2) + 10 \log \frac{W}{W_{ref}} \qquad (6.66)$$

With the data given above we find

$$L_p(R) = L_W - 10 \log(4\pi R^2) + 0.14 \qquad (6.67)$$

where the small number 0.14 has already been explained.

The first point to notice is that as far as Table 6.1 applies to point sources one has to indicate the distance to the source. This indeed has been done for the jet engine and using eq. (6.63) to compare both tables one finds agreement within a few dB.

Another point to consider is the decrease of sound pressure level L_p with increasing distance R. From eq. (6.67) for a point source it follows that a doubling of distance (substituting $R \to 2R$) would give a reduction of the level L_p by 6 dB.

For a *line source*, representing, for example, the constant hum of a motorway, one has to return to the wave equation (6.33) and solve it in cylindrical coordinates. For simplicity we consider a single frequency ω which leads to the Helmholtz eq. (6.32) and has to be simplified by using eq. (B.2). The solution leads to Bessel functions and is left as an exercise. We just mention that for $kr \gg 1$ the asymptotic solution becomes

$$p(r) \approx \frac{A}{\sqrt{r}} e^{-jkr} \qquad (6.68)$$

This equation may be substituted in the Helmholtz equation where it will be found that it satisfies that equation when one ignores terms of the order r^{-2} with respect to terms of the order r^{-1}. The total power output W_1 per unit of length may be found by integrating over a cylinder. This gives

$$W_1 = \frac{2\pi R |A|^2}{2\rho_0 c_0} \tag{6.69}$$

One would define the sound power level as

$$L_{W_1} = 10 \log \frac{W_1}{W_{\text{ref}}} \tag{6.70}$$

with the same $W_{\text{ref}} = 10^{-12}$ W as before, which leads to an equation similar to the one for the point source (6.67):

$$L_p = L_{W_1} - 10 \log(2\pi R) + 0.14 \tag{6.71}$$

In this case a doubling of the distance R to the source leads to a decrease in sound pressure level with only 3 dB.

We note that eqs. (6.57) and (6.71) apply to expansion of waves in free space. Sources just above the ground radiate in half-space only. One therefore could effectively double the power or halve the surface area considered. In both ways the sound pressure level would increase by 3 dB.

Adding Independent Sound (Pressure) Levels

We have already discussed the addition of two sources with the same frequency in eq. (6.33). For sources with different frequencies one would write

$$p_{\text{rms}}^2 = \lim_{T \to \infty} \frac{1}{T} \int_{-T/2}^{T/2} [p_1^2(\mathbf{r}, t) + p_2^2(\mathbf{r}, t) + 2p_1(\mathbf{r}, t)p_2(\mathbf{r}, t)] dt \tag{6.72}$$

with real functions $p_1(\mathbf{r}, t)$ and $p_2(\mathbf{r}, t)$. When both sound sources are uncorrelated, the cross term $2p_1 p_2$ will be as often positive as negative, giving a vanishing time integral. Thus,

$$p_{\text{rms}}^2 = p_{1,\text{rms}}^2 + p_{2,\text{rms}}^2 \tag{6.73}$$

or

$$L_p = 10 \log \frac{p_{1,\text{rms}}^2 + p_{2,\text{rms}}^2}{p_{\text{ref}}^2} \tag{6.74}$$

If both sources are equally strong at position $\mathbf{r}$ it follows that

$$L_p = L_1 + 3 \, \text{dB} \tag{6.75}$$

Reversibly, if one takes one of the sources away the sound pressure level decreases by 3 dB.

We note that the sound from two radios tuned to the same station in the same room will be strongly correlated. One then has to go back to Fig. 6.3 to describe the interference pattern.

The dB scales for L_p, L_I and L_W were defined for arbitrary time-varying fields such as in eq. (6.4). Of course one may always apply a Fourier transformation to $p(\mathbf{r},t)$ and $\mathbf{u}(\mathbf{r},t)$ and discuss single frequencies $f = \omega/(2\pi)$. The total pressure field would be a sum of Fourier components, a generalized form of eq. (6.72). Again the cross terms belonging to different frequencies would average to zero and one could generalize eqs. (6.5), (6.59) and (6.64) to

$$L_p = 10\log\frac{p_{1,\text{rms}}^2 + p_{2,\text{rms}}^2 + p_{3,\text{rms}}^3 + \cdots}{p_{\text{ref}}^2} \tag{6.76}$$

$$L_I = 10\log\frac{I_1 + I_2 + I_3 + \cdots}{I_{\text{ref}}} \tag{6.77}$$

$$L_W = 10\log\frac{W_1 + W_2 + W_3 + \cdots}{W_{\text{ref}}} \tag{6.78}$$

If one has to add sound pressure levels from a few sources, say two, than one first has to go back to their $p_{1,\text{rms}}^2$ and $p_{2,\text{rms}}^2$ and then apply eq. (6.76). This gives

$$L_p = 10\log(10^{L_{p1}/10} + 10^{L_{p2}/10}) \tag{6.79}$$

and similarly for L_I and L_W. In Table 6.3 we summarize the dB scales and add two quantities to be used in Section 6.2.

Diffuse Sound Field

A *diffuse* sound field will consist of waves crossing a unit area from all directions. It also may comprise many frequencies. It is clear that the intensity corresponding to a certain sound pressure becomes smaller than for a plane wave. The relation

Table 6.3 Summary of dB scales

Variable	Quantity	Re value	Definition	Note
Pressure p_{rms}	L_p	2×10^{-5} Pa	(6.5)	All frequencies summed
Intensity I	L_I	10^{-2} W m^{-2}	(6.59)	All frequencies summed
Power W	L_W	10^{-12} W	(6.64)	All frequencies summed
Pressure density $p_{\text{rms}}(f)$	$L_p(f)$	2×10^{-5} Pa	(6.92)	Bandwidth $\Delta f = 1$ Hz
Intensity density $I(f)$	$L_I(f)$	10^{-12} W m^{-2}	(6.94)	Bandwidth $\Delta f = 1$ Hz

is

$$I = \frac{p^2_{\mathrm{rms}}}{4\rho_0 c_0} \tag{6.80}$$

and correspondingly instead of eq. (6.61)

$$L_p = L_I + 6.14 \tag{6.81}$$

The proof of eq. (6.80) is illustrated in Fig. 6.4. The intensity I is the energy crossing a unit area from one side per second. In Fig. 6.4 we therefore have to add contributions from all space angles $d\Omega$, making an angle ϕ with the normal. The power density per steradian will be called g and is by definition a constant for a diffuse field. The incoming power density $g\,d\Omega$ has to be multiplied by $\cos \phi$ to find the power per unit area in the 'horizontal' plane in Fig. 6.4. The total energy I crossing the unit area per second therefore may be written as

$$I = \int_{\mathrm{semisphere}} g \cos \phi \, d\Omega = g \int_0^{\pi/2} \cos \phi (2\pi \sin \phi) \, d\phi = \pi g \tag{6.82}$$

The incoming waves entering in the small angular region $d\Omega$ may be regarded as plane. For their contribution $d(p^2_{\mathrm{rms}})$ to the total p^2_{rms} we may write

$$g\,d\Omega = \frac{d(p^2_{\mathrm{rms}})}{\rho_0 c_0} \tag{6.83}$$

where we used eq. (6.58) and the fact that the incoming waves from all directions may be regarded as independent. The pressure level is determined by waves from all 4π angles. Therefore by integration of eq. (6.83) we find

$$4\pi g = \frac{p^2_{\mathrm{rms}}}{\rho_0 c_0} \tag{6.84}$$

Substitution of eq. (6.84) into eq. (6.82) gives the required relation (6.80). We note

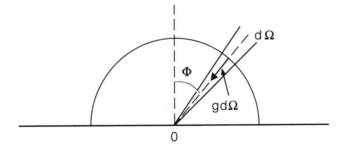

Figure 6.4 Diffuse sound field with (constant) power density g per steradian

in passing that the sound power density per steradian g is analogous to the luminosity from diffuse light sources.

6.2 HUMAN PERCEPTIONS AND NOISE CRITERIA

The human perception of sound and noise is determined by the signal processing in the ears and in the brain. This leads to the capabilities to enjoy music and to communicate by speech, but also to the experience of noise nuisance. In this section we describe some aspects of the human sound perception and speak of the 'ear' while implying the brain functions as well.

We start with the fact that the sensitivity of the human ear to sound depends on its frequency. In Fig. 6.5 we show the response of the human ear to sounds of a single frequency f. The sound intensity L_I is used as defined in eq. (6.59). One notices the *threshold of audibility*, the dotted line in Fig. 6.5, which refers to 'well hearing young people'. In practice, for some 95% of the population the thresholds lie much higher .[*] The reference value $I_{ref} = 10^{-12}$ W m^{-2} defined in eq. (6.59) is used in Fig. 6.5 as well. It can be seen that at about 4 kHz the threshold is around 0 dB, which is the minimum of the curve. For higher frequencies it rises sharply to a cut-off at some 20 kHz. This higher end of the audibility spectrum decreases with age to some 15 or even 10 kHz.

The curves in Fig. 6.5 connect points of equal *loudness level L_N*. Two signals at two different frequencies on the same curve were experienced as being equally loud in a series of test experiments performed in the 1950's. They are since then taken as standard by the International Standardization Organization (ISO), although, once again, individual persons will have a 'personal' curve.

The *loudness level L_N* is given in units of *phon*, which is the intensity level of the contour in dB at a frequency of 1 kHz. They are indicated near the curves in Fig. 6.5. At the top of the graph one notices a threshold of feeling, which is a trickling sensation in the ears, which changes into pain at an intensity level of some 140 dB.

Loudness

The *loudness N* of a signal at a certain frequency f should be defined in such a way that a doubling of loudness results in a doubling of the value of N. This has of course to be determined by experimental subjects. The value $N = 1$ *sone* is defined by the loudness level of 40 phon. Note that the loudness is independent of the frequency.

---

[*]In Chedd ([1], p. 17) one may find curves indicating the percentage of the population that have thresholds of audibility at a certain frequency f below that curve. It appears that the 50% curve lies some 14 dB above the 5% curve shown in Fig. 6.5.

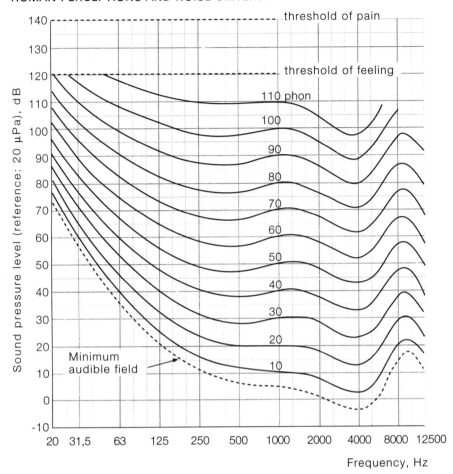

Figure 6.5 Curves of equal loudness. The dashed threshold of audibility applies to the best hearing 5% of the population (modal value of the binaural hearing threshold of otologically normal persons within the age limits from 18 to 30 years). The loudness at 1 kHz is indicated near the curves and defines the phon. (Permission to reproduce the figure in annex A, p. 4, of ISO 226: 1987E has been given by the International Organization for from Standardization, ISO. The complete standard may be purchased from the ISO member bodies or directly from the ISO central secretariat, Case Postale 56, 1211 Geneva 20, Switzerland)

The relation between loudness N and loudness level L_N is indicated in Fig. 6.6. On the semi-logarithmic scale the relation is linear for $L_N > 40$ phone corresponding to

$$N = 0.046 \times 10^{L_N/30} \tag{6.85}$$

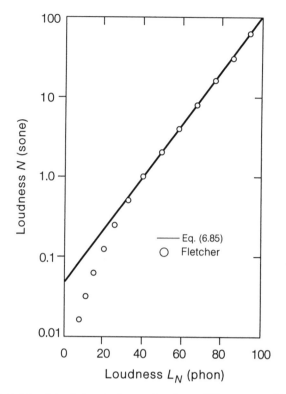

Figure 6.6 Experimental relation between loudness N in sone and loudness level L_N in phon. (Reproduced with permission from Lawrence E. Kinsler, Austin R. Frey, Alan B. Coppers and James V. Sanders, *Fundamentals of Acoustics*, 3rd edn, Wiley, New York, 1982, Fig. 11.13, p. 271)

As for a frequency of $f = 1\,\text{kHz}$ the loudness level L_N is proportional to the logarithm of the intensity I (cf. eq. (6.59)); there is a log–log relation between N and I which implies a power relation between the subjective experience of loudness N and the physical intensity I of the sound signal. From Fig. 6.6 one may deduce

$$N \approx 460F(f)I^{1/3} \tag{6.86}$$

where $F(f) = 1$ for the frequency $f = 1\,\text{kHz}$ (cf. Exercise 6.7).

Equation (6.86) is an example of the *psycho-physical power law*, first formulated by Stevens [2]. It says that 'equal objective ratios I_2/I_1 and I_4/I_3 of physical stimuli I produce equal subjective ratios N_2/N_1 and N_4/N_3 of effects N'. This would hold for a.o. brightness, electric shocks, pressure on the palm and loudness.

This implies that

$$N = cI^\kappa \qquad (6.87)$$

where for loudness the particular form (6.86) holds. The exponent may be bigger than one (such as for electric shocks with $\kappa \sim 3.5$) or smaller than one, as for sound.

In the case that $\kappa < 1$ in eq. (6.87) it appears that the curve of N versus I somewhat resembles the logarithmic curve of L_I versus I if one only plots less than two orders of magnitude (cf. Exercise 6.8). This fact erroneously led the German physicist Fechner around 1860 to postulate a logarithmic rule

$$N \approx 10 \log I + \text{constant} \qquad \text{(incorrect)} \qquad (6.88)$$

This would imply that the decibel scale (6.59) for acoustic intensity is precisely proportional to the perceived loudness. This assumption, now proven false, has led to the introduction of the dB scale.

Two remarks seem appropriate. First, it is clear from Fig. 6.6 that eqs. (6.85), (6.86) and (6.87) only hold for $L_N > 40$ phon. Second, it is remarkable that the human ear is able to process sound intensities L_I or pressure levels L_p that cover the many powers of ten, displayed in Fig. 6.5 and Table 6.1. On the quiet side man is able to detect the rustling of leaves caused by a wild animal and on the loud side he may stand the noise of the Great Falls to catch a fish.

Octave Bands

Perhaps more amazing still is the way in which we perceive frequencies. Consider two frequencies f_1 and f_2, close to one another. We know from elementary mechanics that the sum signal exhibits *beats* with frequency $|f_1 - f_2|$. The interesting point is that the human brain perceives similar beats if the ratio of frequencies f_2/f_1 is close to a simple ratio of integers: 2/1 or 3/2 etc. The beating becomes more subtle if the integers become greater and it requires more experience to hear them. If we experience *consonance* the ratio f_2/f_1 is precisely that of two integers. This also happens if the two signals are offered to different ears, so this phenomenon is a property of the human brain and the basis of our enjoyment of music. We note in passing that there are some integer ratios which some people experience as dissonance, e.g. 4/3.

The ratio $f_2/f_1 = 2/1$ defines an *octave*. The ratio $f_2/f_1 = 2^{1/3}$ is called 1/3-octave or in music the *tierce*. The properties of the human ear, just discussed, have led to the division of the audible spectrum of Fig. 6.5 in *octave bands* with a subdivision in 1/3-octave bands.

The centre f_c of a band is defined by

$$f_c = \sqrt{f_1 f_2} \qquad (6.89)$$

The octave band centres f_c are given on top of Fig. 6.7 following international

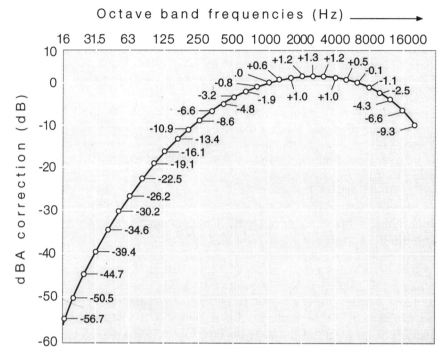

Figure 6.7 dBA-correction curve. Octave band centre frequencies f_c are shown at the top. Corrections Δ required for finding dBA values are added to the curves, both for the octave centre frequencies and for the 1/3 octave centre frequencies

conventions. For precision measurements it is often required to subdivide the octave bands into 1/3-octave bands or even 1/10-octave bands. Acoustic filters then determine the levels L_I or L_p in each band and the total levels, if required, may be found by means of eqs. (6.76) and (6.77).

dBA Scale

When one knows the distribution of L_I or L_p over the bands one might compare the data with the curves of Fig. 6.5 in order to deduce a single number for an overall sound level. It is easiest to take the difference between the actual level L_I and the threshold of audibility at a certain frequency. The complications of loudness and loudness levels are then ignored.

In practice one uses a smoothened version of the audibility curve, which resembles the 40 phon curve of Fig. 6.5 without the structure at high frequencies.

That curve is shown in Fig. 6.7 and it leads to corrections Δ to be added to the L_p values of the octave bands indicated at the top with centre frequencies 16, 31.5, 63, etc. It is clear that the correction values change considerably over a bandwidth. Therefore, the corrections are also given for the more accurate 1/3-octave band distribution. After the corrections Δ have been applied, one should add pressure levels according to eq. (6.76) and find L_A, sometimes indicated by $L_{A,eq}$, which is expressed in units dBA. This quantity is used to define noise criteria, to be discussed below.

Tones

Measurements in octave bands tend to obscure the presence of *tones*. Musical instruments, for example, generate a series of discrete frequencies, each with their own pressure level L_p. Examples are shown in Fig. 6.8. One notices that the pure tones (g_0 of the violin and c_0 of the clarinet) consist of the fundamental and a series of overtones. The set of pressure levels shown determines the sound of the musical instrument.

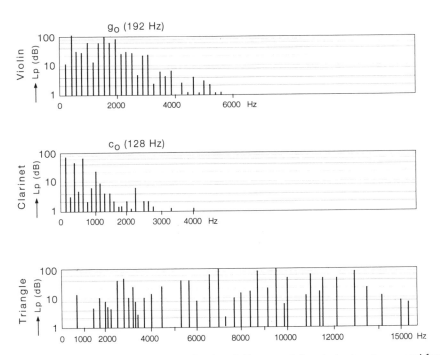

Figure 6.8 Sound pressure levels for the violin g_0 and the clarinet c_0 tones and for the triangle taken by a frequency analyser

It is convenient to introduce pressure densities $p_{rms}^2(f)$ for frequency f, and similarly intensity densities $I(f)$. They represent the partial p_{rms}^2 or the partial intensity I over an interval $\Delta f = 1$ Hz, defined as

$$p_{rms}^2(f) = \frac{\Delta p_{rms}^2(f)}{\Delta f} \tag{6.90}$$

$$I(f) = \frac{\Delta I}{\Delta f} \tag{6.91}$$

They may also be expressed in a dB scale by

$$L_p(f) = 10 \log \frac{p_{rms}^2(f)\Delta f}{p_{ref}^2} \tag{6.92}$$

and

$$L_p = 10 \log \frac{\int p_{rms}^2(f)\,\mathrm{d}f}{p_{ref}^2} \tag{6.93}$$

One again should realize that the value of (6.93) may be found from (6.92) by the detour of eq. (6.76) in integral form. For the intensities one defines similarly

$$L_I(f) = 10 \log \frac{I(f)\Delta f}{I_{ref}} \tag{6.94}$$

and

$$L_I = 10 \log \frac{\int I(f)\,\mathrm{d}f}{I_{ref}} \tag{6.95}$$

The reference values are the same as before. The equations are summarized in Table 6.3. An example of tones against a background is shown in Fig. 6.9. The continuous distribution of $L_p(f)$ in dB may be reduced to the values of L_p over an octave band by integrating in eq. (6.93) over the bandwidth giving the histogram shown. Without the tones the histogram would have been lower, but from the histogram one could not deduce that tones were present.

It is instructive to look more closely at the histogram concept, using Fig. 6.10, where we indicated a single tone at 250 Hz against a constant background over the octave bandwidth. Let us first look at the background alone. For a width w and a constant pressure density $p_{rms}^2(f) = p_{rms}^2$ we find by eq. (6.93) that

$$L_p = L_p(f) + 10 \log w \tag{6.96}$$

The single tone is defined on a width $\Delta f = w = 1$ Hz, so its pressure density level $L_p(f)$ equals its pressure level L_p. The background pressure has to be added to the value L_{p_2} of the tone by eq. (6.79). With the numbers indicated in Fig. 6.10 one would find $L_p = 62.5$ dB for the background. Adding to 60 dB for the tone

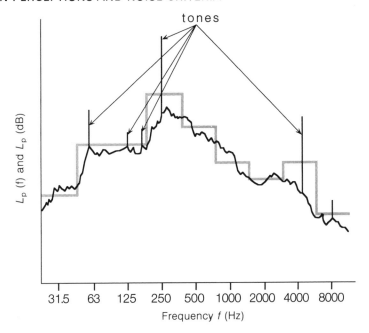

Figure 6.9 Tones as narrow peaks against a continuous background. The octave band histogram includes the tones. (Reproduced with permission from Lawrence E. Kinsler, Austin R. Frey, Alan B. Coppers and James V. Sanders, *Fundamentals of Acoustics*, 3rd edn, Wiley, New York, 1982, Fig. 11.1, p. 250)

would give 64.5 dB. In order to represent both together in a single histogram we use eq. (6.96) and find $L_p(f) = 42$ dB, indicated in Fig. 6.10 as well.

The contribution of this band to the A-weighted pressure level L_A is found by Fig. 6.7 as $64.5 - 8.6$ dB $= 55.9$ dB. When there are more bands present, as in Fig. 6.9 one has to add their contributions according to eq. (6.79).

Noise Criteria

Speech is a way of communication by sound. The intelligibility of speech signals against background noise is given in Fig. 6.11. It is apparent that sentences are more easily understood than words without a context, as one would expect. The curves of Fig. 6.11 may be used to determine the amount of sound insulation required between rooms in offices and dwellings. Alternatively, they could be used to determine the amount of background noise to be added in order to make speech unintelligible.

Background noise may be a nuisance as well. Depending on location and use some '*acceptable noise levels*' have been defined and are shown in Table 6.4. The

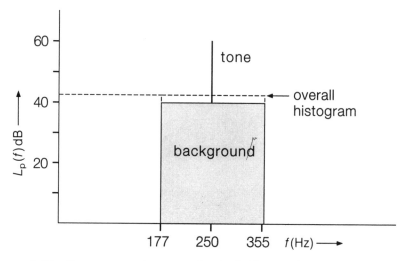

Figure 6.10 Constant sound pressure density $L_p(f)$ in the 250 Hz octave band with a single tone. The equivalent overall histogram has been dotted

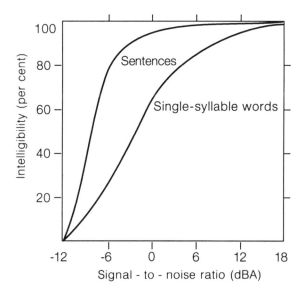

Figure 6.11 Percentage of words and sentences correctly identified in the presence of background noise. (Reproduced with permission from Lawrence E. Kinsler, Austin R. Frey, Alan B. Coppers and James V. Sanders, *Fundamentals of Acoustics*, 3rd edn, Wiley, New York, 1982, Fig. 12.3, p. 284)

Table 6.4 Acceptable noise levels in unoccupied rooms. (Reproduced with permission from L. E. Kinsler, A. R. Frey, A. B. Coppers and J. V. Sanders, *Fundamentals by Acoustics*, 3rd edn, Wiley, New York, 1982, Table 12.3, p. 287)

Location	Noise criteria (NC)
Concert hall, recording studio	5–20
Music room, legitimate theatre, classroom	20–25
Church, courtroom, conference room, hospital, bedroom	25–30
Library, business office, living room	30–35
Restaurant, movie theatre	35–40
Retail shop, bank	40–45
Gymnasium, clerical office	45–50
Shops and garages	50–55

noise criteria are determined in a way described by Fig. 6.12. This figure shows a dotted line representing the measured sound pressure levels for the octave bands, which are not *A*-weighted. Drawn curves help to define the *preferred noise criteria* (PNC). The intersection, which gives the highest (PNC) value is defined as the PNC of the room.

Table 6.4 refers to public buildings. For dwellings governments have defined norms for acceptable sound pressure levels. They are used in planning highways and deciding on their sound insulation. They are also used to determine whether a tool shop or a discotheque within a community may be accepted, and if so, what insulation measures the proprietor should take.

As the experienced noise nuisance will depend on the time of the day and on the duration of the nuisance a single number does not suffice and one needs a set of criteria. They usually take *A*-weighted noise levels as the starting point. We mention the time averaged pressure level L_A.

The time average may be taken over the day (7.00–19.00 h), the evening (19.00–23.00 h) and the night (23.00–7.00 h) separately. In order to obtain a single number one may define an overall day–evening–night sound level L_{den} as the highest of the following three numbers: the day sound level in dBA, the evening sound level plus 5 dBA and the night level plus 10 dBA.

In a country like the Netherlands the norms are put as $L_{den} = 50$ dBA outside the house and $L_{den} = 35$ dBA inside with closed windows. Exceptions are made for zones in the neighbourhood of roads and industries, where the boundaries of the zones are determined by political decision making in local councils. An

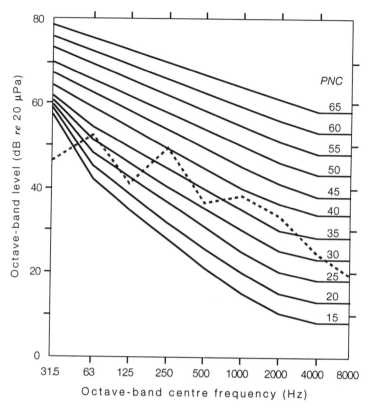

Figure 6.12 Octave band levels used to determine preferred noise criteria (PNC) for a room by intersection of a measured curve with the highest defined one. In this case PNC = 40. (Reproduced with permission from Lawrence E. Kinsler, Austin R. Frey, Alan B. Coppers and James V. Sanders, *Fundamentals of Acoustics*, 3rd edn, Wiley, New York, 1982, Fig. 12.5, p. 286)

example of the regulations is given in Table 6.5.* It should be mentioned that the regulations may vary from country to country. Also one could or should take into account short lasting events by defining an x-percentile-exceeded sound level L_{Ax}, where L_{A10} would be exceeded 10% of the time.

In all cases, the walls and windows should reduce the outside noise levels to much lower inside levels, implying a reduction by 15 dBA to 20 dBA. Of course, walls between apartments should give a similar reduction. This is the subject of the next section.

---

*The example was taken from Dutch Law (Wet Geluidshinder, 1979). Compared with some other countries the limits may seem rather tight. However, temporarily, limits of 5 dBA higher are allowed.

Table 6.5 Example of noise level limits L_{den} (in dBA) outside dwellings

Norm	50
Acceptable in rural areas	55
Acceptable in cities	60

6.3 REDUCING THE TRANSMISSION OF SOUND

The sound level L_0 in a room will be determined by a number of factors, depicted in Fig. 6.13. Suppose that there are no internal sources of sound; then there is *direct transmission* from the neighbours (flow 1 in Fig. 6.13) and from the outside (flow 6). There is *flanking transmission* by the walls, floors and ceilings (flows 2, 3 and 4) and *contact noise* from children playing (flow 5).

Sound Insulation

The direct transmission between two rooms is determined by the insulation properties of the separation wall. Suppose that an intensity I_i of sound is hitting the wall. Then a part $I_t = tI_i$ will be transmitted. The *air insulation* of the wall is defined as

$$R = 10 \log \frac{I_i}{I_t} \tag{6.97}$$

This quantity is called air insulation as opposed to contact insulation. If the sound field is diffuse, both in the emitting room and in the receiving room, one may write, using eq. (6.80),

$$I_i = \frac{p^2_{rms,e}}{4\rho_0 c_0} \tag{6.98}$$

where the extra subscript e refers to the emitting room and an index 0 will refer to the receiving room. The acoustic power P entering the receiving room by the wall with surface area S may be written as

$$P = I_t S \tag{6.99}$$

In a stationary situation this power is absorbed in the receiving room.

In any room, sound may be absorbed by the walls, the carpeting, etc.; it may be reflected and it may be transmitted. The last fraction is small, so within a room absorption and reflection dominate. For an open window the absorption coefficient is taken as one. For other surfaces the absorption may be found by

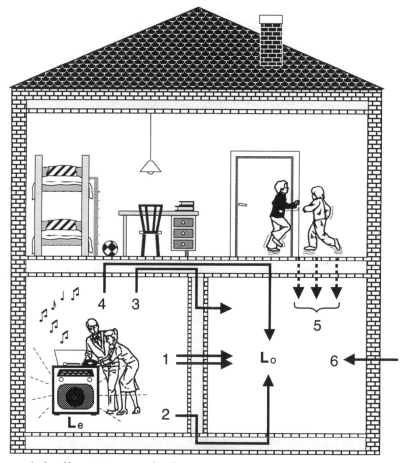

1,6=direct transmission
5=contact noise
2,3,4=flanking transmission

Figure 6.13   Categories of sound transmission

multiplying the surface area and absorption coefficient. By adding all contributions one finds an area A, which is called the *effective open window area*. The power P entering the room must be absorbed. For a stationary situation this implies

$$P = I_0 A = \frac{p_{\text{rms},0}^2}{4\rho_0 c_0} A \qquad (6.100)$$

where I_0 is the sound intensity in the receiving room. From eqs. (6.100) and (6.99) we find I_t to be substituted into eq. (6.97) together with eq. (6.98). This gives

$$R = 10 \log \left(\frac{p_{\text{rms},e}^2}{4\rho_0 c_0} \right) - 10 \log \left(\frac{p_{\text{rms},0}^2}{4\rho_0 c_0} \frac{A}{S} \right) \qquad (6.101)$$

With eq. (6.5) we may write

$$R = L_e - L_0 + 10 \log \frac{S}{A} \qquad \text{dB} \qquad (6.102)$$

where L_e is the sound pressure level in the emitting room and L_0 is the same in the receiving room. We note that S is the surface area of the separation wall and A the total absorption in the receiving room. When one measures the sound insulation R of a wall in the laboratory and then puts the wall in a building the resulting isolation R will be smaller because of the flanking transmissions.

The sound insulation R appears to be a function of the frequency f of the sound. One therefore has to add an index f as in R_f to denote the frequency. The value R_{500} gives a practical measure for the average insulation (cf. Exercise 6.13)

Sound Leaks

It appears that small holes in a separation wall give rise to a considerable reduction in sound insulation. The same is true for narrow openings underneath doors. To show this we consider one leak with surface area S_1 and sound insulation R_1 and transmission coefficient t_1. The rest of the surface will be $S - S_1$ with sound insulation R and transmission coefficient t. The total area S will have a mean transmission coefficient t_r and insulation R_{eff}.

The transmission coefficients represent the transmission per unit area, so eq. (6.97) gives the simple relation

$$R = 10 \log \frac{I_i}{I_t} = 10 \log \frac{I_i}{t I_i} = 10 \log \frac{1}{t} \qquad (6.103)$$

or

$$t = 10^{-R/10} \qquad (6.104)$$

for the main surface. Similarly $t_1 = 10^{-R_1/10}$ for the leak and $t_r = 10^{-R_{\text{eff}}}$ for the mean transmission. The transmission coefficients obey the relation

$$t_r S = t_1 S_1 + t(S - S_1) \qquad (6.105)$$

We ignore the flanking transmission. The power P entering the receiving room then is determined by $t_r S$ and the incoming intensity I_i from the emitting room:

$$P = I_i t_r S = I_i t_1 S_1 + I_i t(S - S_1) \qquad (6.106)$$

The effective air insulation R_{eff} follows as

$$R_{eff} = 10 \log \frac{1}{t_r} = -10 \log \left\{ t \left[1 + \frac{S_1}{S} \left(\frac{t_1}{t} - 1 \right) \right] \right\} \qquad (6.107)$$

Using eq. (6.104) and its analogues to go back to insulations R and R_1 one finds easily that

$$R_{eff} = R - 10 \log \left(1 - \frac{S_1}{S} + \frac{S_1}{S} 10^{R - R_1/10} \right) \qquad (6.108)$$

As an example we take a wall with $R = 50\,dB$ and an air leak of $S_1/S = 10^{-3}$ with $R_1 = 0$. One finds $R_{eff} = 30\,dB$, a reduction of sound resistance with 20 dB. Thus, in problems with sound insulation one should close all holes, including those underneath doors.

Mass Laws for Sound Insulation

Let us consider a plane layer such as a flat wall with thickness d and mass m per m^2. Its boundaries are ignored, so it is considered as infinitely extended. We assume that a plane harmonic wave of frequency f is incident perpendicular on the surface. We may find the sound insulation R of the layer by using Fig. 6.14.

The plane waves obey eq. (6.7), giving $p_i = \rho_0 c_0 u_i$ for the incident wave, $p_r = \rho_0 c_0 u_r$ for the reflected wave and $p_t = \rho_0 c_0 u_t$ for the transmitted wave. The construction will have a (very small) displacement s which is assumed to be the same at both sides of the rigid layer (implying that the sound velocity in the layer is much larger than in air). As with the piston in Section 6.1 the velocity ds/dt equals the velocity of the air particles. Therefore on the right-hand side of the layer

$$u_t = \frac{p_t}{\rho_0 c_0} = \frac{ds}{dt} \qquad (6.109)$$

On the left-hand side the absolute velocity change $u_i - u_r$ would vanish for a rigid wall with $ds/dt = 0$. In our case we have instead

$$u_i - u_r = \frac{p_i - p_r}{\rho_0 c_0} = \frac{ds}{dt} \qquad (6.110)$$

where u_r corresponds to a leftward moving wave. In order to find the pressure p_1 at the wall one has to add the pressures of the waves

$$p_1 = p_i + p_r \qquad (6.111)$$

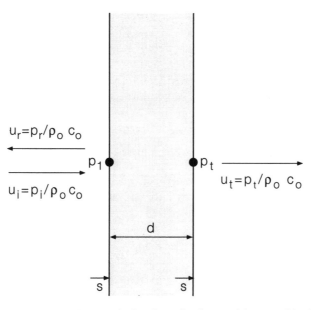

Figure 6.14 Sound transmission for a flat layer with normal incidence

For a unit area of the layer the pressure difference $p_1 - p_t$ acts as a force, so

$$p_1 - p_t = m\frac{d^2s}{dt^2} = m\frac{du_t}{dt} \tag{6.112}$$

where we used eq. (6.109). We eliminate p_1 and p_r from eqs. (6.109), (6.110), (6.111) and (6.112) and find

$$p_i = p_t + \frac{m}{2\rho_0 c_0}\frac{dp_t}{dt} \tag{6.113}$$

We remember that p_i and p_t refer to pressures at fixed positions, the boundaries of the layer. The easiest way to solve eq. (6.113) is to use the complex notation

$$p_i = A\,e^{j\omega t} \tag{6.114}$$

$$p_t = B\,\omega^{j\omega t} \tag{6.115}$$

and find

$$A = B\left(1 + \frac{m\omega}{2\rho_0 c_0}j\right) \tag{6.116}$$

Consequently the amplitudes of p_i and p_t relate as

$$\frac{|p_i|^2}{|p_t|^2} = 1 + \left(\frac{m\omega}{2\rho_0 c_0}\right)^2 \tag{6.117}$$

and there is also a phase difference of arctan $[m\omega/(2\rho_0 c_0)$). The intensities of the incident and transmitted waves are proportional to p_{rms}^2 (cf. eq. (6.58)) so the definition (6.97) for air insulation gives

$$R = 10\log\frac{I_i}{I_t} = 10\log\left[1 + \left(\frac{\pi m f}{\rho_0 c_0}\right)^2\right] \tag{6.118}$$

where $\omega = 2\pi f$ was used. We remember that for air $\rho_0 c_0 \approx 413\,\text{kg}\,\text{m}^{-2}\,\text{s}^{-1}$. Therefore the second term underneath the logarithm dominates when $mf > 500$. Then we may simplify the mass law as

$$R = 20\log m + 20\log\frac{f}{500} + 12\,\text{dB} \tag{6.119}$$

A doubling of mass m per unit area or of frequency f should give an increase in insulation R with 6 dB. In practice this number is 5 dB. Also the coefficients in eq. (6.119) are somewhat lower:

$$R_{\text{practical}} \approx 17.5\log m + 17.5\log\frac{f}{500} + 3\,\text{dB} \tag{6.120}$$

Hollow Wall

In building a hollow wall is often used for thermal insulation. It consists of two walls or two glass panels with air in between of thickness D and pressure p_0. Figure 6.15 shows the situation. A plane wave is incident from the left with $p_i = \rho_0 c_0 u_i$ and reflected with $p_r = \rho_0 c_0 u_r$. This produces a pressure p_1 at the left wall according to

$$p_1 = p_i + p_r \tag{6.121}$$

The left wall has a displacement s_1 and the right wall a displacement s_2. The pressure in the cavity is p_2. Again the velocity change of the air particles equals ds_1/dt, giving

$$u_i - u_r = \frac{p_i - p_r}{\rho_0 c_0} = \frac{ds_1}{dt} \tag{6.122}$$

Similarly, at the right wall

$$u_t = \frac{p_t}{\rho_0 c_0} = \frac{ds_2}{dt} \tag{6.123}$$

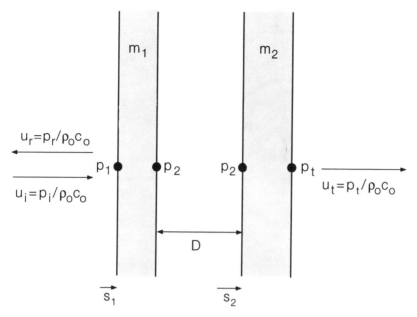

Figure 6.15 Sound transmission by a hollow wall. (Reproduced by permission of Delft University of Technology, Faculty of Civil Engineering from lecture notes gc45 vakgroep bouwfysica', Professor A. J. Verhoeven, Technological University, Delft, Netherlands, p. 3.68)

The pressure p_2 in the cavity is found by assuming that the air inside behaves like a string with strength parameter C^{-1} defined by Hooke's law:

$$\Delta p = -C^{-1}\Delta D \tag{6.124}$$

Here $\Delta p = p_2$ is the deviation from the atmospheric pressure p_0 and ΔD is the deviation from the equilibrium distance D_0. We write C^{-1} in order to simplify future equations. As the air in the cavity will change adiabatically we apply Poisson's law to $1 \, \text{m}^2$ of area:

$$(p_0 + p_2)D^\kappa = p_0 D_0^\kappa \tag{6.125}$$

We find C from eq. (6.124) as $-dD/dp$ taken at equilibrium with $p_2 = 0$. Therefore, from eq. (6.125) we have

$$D = \left(\frac{p_0}{p_0 + p_2}\right)^{1/\kappa} D_0 \tag{6.126}$$

and finally

$$C = -\frac{dD}{dp_2}\bigg|_0 = \frac{1}{\kappa}\frac{D_0}{p_0} \qquad (6.127)$$

Let us now apply eq. (6.124) to Fig. 6.15. We find

$$p_2 = -C^{-1}(s_2 - s_1) \qquad (6.128)$$

$$s_1 = s_2 + p_2 C \qquad (6.129)$$

Both walls obey Newton's laws, resulting in

$$p_1 - p_2 = m_1 \frac{d^2 s_1}{dt^2} \qquad (6.130)$$

$$p_2 - p_t = m_2 \frac{d^2 s_2}{dt^2} \qquad (6.131)$$

From the six equations (6.121), (6.122), (6.123), (6.129), (6.130) and (6.131) we deduce a relationship between p_i and p_t by eliminating all other variables one by one. This gives

$$p_i = p_t + \frac{m_1 + m_2 + C(\rho_0 c_0)^2}{2\rho_0 c_0}\frac{dp_t}{dt} + \frac{(m_1 + m_2)C}{2}\frac{d^2 p_t}{dt^2} + \frac{Cm_1 m_2}{2\rho_0 c_0}\frac{d^3 p_t}{dt^3} \qquad (6.132)$$

We again substitute the simple complex exponentials (6.114) and (6.115) and find

$$A = B\left(\left[1 - \frac{(m_1 + m_2)C\omega^2}{2}\right] + j\left\{\frac{[m_1 + m_2 + C(\rho_0 c_0)^2]\omega}{2\rho_0 c_0} - \frac{Cm_1 m_2 \omega^3}{2\rho_0 c_0}\right\}\right) \qquad (6.133)$$

We ignore the phase difference as we are only interested in the air sound insulation R determined from eq. (6.97) as

$$R = 10\log\frac{|A|^2}{|B|^2} \qquad (6.134)$$

$$R = 10\log\left\{\left[1 - \frac{(m_1 + m_2)C\omega^2}{2}\right]^2 \right.$$
$$\left. + \left(\frac{\omega}{2\rho_0 c_0}\right)^2 [m_1 + m_2 + C(\rho_0 c_0)^2 - Cm_1 m_2 \omega^2]^2\right\} \qquad (6.135)$$

In analysing insulation R as a function of frequency $f = \omega/(2\pi)$ we should realize that the wall in Fig. 6.15 may be interpreted as a mass–string–mass system. Therefore, even without external driving forces like sound it has normal vibrations with a frequency (cf. Exercise 6.15)

$$f_r = \frac{1}{2\pi} \sqrt{\frac{m_1 + m_2}{m_1 m_2 C}} \approx 60 \sqrt{\frac{m_1 + m_2}{m_1 m_2 D_0}} \qquad (6.136)$$

where in the last equality we used eq. (6.127) with $\kappa = 1.4$ and atmospheric pressure p_0.

It is clear that sound will easily pass at resonance, so the air insulation R should exhibit a minimum at the resonance frequency $f = f_r$. This is indeed the case, as is illustrated for a double-glazing construction in Fig. 6.16. One notes the minimum at $f_r = 82$ Hz according to the data given.

For low frequencies $f \ll f_r$ the curve $R(f)$ approximates a straight line with a slope of 6 dB/octave. This reminds us of the mass law (6.119) and corresponds to that law for $m = m_1 + m_2$. For high frequencies $f \gg f_r$ the term with ω^6 in eq.

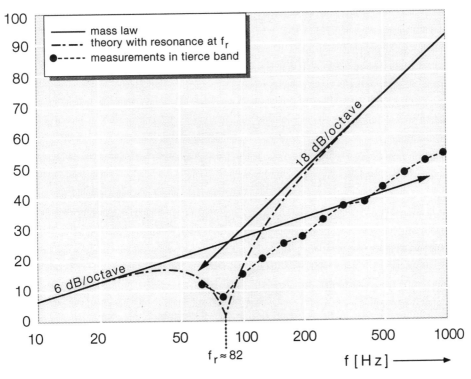

Figure 6.16 Double glazing with $D_0 = 0.08$ m, $m_1 = 20$ kg m^{-2}, $m_2 = 10$ kg m^{-2} and $f_r = 82$ Hz. (Reproduced by permission of Delft University of Technology, Faculty of Civil Engineering from lecture notes gc45 'vakgroep bouwfysica', Professor A. J. Verhoeven, Technological University, Delft, Netherlands, Fig. 3.21, p. 3.43)

(6.135) will dominate and the curve should approach a straight line with a slope of 18 dB/octave, as is the case.

In Fig. 6.16 experimental points are also shown. The insulation at higher frequencies is much lower than predicted. It seems that the weakest points of the construction, the edges, will then dominate the transmission. They in fact behave like a single layer with mass $m = m_1 + m_2$ and roughly continue the tendency at lower frequencies.

The theoretical insulation $R(f)$ at resonance $f = f_r$ may be found by substituting eq. (6.136) into eq. (6.135). This gives

$$R(f_r) = 10 \log \left\{ \left[1 - \frac{(m_1 + m_2)^2}{2m_1 m_2} \right]^2 + \frac{m_1 + m_2}{4m_1 m_2} C(\rho_0 c_0)^2 \right\} \qquad (6.137)$$

Therefore, for $m_1 = m_2$ the insulation at resonance will be very small. It may be remarked that much street noise has low frequencies so double glazing will not help much to prevent it entering the home.

Coincidence

In the preceding part we discussed perpendicular incident waves. Let us now consider the oblique incidence of plane waves. The situation is sketched in Fig. 6.17 for a flat plate where the planes of equal sound phase make an angle of θ with the undisturbed plate. From Fig. 6.17 one observes that the high and low sound pressures will produce a fall and rise in the plate. The wavelength λ_{plate} of the induced waves in the plate and the wavelength λ_{air} of the sound waves relate as

$$\lambda_{\text{plate}} = \frac{\lambda_{\text{air}}}{\sin \theta} \qquad (6.138)$$

As the frequency f will be the same in air and plate the propagation velocities relate similarly:

$$c_{\text{plate}} = \frac{c_0}{\sin \theta} \qquad (6.139)$$

Flat plates may perform transverse vibrations without external force as well. The relation between frequency f and speed c'_{plate} for normal vibrations turns out to be

$$c'_{\text{plate}} = \left[\frac{Yt^2(2\pi f)^2}{12\rho} \right]^{1/4} \qquad (6.140)$$

Here Y is the Young modulus of the plate (N m^{-2}), describing its elastic properties, t is its thickness (m) and ρ is the mass density (kg m^{-3}). One may check that eq. (6.140) gives the right dimension (m s^{-1}) for the velocity c'_{plate}.

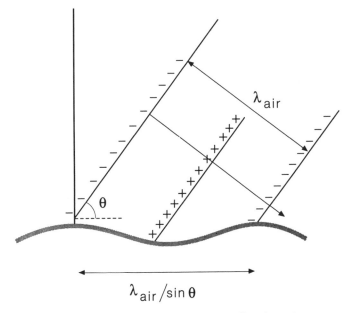

Figure 6.17 Oblique incidence of plane waves on a flat plate. A transverse wave is induced with velocity $c_0/\sin\theta$. The lower points of the plate correspond with maximum air pressure

The proof of eq. (6.140) is too elaborate to reproduce here.* We just remark that the quarter-power on the right-hand side originates from the bending of the plate. Its wave equation therefore is more complicated than, for example, eq. (6.23) and contains $\nabla^2\nabla^2$ terms. The consequence is that the velocity c'_{plate} of the transverse waves depends on the frequency f.

It will be clear that when the incident waves of frequency f induce a propagation velocity (6.139) in the plate that equals the normal propagation speed (6.140) the plate as a body will resonate. Its back emits the waves that are forced at its front. The sound field therefore propagates as if the plate were not there. This phenomenon is called *coincidence*. From eqs. (6.139) and (6.140) we find that it occurs for a frequency f_c where

$$f_c = \frac{1}{2\pi t}\left(\frac{c_0}{\sin\theta}\right)^2\sqrt{\frac{12\rho}{Y}} \qquad (6.141)$$

For normal incidence $\sin\theta = 0$, so there is no coincidence; of course there are no

---

*One may find the proof in books on elasticity which discuss the transverse vibrations of a bar. In fact, Fig. 6.17 extends infinitely perpendicular to the drawing and therefore its differential equation equals that of a bar. (See, for example, Ref. 3, pp. 68–71.)

Table 6.6 Elasticity properties for some building materials

	$Y\,(10^9\,\mathrm{N\,m^{-2}})$	$\rho\,(\mathrm{kg\,m^{-3}})$	$\sqrt{Y/\rho}\,(\mathrm{m\,s^{-1}})$
Brick	25	2100	3450
Gypsum plates	4	800	2236
Aluminium	70	2800	5000
Window glass	70	2500	5292
Pine wood	11	520	4599
Soft board	3	650	2148

induced transversal waves either. For $\theta > 0$ the frequency becomes lower and for $\theta = \pi/2$, the so-called *grazing incidence*, one has the lowest coincidence frequency, the critical frequency f_{cr}:

$$f_{cr} = \frac{c_0^2}{2\pi t}\sqrt{\frac{12\rho}{Y}} \approx \frac{65 \times 10^3}{t\sqrt{Y/\rho}} \qquad (6.142)$$

where at the extreme right-hand side data for standard air were used.

In a diffuse sound field waves will hit a wall from all directions. If, however, its frequency $f < f_{cr}$ no coincidence will occur and therefore no loss of sound insulation. Figure 6.5 therefore suggests that a wall should be constructed such that f_{cr} becomes higher than 10 kHz.

The values of $\sqrt{Y/\rho}$ which occur in the denominator of eq. (6.142) are shown in Table 6.6 for some common building materials. It will be found that the critical frequency for a brick wall with a thickness of 10 cm becomes $f_{cr} \sim 190$ Hz and for 4 mm window glass it becomes $f_{cr} \sim 3000$ Hz. In both cases coincidence will cause loss of sound insulation. One sometimes tries to prevent this by adding a thin (small t) and soft (small $\sqrt{Y/\rho}$) wall to the major wall.

Avoiding Contact Noise

Transmission of the contact noise shown in Fig. 6.13 can best be diminished by using sprung floors. More generally, one could put noisy vibrating machines on springs. The simplest example is shown in Fig. 6.18. A mass m_1 is put on a spring with constant C^{-1} on a floor with a total mass m_2. As $m_2 \gg m_1$ one may either use eq. (6.136) in the limit $m_2 \to \infty$ or ignore the second mass altogether. In both cases one finds the resonance frequency

$$f_r = \frac{1}{2\pi}\sqrt{\frac{1}{m_1 C}} \qquad (6.143)$$

Suppose that the mass of the spring is compressed over a distance s by the weight $m_1 g$ of the mass. Then Hooke's law gives for the equilibrium compression

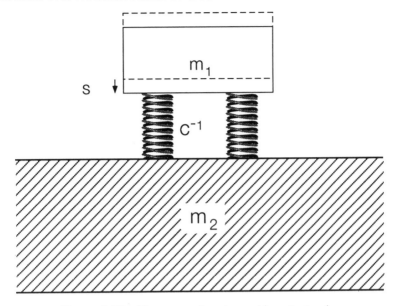

Figure 6.18   Mass on springs to avoid contact noise

$m_1 g = C^{-1}s$ or $m_1 g C = s$. Multiplying numerator and denominator in eq. (6.143) by g gives

$$f_r = \frac{1}{2\pi}\sqrt{\frac{g}{m_1 C g}} = \frac{1}{2\pi}\sqrt{\frac{g}{s}} \qquad (6.144)$$

Thus the resonance frequency f_r is simply found by the equilibrium compression s of the spring. When the mass m_1 contains machines producing vibrations with frequency f, elementary mechanics shows that for $f \gg f_r$ the amplitudes of the forced vibrations of the spring go to zero.

Refraction

In the open air it is only a rough approximation to represent sound waves by straight lines for it is well known that sound barriers between a highway and a garden do not reduce the sound level completely. In fact, according to Huygens' principle, the top of the barrier will again radiate sound in all directions.

   We do not discuss barriers in this text, but rather pay some attention to the curving of sound transmission by (vertical) temperature and air velocity gradients. As a preliminary we regard the refraction of sound waves on the interface of two media, sketched on the left-hand side of Fig. 6.19.

   A plane harmonic wave in an arbitrary direction **n** is given by its wave vector

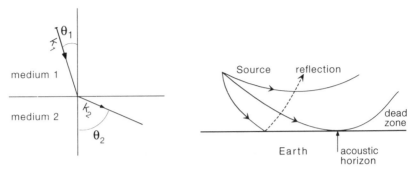

Figure 6.19 Refraction of sound waves at the interface of two media (left) and in the atmosphere (right)

$\mathbf{k} = k\mathbf{n}$ in that direction. Generalization of eq. (6.28) gives, for medium 1,

$$p_1(\mathbf{r}, t) = A\, e^{j(\omega t - \mathbf{k}_1 \cdot \mathbf{r})} \tag{6.145}$$

We will consider a single frequency ω and omit the time dependence. We also use the geometry of Fig. 6.19 which gives

$$p_1(\mathbf{r}, t) = A\, e^{-jk_1 x \sin\theta_1 + jk_1 z \cos\theta_1} \tag{6.146}$$

The pressure in medium 2 is given similarly by

$$p_2(\mathbf{r}, t) = B\, e^{-jk_2 x \sin\theta_2 + jk_2 z \cos\theta_2} \tag{6.147}$$

At the interface $z = 0$ the x dependence of both pressures should be the same, although the amplitude B in the second medium may be smaller. Therefore we find *Snell's law*

$$k_1 \sin\theta_1 = k_2 \sin\theta_2 \tag{6.148}$$

or

$$\frac{\sin\theta_2}{\sin\theta_1} = \frac{k_1}{k_2} = \frac{c_2}{c_1} \tag{6.149}$$

where we used eq. (6.26) and the fact that the frequency f is the same in both media. If $c_2 > c_1$ as is the case for the air–water transition it follows that $\theta_2 > \theta_1$. The boundary angle will be found for $\theta_2 = 90°$ which gives for air–water with $c_2 \approx 4c_1$ that $\theta_1 \leqslant 14.5°$.

Acoustic Horizon

We know from eq. (3.28) for the dry adiabat that the temperature of air will decrease with height. According to eq. (6.12) and eq. (3.22),

$$c_0^2 = \frac{p}{\rho} = RT \tag{6.150}$$

with R the corrected gas constant for dry air. The velocity of sound also decreases with altitude. The resulting sound 'rays' are shown on the right-hand side of Fig. 6.19. Instead of two straight rays $\mathbf{k}_1$ and $\mathbf{k}_2$ we have a continuous tendency of slowly increasing angle θ. From a source high in the atmosphere there will be one ray that just touches the ground: its *acoustic horizon*. Points on the earth to the right of the acoustic horizon cannot be reached by sound: a *dead zone*.

This phenomenon may be noticed when an aeroplane is passing overhead. Its sound first gradually decreases and then suddenly disappears. It must be mentioned, however, that the wind velocity gradient is the major contributor to this effect.

6.4 ACTIVE CONTROL OF SOUND

When measures to reduce sound at its origin are exhausted one may try to add sound of opposite phase to the noise, as was written down in eq. (6.37) and illustrated in Fig. 6.3. We will discuss this active control of sound in the simplest case of a monopole sound source in an infinite duct.

The principle is sketched in Fig. 6.20. One notices the primary source at $x = 0$ and the secondary source at $x = L$. An electronics device V should control the secondary source such that the required phase relations are maintained. We note the definitions of downstream ($x \to \infty$) and upstream ($x \to -\infty$). The figure gives the essentials of the US patent of 1936 granted to the inventor Paul Lueg. It was

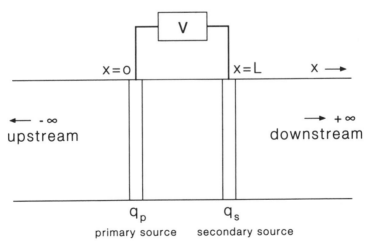

Figure 6.20 Two sound sources in an infinite duct with an electronic control V. (Reproduced, with permission, from *Active Control of Sound*, P. A. Nelson and S. J. Elliott, Academic Press, 1992, Fig. 5.5, p. 123)

only in the 1980s that the electronics became sophisticated enough to apply the idea to various practical designs. Inside cars, for example, one may achieve a noise reduction of 5–7 dB over the frequency range of 0–250 Hz by analysing six reference signals from detectors on the underside of a car body (see Ref. 4, p. 406)

In the present text we do not discuss electronics. We will also restrict ourselves to a single primary monopole source and a single secondary source with sound propagating in the x direction only. In a duct the approximation is justified if the wavelength is at least twice as large as the cross size of the duct. For frequencies up to 170 Hz the wavelengths are larger than 2 m, so for low frequencies and not too wide ducts the discussion below will be valid.

In an ideal case the secondary source should cancel the pressure field downstream and moreover should not disturb the primary field upstream. The latter requirement makes it easier to control the primary field. We shall show that the combination of a dipole and a monopole as a secondary source may do the trick. We will start by defining and discussing monopole sources and at the end turn towards a dipole.

Monopole Source

A (hypothetical) monopole source at $x = x_0$ may be defined as a combination of two massless pistons between which air is compressed and decompressed with a frequency f. As the pistons will have opposite velocities all the time, the air velocities $U(x_0^-)$ just to the left of the pistons and $U(x_0^+)$ just to the right will be precisely opposite:

$$U(x_0^+) = - U(x_0^-) \qquad (6.151)$$

The two pistons are supposed to be very close to each other. Moreover, they may be penetrated by an external sound field without losses.

The situation is sketched in Fig. 6.21. For $x > x_0$ the right piston produces an upstream harmonic sound wave with frequency f, angular frequency ω, wave number k and propagation velocity c_0. In eqs. (6.24) to (6.29) the properties of the wave were given. We further recall eq. (6.7) saying that $p = \rho_0 c_0 u$. We now use the symbols $U(x_0^+)$ and $U(x_0^-)$ for the amplitudes of the velocity but keep the minus sign as in eq. (6.151). It follows that

$$p(x) = \rho_0 c_0 U(x_0^+) e^{-jk(x - x_0)} \qquad (x > x_0) \qquad (6.152)$$

$$u(x) = U(x_0^+) e^{-jk(x - x_0)} \qquad (x > x_0) \qquad (6.153)$$

Here we used the complex notation of eqs. (6.28) and (6.38), implying complex velocities $U(x_0^+)$ and $U(x_0^-)$. We put the origin of the wave at $x = x_0$ in order to simplify the discussion of interference later on. Finally we recall that the time dependence $e^{j\omega t}$, which has been omitted, implies that eqs. (6.152) and (6.153) represent waves going to the right.

The left-hand side piston in Fig. 6.152 produces an upstream going sound

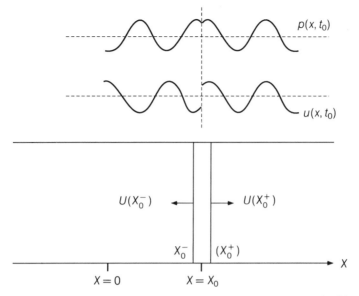

Figure 6.21 Monopole source at $x = x_0$ with the air particle velocities $u(x, t)$ indicated at a certain time $t = t_0$. Note that in reality the wavelength should be larger than the size of the duct. (Reproduced, with permission, from *Active Control of Sound*, P. A. Nelson and S. J. Elliott, Academic Press, 1992, Fig. 5.4, p. 121)

wave. Its particle velocities $U(x_0^-)$ just to the left of $x = x_0$ are defined with respect to a positive axis pointing to the right. The relation $p = \rho_0 c_0 u$ was derived for the case that the positive direction was the propagation of the wave. Therefore we find the equations analogous to (6.152) and (6.153) by inserting an extra minus sign:

$$p(x) = -\rho_0 c_0 U(x_0^-)e^{jk(x-x_0)} \qquad (x < x_0) \qquad (6.154)$$

$$u(x) = U(x_0^-)e^{jk(x-x_0)} \qquad (x < x_0) \qquad (6.155)$$

where the exponential now guarantees a leftwards going wave. Equations (6.151), (6.153) and (6.155) imply a discontinuity of particle velocity at $x = x_0$. Because of the minus sign in eq. (6.154) the pressure joins continuously. This is illustrated at the top of Fig. 6.21.

Let us define a *source strength* $q(x_0)$ as the volume velocity introduced into the duct by the monopole. Then

$$q(x_0) = SU(x_0^+) - SU(x_0^-) = 2SU(x_0^+) = -2SU(x_0^-) \qquad (6.156)$$

The pressures (6.152) and (6.154) may then be comprised

$$p(x) = \frac{\rho_0 c_0}{2S} q(x_0)e^{-jk|x-x_0|} \qquad (6.157)$$

Note that in the exponential the absolute distance $|x - x_0|$ from the source shows up.

Two Monopole Sources at Distance L

Let us return to the problem of Fig. 6.20 and take both the primary source at $x = 0$ and the secondary at $x = L$ to be monopole sources. They will both follow eq. (6.157) with strengths q_p for the primary source and q_s for the secondary. This gives for the pressures from the sources

$$p_p(x) = \frac{\rho_0 c_0}{2S} q_p e^{-jk|x|} \tag{6.158}$$

$$p_s(x) = \frac{\rho_0 c_0}{2S} q_s e^{-jk|x-L|} \tag{6.159}$$

The total pressure $p(x)$ will become the sum

$$p(x) = p_p(x) + p_s(x) \tag{6.160}$$

We have to choose the parameters such that downstream from the secondary source $(x \geqslant L)$ the total sound pressure vanishes:

$$q_p e^{-jkx} + q_s e^{-jk(x-L)} = 0 \qquad (x \geqslant L) \tag{6.161}$$

where we omitted the common factor $\rho_0 c_0/(2S)$. Equation (6.161) gives as a requirement for the secondary source strength

$$q_s = -q_p e^{-jkL} \tag{6.162}$$

The secondary source therefore should be as strong as the primary one as $|q_s| = |q_p|$. The exponential e^{-jkL} implies a phase difference kL corresponding with the time L/c_0 that it takes the primary wave to travel the distance L.*

In the region between both sources one obtains a standing wave (cf. Exercise 6.16) and upstream the primary source one finds from eqs. (6.158), (6.159), (6.160) and (6.162) that

$$p(x) = \frac{\rho_0 c_0}{2S} q_p (1 - e^{-j2kL}) e^{jkx} \qquad (x \leqslant 0) \tag{6.163}$$

This implies that for $2kL = n(2\pi)$ with integer n the pressure $p(x)$ vanishes upstream. Therefore, for primary sound of a single frequency one can adjust the distance between the sources such that the pressure vanishes both for $x < 0$ and

---

*This is found by adding the time dependence $e^{j\omega t}$ resulting in $t = kL/\omega = L/c_0$. More generally one may perform Fourier transformations to go back and forth between the frequency domain and the time domain.

$x > L$. In practice one will have a frequency range and not a single frequency, so adjusting the distance L will not help much. Before discussing more complicated set-ups we investigate what happens with the acoustic energy from both sources.

Energy Intensity

The acoustic intensity $\mathbf{I}$ was defined in eqs. (6.53) and (6.54) as a vector representing the power flow (W m^{-2}). The power flows in one dimension will count positive when they go to the right. When we compare the power flows just to the left and the right of a monopole source the difference should be the power added by the source

$$W = S[I(x_0^+) - I(x_0^-)] \qquad (6.164)$$

as the background field will go continuously past $x = x_0$ and the power from the source going to the left will count negative. We will work with the complex notation and a single frequency ω. We may therefore use eq. (6.56) to express the acoustic intensity. We write down the components of eq. (6.59) in the positive x direction at points x_0^+ and x_0^-:

$$I(x_0^+) = \tfrac{1}{2}\text{Re}[p^*(x_0^+)u(x_0^+)] \qquad (6.165)$$

$$I(x_0^-) = \tfrac{1}{2}\text{Re}[p^*(x_0^-)u(x_0^-)] \qquad (6.166)$$

We consider a source at $x = x_0$ in an external acoustic field. The external field will be continuous at $x = x_0$ both for $p(x)$ and for $u(x)$. Therefore any difference in air particle velocity between $x = x_0^-$ and $x = x_0^+$ should be ascribed to the source. Similar to eq. (6.156) we therefore write

$$q(x_0) = S[u(x_0^+) - u(x_0^-)] \qquad (6.167)$$

Inserting eqs. (6.165) and (6.166) into (6.164) and using the continuity of the pressure field shown in Fig. 6.21 gives with eq. (6.167) for the source power

$$W = \tfrac{1}{2}\text{Re}[p^*(x_0)q(x_0)] \qquad (6.168)$$

In the case of the primary monopole source at $x = 0$ and the secondary source at $x = L$ relation (6.162) was deduced by demanding that for $x \geq L$ the pressure vanishes. Therefore $p(L) = 0$ and eq. (6.168) shows that the secondary source does not radiate net power. During the period of time $T = 1/f$ it apparently absorbs as much power as it emits.

A Pair of Secondary Monopole Sources

Let us now consider two secondary monopole sources, one with strength q_{s1} located at $x = L$ and a second one with strength q_{s2} at position $x = L + d$, as depicted in Fig. 6.22. Physically the situation is identical to the two monopole

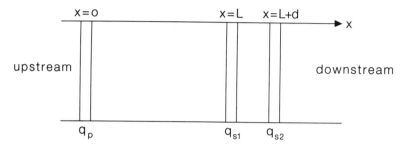

Figure 6.22 Two plane monopole sources q_{s1} and q_{s2} cancelling downstream primary sound whereas no upstream disturbance results. (Reproduced, with permission, from *Active Control of Sound*, P. A. Nelson and S. J. Elliott, Academic Press, 1982, Fig. 5.12, p. 135)

sources encountered in eqs. (6.158) and (6.159). We write

$$p_{s1}(x) = \frac{\rho_0 c_0}{2S} q_{s1} e^{-jk|x-L|} \tag{6.169}$$

$$p_{s2}(x) = \frac{\rho_0 c_0}{2S} q_{s2} e^{-jk|x-L-d|} \tag{6.170}$$

We can easily ensure zero radiation in the upstream direction by demanding the parallel of eq. (6.162) to hold:

$$q_{s2} = -q_{s1} e^{jkd} \tag{6.171}$$

The plus sign in the exponent on the right guarantees a vanishing sound field upstream. This relation implies that source q_{s2} will be precisely opposite to source q_{s1} at a time d/c_0 later. The downstream complex pressure $p_s(x)$ due to the two sources together will be found by adding eqs. (6.169) and (6.170) while using eq. (6.171):

$$p_s(x) = \frac{\rho_0 c_0}{2S} q_{s2}(-e^{-jk(x-L)-jkd} + e^{-jk(x-L-d)}) \qquad (x \geqslant L+d) \quad (6.172)$$

$$p_s(x) = \frac{\rho_0 c_0}{2S} q_{s2} e^{-jk(x-L)}[2j \sin(kd)] \qquad (x \geqslant L+d) \quad (6.173)$$

We now add a primary source with strength q_p at $x = 0$, and add eq. (6.157) to eq. (6.173) to find the total pressure downstream as

$$p(x) = \frac{\rho_0 c_0}{2S}\{q_p e^{-jkx} + q_{s2} e^{-jk(x-L)}[2j \sin(kd)]\} \qquad (x \geqslant L+d) \quad (6.174)$$

The pressure $p(x)$ downstream will vanish if the secondary strength q_{s2} obeys

$$q_{s2} = \frac{-q_p e^{-jkL}}{2j \sin(kd)} \tag{6.175}$$

together with eq. (6.171). For frequencies where $kd = n\pi$ with integer n we have $\sin(kd) = 0$ and q_{s2} becomes infinite. This is understandable as eq. (6.173) shows that for that case the secondary source produces no downstream radiation at all.

The Plane Dipole Source

Let us consider two monopoles with definitions (6.169) and (6.170) and put $L = 0$. This is sketched in Fig. 6.23. We have omitted the indices and write, contrary to eq. (6.171),

$$q_2 = -q_1 = q \tag{6.176}$$

With the definitions (6.176), eqs. (6.169) and (6.170) reduce to

$$p_1(x) = -\frac{\rho_0 c_0}{2S} q\, e^{-jk|x|} \tag{6.177}$$

$$p_2(x) = \frac{\rho_0 c_0}{2S} q\, e^{-jk|x-d|} \tag{6.178}$$

We now calculate the pressure $p(x)$ by assuming that $kd \ll 1$, which means that the distance d between the two monopoles is much smaller than the wavelength λ of the sound radiation:

$$p(x) = \frac{\rho_0 c_0}{2S} q(e^{-jk(x-d)} - e^{-jkx}) \qquad (x > d) \tag{6.179}$$

$$p(x) = \frac{\rho_0 c_0}{2S} q\, e^{-jk(x-d)}(1 - e^{-jkd}) \qquad (x > d) \tag{6.180}$$

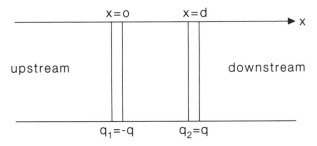

Figure 6.23 The dipole source. (Reproduced, with permission, from *Active Control of Sound*, P. A. Nelson and S. J. Elliott, Academic Press, 1992, Fig. 5.19, p. 145)

$$p(x) \approx \frac{j\omega\rho_0 qd}{2S} e^{-jk(x-d)} = \frac{f}{2S} e^{-jk(x-d)} \qquad (x > d) \qquad (6.181)$$

Here we used a series expansion for the exponential e^{-jkd} and the relation $\omega = c_0 k$. We also introduced a strength

$$f = j\omega\rho_0 qd \qquad (6.182)$$

Upstream we have

$$p(x) = \frac{\rho_0 c_0}{2S} q(e^{jk(x-d)} - e^{jkx}] \qquad (x < 0) \qquad (6.183)$$

$$p(x) \approx \frac{j\omega\rho_0 qd}{2S} e^{jkx} = -\frac{f}{2S} e^{jkx} \qquad (x < 0) \qquad (6.184)$$

A *dipole* is defined by the limit $d \to 0$ while keeping the product qd and therefore the strength f in eq. (6.182) constant. Let us locate the dipole at $x = x_0$ and adjust the equations. Equation (6.181) becomes

$$p(x) = \frac{f}{2S} e^{-jk(x-x_0)} \qquad (x > x_0) \qquad (6.185)$$

while eq. (6.184) becomes

$$p(x) = -\frac{f}{2S} e^{jk(x-x_0)} \qquad (x < x_0) \qquad (6.186)$$

The Monopole–Dipole Combination

We recall that a combination of two monopoles as secondary sources could give a cancelling pressure contribution upstream if eq. (6.171) holds. Therefore the distance and wavelength are coupled. We will now show that such a restriction no longer applies if one takes a dipole with strength $f(x_0)$ and a monopole with strength $q(x_0)$ at $x = x_0$. Downstream we use eqs. (6.157) and (6.185) and find

$$p(x) = \frac{\rho_0 c_0 q(x_0)}{2S} e^{-jk(x-x_0)} + \frac{f(x_0)}{2S} e^{-jk(x-x_0)} \qquad (x > x_0) \qquad (6.187)$$

Upstream we use eqs. (6.157) and (6.186) and find

$$p(x) = \frac{\rho_0 c_0 q(x_0)}{2S} e^{jk(x-x_0)} - \frac{f(x_0)}{2S} e^{jk(x-x_0)} \qquad (x > x_0) \qquad (6.188)$$

The upstream pressure caused by this set of secondary sources vanishes when

$$f(x_0) = \rho_0 c_0 q(x_0) \qquad (6.189)$$

a relation which only contains strengths and phases but no geometry. The downstream sound pressure (6.187) then becomes

$$p(x) = \frac{f(x_0)}{S} e^{-jk(x-x_0)} \qquad (x > x_0) \qquad (6.190)$$

If we now have a primary monopole source at $x = 0$ with source strength q_p and the secondary set at $x_0 = L$ we find its downstream contribution from eqs. (6.157) and (6.190) as

$$p(x) = \frac{\rho_0 c_0}{2S} q_p e^{-jkx} + \frac{f}{S} e^{-jk(x-L)} \qquad (x > L) \qquad (6.191)$$

This total downstream sound pressure will vanish if

$$f = -\frac{\rho_0 c_0 q_p}{2} e^{-jkL} \qquad (6.192)$$

This determines the secondary dipole strength f and also the monopole strength by eq. (6.189). It solves the problem of this section as it cancels the downstream sound and leaves the primary field upstream undisturbed.

Exercises

6.1 Newton calculated the velocity of sound in air with Boyle's law $pV = $ constant. What velocity did he find?

6.2 Check Fig. 6.3 by calculating a few points, using eq. (6.37).

6.3 From tables of air density find the dry air temperature where the last term in eq. (6.53) would vanishes.

6.4 Calculate for the two values $p_{ref} = 2 \times 10^{-5}$ Pa and $p_{rms} = 0.6$ Pa the maximum velocities u of air parcels. Calculate the maximum displacements from equilibrium and the maximum accelerations for frequencies $f = 20$ Hz and $f = 1000$ Hz. Note that accelerations may be in the order of the gravity acceleration g or even larger. Compare the spatial amplitudes for p_{ref} and $f = 1000$ Hz with the Bohr radius of the hydrogen atom. Note that the ear can just identify these vibrations.

6.5 Show that in cylindrical coordinates the Bessel functions $J_0(kr)$ and Neumann functions $N_0(kr)$ are the cylindrical symmetric solutions of the wave equation with asymptotic solutions (6.60).

6.6 Show that eqs. (6.43), (6.68) and (6.28) can be understood qualitatively by looking at energy conservation. Consider a high apartment building in the middle of a busy town. Does it help to move from the lower to the higher floors to reduce the noise nuisance from the town?

6.7 Show that the linear part of Fig. 6.6 may be fitted by eq. (6.85).

6.8 Make a plot of eq. (6.86) for $F(f) = 1$ for the region $10^{-7} < I < 10^{-5}$ on a linear scale for I. Also plot $L_I + c_2$ and fit c_2 so that both curves intersect at $I = 5 \times 10^{-6}$. Check that both plots are rather similar around the intersection point.

6.9 At the six frequencies 125, 250, 500, 1000, 2000 and 4000 Hz a signal with $L_I = 60\,dB$ is offered. Deduce from Fig. 6.5 at each frequency the loudness level L_N in phon and from eq. (6.85) the loudness N in sone. What is the total loudness N in sone? What is the corresponding L_N using eq. (6.85)? Compare this with the total number L_I of dB and comment.

6.10 From eq. (6.97) one might conclude that the resistance of a wall may be increased by decreasing A. Show why this is wrong.

6.11 Consider a power source in a single room producing an intensity I_1 of a diffuse sound field with absorption A_1. After a change of absorption to A_2 the intensity becomes I_2. Show that for the sound pressure levels $L_2 - L_1 = 10 \log(A_1/A_2)$.

6.12 Two rooms are separated by a wall with $S = 12\,m^2$. If both rooms are unfurnished the emitting room has a sound pressure level $L_e = 110\,dB$ and the receiving room $L_0 = 63\,dB$. In the emitting room the absorption $A_e = 2.5\,m^2$ and in the receiving room $A_0 = 3\,m^2$. (a) Determine the sound resistance of the wall. Next both rooms are furnished, giving $A_e = 8\,m^2$ and $A_0 = 20\,m^2$. (b) Calculate L_e and L_0 assuming the same sound power as before (use the results of Exercise 6.11). (c) Will it be easy to get another decrease in L_0 with 10 dB by increasing absorption?

6.13 Define an average air insulation as the mean of R of eq. (6.119) for the frequencies $f = 100, 200, 400, 800, 1600$ and 3200 Hz. Show that $f = 565$ Hz would represent the average. Note that this result also holds for the practical mass law (6.120).

6.14 Consider a wall consisting of a brick layer of 22 cm thickness with 1 cm of gypsum plaster on both sides. Calculate the air sound insulation according to the mass law eq. (6.117) and the practical law (6.120). Use Table 4.1.

6.15 Interpret the hollow wall in Fig. 6.15 as two masses m_1 and m_2 connected by a string with strength constant C^{-1}. Calculate the eigenfrequency (6.136).

6.16 Study the case of two monopole sources at distance L, given in eqs. (6.158) to (6.162). Find the standing wave for $0 \leqslant x \leqslant L$. Make a graph of the intensity $|p(x)|$ for $-\infty \leqslant x \leqslant L$ if $kL = \frac{5}{8}(2\pi)$.

6.17 Use eqs. (6.168) and (6.163) to find the power output of the primary monopole source at $x = 0$ when there is a single secondary source at $x = L$. Compare it with the power output of the primary source in the absence of the secondary.

References

1. G. Chedd, Sound, Alders Books, London, 1970.
2. Roger Brown and Richard J. Hernstein, *Phychology*, Methuen, London, 1975. See the interesting discussion in Figs. 7-52 to 7-56.

3. Lawrence E. Kinsler, Austin R. Frey, Alan B. Coppers and James V. Sanders, *Fundamentals of Acoustics*, 3rd edn, John Wiley, New York, 1982. A general text on the same level as the present one, with special chapters on environmental acoustics and architectural acoustics.
4. P. A. Nelson and S. J. Elliott, *Active Control of Sound*, Academic Press, 1992. Written on a somewhat more advanced level than the present text, it is finely and precisely detailed with much relevant information. May be used for Section 6.1 and in particular Section 6.4, where it gives many more applications.

Bibliography

Morse, Philip M., *Vibration and Sound*, American Institute of Physics, 1983. A classic text, widely appreciated since the first edition in 1936. It makes ample use of acoustic impedance and equivalent electric circuits.
Möser, M. *Vorlesungen Technische Akustik*, T.U. Berlin, Fachbereich 21, Umwelttechnik, 1991.

7

Environmental Spectroscopy: Some Examples

Accurate quantitative analysis of the composition of the soil, the surface water or the atmosphere is crucial in our assessment of the quality of the environment and our judgement of the relative success of measures taken to reduce pollution. Many of the techniques to analyse the environment are based on spectroscopy. This is mainly due to the fact that each atom, molecule (small or large) or molecular aggregate is uniquely characterized by a set of energy levels. Transitions between levels by the absorption or emission of electromagnetic radiation result in highly specific spectroscopic features. Moreover, since each atom or molecule directly interacts with its close environment, the relevant energy levels and transition intensities may be perturbed, so each atom or molecule acts as a sensitive probe for its surroundings. These properties allow both the identification and the quantification of trace amounts of specific elements or molecules and an accurate assessment of their environment.

Important examples of the application of spectroscopic techniques in environmental analysis concern the monitoring of the atmosphere by laser remote sensing, the analysis of soil or water by high-resolution laser spectroscopy or by fluorescence and the identification of specific elements using the absorption or scattering of X-rays.

7.1 OVERVIEW OF SPECTROSCOPY

For atoms and molecules all states and their corresponding energies are given by the solution of the stationary Schrödinger equation for that system. For atoms this yields a set of electronic states, each characterized by their own combination of quantum numbers. Electronic transitions may occur between these states, giving absorption (transitions up in energy) or emission (down in energy) of electromagnetic radiation. For molecules the states are calculated by the assumption that in first order the motion of the electrons can be separated from the motion of the nuclei (the latter is assumed to be much slower: the Born–Oppenheimer approximation). In addition, the rotation of the molecule is separated from the relative vibrations of its nuclei. Thus, in general, the energy of a molecule is written as

$$E_{MOL} = E_{EL} + E_{VIB} + E_{ROT} \tag{7.1}$$

Consequently, molecules may undergo transitions, not only between electronic states but also between different vibrational and rotational states.

In a typical *absorption experiment* the amount of light transmitted by a sample is monitored by a detector as the frequency of the light source is swept over the desired frequency range. An atom or molecule may absorb electromagnetic radiation if the following condition is met:

$$E_f - E_i = \hbar\omega \tag{7.2}$$

where $E_f - E_i$ is the difference in energy between the initial (i) and final (f) states involved and ω the frequency of the incident light beam. In *emission spectroscopy* the electromagnetic radiation emitted by the molecules/atoms is detected as a function of the frequency of the radiation. Since in principle the same energy levels are involved in the absorption and in the emission process, the same spectroscopic information is obtained from both types of experiments and it will depend on the specific conditions which type is to be preferred.

The absorption/emission spectra thus obtained exhibit a number of lines or bands; some will be intense, some weak, while expected lines may be totally absent. Most lines will have a certain width. It is the purpose of this section to explain in general terms why this is so and how the measured spectra are related to atomic and molecular structures.

In Fig. 7.1 we summarize the various types of spectroscopy that are available for these studies. We distinguish these types by the energy of the photon involved in the transition, represented by its wavelength or frequency.

In *nuclear magnetic resonance* (NMR) spectroscopy the transitions between nucleon spin states in an applied magnetic field are detected. Typical frequencies are in the range of $v \approx 100\,\mathrm{MHz}$ (the largest NMR machines today available use a frequency $v \approx 700\,\mathrm{MHz}$), corresponding to roughly $0.0033\,\mathrm{cm}^{-1}$. The wavenumber $\bar{v}$ is given by $\bar{v} = v/c = \lambda^{-1}$ and is for historic reasons expressed in cm^{-1}. In the following the bar on top of the wave number is sometimes omitted and the dimension cm^{-1} added in order to avoid confusion. NMR spectroscopy is a powerful tool for detecting magnetic interactions between nuclei and the spectra are very sensitive to the molecular structure and chemical environment in which these nuclei are placed. Consequently, NMR can be applied to identify and even resolve three-dimensional structures of complex molecules in solution and in the solid phase.

In *electron spin resonance* (ESR) spectroscopy, atomic or molecular species

NMR
(rf)

342

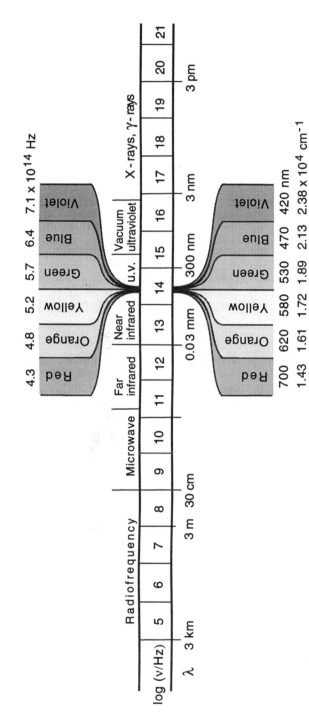

Figure 7.1 The electromagnetic spectrum and the classification by permission of Oxford University Press of the various spectral regions. (Adapted and reproduced from P. W. Atkins, *Physical Chemistry*, 3rd edn, Oxford University Press, 1986, Fig. 18.1, p. 432)

containing unpaired electrons, e.g. organic free radicals, are placed in an external *ESR (µWave)* magnetic field. Transitions between electron spin states can then be induced by radiation in the microwave region at about 10 GHz (9500 MHz or X-band ESR; 12 000 MHz or Q-band ESR). Note that in both NMR and ESR spectroscopies the relevant energy differences are much smaller than kT (which by $kT = h\nu$ corresponds with a wave number $\bar{\nu} \approx 200\,cm^{-1}$ at room temperature), implying that the signals are often very weak due to the extremely small population difference between the states involved (see below).

In *rotational* spectroscopy transitions between different rotational states of a *rotatnl (µWave)* molecule are observed. Most of these transitions occur in the microwave region, except those of very light molecules, which take place in the far-infrared region.

Transitions between *vibrational states* in molecules occur in the infrared region *vibratnl (IR)* of the spectrum, approximately between wavelengths of 50 and 2.5 µm. The energy difference between different vibrational states is larger than kT (corresponding with $\lambda = \bar{\nu}^{-1} \approx 50\,\mu m$ at room temperature) and consequently transitions involving the lowest vibrational level of a molecule are easily observed. Often 'overtones', i.e. transitions to high vibrational states, are possible in molecules and the corresponding wavelengths may extend into the 2.5 µm to 800 nm region of the spectrum.

Raman spectroscopy explores vibrational and rotational energy levels of *Raman* molecules by studying the frequency of light scattered by them. As a consequence of the scattering process, the molecule may end up in a higher vibrational or rotational state and the scattered light emerges with a lower frequency than the original excitation beam (and in a different direction). These low-energy scattered photons give rise to 'Stokes Raman scattering'. Alternatively, if molecules are *Stokes-Raman* present in excited vibrational/rotational states, the scattered radiation may emerge with a higher frequency than the excitation beam and thus produce 'anti-Stokes Raman scattering'. Note that in general most of the scattered radiation will have the same frequency as the original excitation source (Rayleigh and Mie scattering). Relative to this elastic scattering the Raman effect is very weak and requires highly sophisticated equipment to be observed accurately.

Electronic transitions between electronic states of atoms and molecules occur *electronic (IR-UV)* over a wide range of energies extending over the whole near-IR (= near-infrared), visible, ultraviolet (= UV) and far-UV regions. *Electronic spectra* give rise to the line spectra of atoms (such as the well-known spectral series of the H atom) and the complex spectra of molecules. In important biological molecules such as chlorophyll, β-carotene, aromatic amino acids or the DNA bases the absorption arises from electronic transitions between delocalized π-electron levels and these transitions are generally very intense.

X-ray spectroscopy is used to study the inner electrons of atoms (also within *X ray* molecules). Such electrons are much more strongly bound than the electrons involved in optical electronic spectroscopy and consequently high-energy photons are involved in absorption or emission. Examples are X-ray absorption, X-ray emission, photoelectron spectroscopy and Auger spectroscopy.

7.1.1 POPULATION OF ENERGY LEVELS AND INTENSITY

The rate of transition from a level with energy E_i and consequently the measured intensity of an absorption line depends on the population of that level. The probability $P(E_i)$ that an atom/molecule at a temperature T is in the level with energy E_i is given by the Boltzmann equation

$$P(E_i) \approx g(E_i)e^{-E_i/(kT)} \tag{7.3}$$

where $g(E_i)$ is the degeneracy of the level E_i (which we ignored in our derivation of the Einstein coefficients in Section 2.2.2, but can simply be included). If two transitions occur in the same atom/molecule, one arising from level E_i, the other from E_i', then with all other factors being equal, the ratio of the intensities of these two transitions will be given by the ratio of the populations of both levels, or

$$\frac{I}{I'} = \frac{g(E_i)}{g(E_i')} e^{-(E_i - E_i')/(kT)} \tag{7.4}$$

The relative intensity of both lines in an absorption spectrum thus depends on the values of $E_i - E_i'$ relative to kT. With $kT \approx 200 \, \text{cm}^{-1}$ at room temperature and a rotational transition in a medium sized molecule we have $E_i - E_i' \ll kT$, so that both lines will be observed with comparable intensities. On the other hand, for electronic and vibrational transitions the separation between two adjacent energy levels is of the order of $10^4 – 10^5 \, \text{cm}^{-1}$ and $10^2 – 10^3 \, \text{cm}^{-1}$ respectively, and consequently only (or mainly) transitions from the lowest level will be observed. Because of the factor kT in eq. (7.4), cooling will have strong effects on rotational spectra, while in vibrational spectra the already weak absorption lines from higher excited states will disappear.

7.1.2 THE TRANSITION DIPOLE MOMENT: SELECTION RULES

As shown in Section 2.2.1, transitions between the levels E_i and E_f induced by the electric field component of the electromagnetic field are governed by the size of the transition electric dipole moment (eq. 2.12))

$$\mu_{if} = \langle i|\mu|f \rangle \tag{7.5}$$

Only if μ_{if} has a value different from zero can transitions between levels i and f be induced by the electromagnetic field. One of the selection rules for optical transitions that immediately follows from eq. (7.5) arises from the conservation of angular momentum of the atom/molecule plus photon system. Since the photon carries one unit of angular momentum, this must be accounted for in the absorption of the photon to an excited state. For this reason the transition between the $|1s\rangle$ and $|2s\rangle$ states in atomic hydrogen cannot be induced by electromagnetic radiation. This transition is not observed in an absorption/emission spectrum and is called 'forbidden'. On the other hand, for the $|1s\rangle$ to $|2p\rangle$ transition in atomic hydrogen the value of the orbital angular quantum number l increases from 0 to 1 ($\Delta l = 1$). This transition is observed as a strong line and is called 'allowed'. These results are easily checked using the appropriate expressions for the H atom wave functions.

The definition $\mu = \sum q_i r_i$ of the dipole operator shows that the two states involved must have opposite parity as the operator changes sign under inversion. For complex molecules the symmetry properties of the electronic and vibrational wave functions play an essential role in deciding whether a particular transition is allowed or forbidden.

For rotational transitions the electric dipole transition moment μ_{if} equals zero, unless the molecule possesses a permanent dipole moment. This implies that the molecule must be polar. For instance, a molecule like H_2O will show a rotational spectrum, while N_2, CO_2 and CH_4 will not. The classical basis for this rule is that a polar rotating molecule represents an in time fluctuating dipole, which can interact with the oscillating electromagnetic field.

For vibrational transitions the electric transition dipole moment is zero, unless the electric dipole of the molecule fluctuates during the vibration that is excited, either in size or in direction. Examples are the asymmetric stretching and bending vibrations of CO_2, H_2O, CH_4, etc. A molecule like N_2 does not possess a vibration that is associated with a change in dipole moment, and, consequently, does not exhibit a vibrational spectrum. Note that the infrared active molecules do not need to have a permanent dipole moment.

7.1.3 LINE WIDTHS

Spectral lines are not infinitely narrow, but show a certain width $\Delta\omega$. A large variety of phenomena and processes may contribute to the observed line widths and here we will discuss a few, assuming a single type of atom or molecule. In principle, we distinguish 'homogeneous broadening' where all the atoms/molecules that contribute to an absorption line suffer from the same broadening of the line and 'inhomogeneous broadening' where different atoms/molecules of the sample absorb at different frequencies.

Homogeneous Broadening

We first discuss lifetime broadening as a form of homogeneous broadening. If the excited state produced by the absorption of electromagnetic radiation has a finite lifetime τ, the correct expression for the time-dependent excited state wave function Ψ_f is

$$\Psi_f(\mathbf{r}, t) = \Psi_f(\mathbf{r}, 0)e^{-iE_f(t/\hbar) - t/2\tau} \tag{7.6}$$

We may express the time-dependent part of $\Psi_f(\mathbf{r}, t)$ as a linear combination of pure oscillating functions of the type $e^{(-iEt)/\hbar}$ by means of a Fourier transformation and obtain

$$e^{-iE_f(t/\hbar) - t/2\tau} = \int L(E)e^{-iE(t/\hbar)} dE \tag{7.7}$$

in which $L(E)$ is the distribution or line shape function. Equation (7.7) demonstrates that the finite excited state lifetime τ implies that we have an 'uncertainty' in the excited state energy, reflected by the distribution function $L(E)$, which can be easily calculated to have the form

$$L(E) = \frac{\hbar/\tau}{(E_f - E)^2 + (\hbar/2\tau)^2} \tag{7.8}$$

The line shape represented by eq. (7.8) is a Lorentzian. The full width at half-maximum (FWHM) of this function is given by

$$\Delta E = \hbar/\tau \tag{7.9}$$

or

$$\Delta\omega = 1/\tau \tag{7.10}$$

Note that eqs. (7.9) and (7.10) are strongly reminiscent of Heisenberg's uncertainty relation and therefore lifetime broadening is often referred to as uncertainty broadening.

For atoms/molecules where the excited state lifetime is only determined by spontaneous emission the lifetime is called the natural or radiative lifetime τ_R, given by

$$\tau_R^{-1} = A \tag{7.11}$$

where A is the Einstein coefficient (cf. eq. (2.20)) for the transition, representing the number of transitions per second. The resulting 'natural' or 'radiative' line width is the minimum line width. For atoms in the gas phase or for molecules at $T \approx 0$, where other decay processes no longer contribute, τ_R^{-1} may be the observed line width. Note that natural line widths depend strongly on the frequency, through the Einstein coefficient (eq. (2.20)). Since they increase with ω^3, electronic transitions will have significant natural line widths, while rotational transitions will be very narrow. Typically, for an allowed electronic transition the natural lifetime may be of the order of 10^{-7}–10^{-8} s, corresponding to a natural line width of 1.5–15 MHz. A typical natural lifetime for a rotational transition is about 10^3 s, corresponding to a natural line width of only 10^{-4} Hz.

Normally, for atoms and molecules many processes may contribute to the shortening of the lifetime τ and hence increase the width of the absorption band. Let us discuss atoms in the gas phase which may undergo collisions. If we assume that the phases of the excited state wave functions before and after the collision are totally uncorrelated then it is not difficult to show that this will result in a

Lorentzian line shape of the transition with the FWHM given by

$$\Delta\omega_{coll} = \frac{2}{\tau_{coll}} \qquad (7.12)$$

where τ_{coll}^{-1} is the rate at which the collisions occur. A reasonable value of the collision time at a gas density corresponding to a pressure of 10^5 Pa and at room temperature is $\tau_{coll} \approx 3 \times 10^{-11}$ s or $\Delta\nu_{coll} \approx 10^{10}$ Hz, which largely exceeds the natural line width.

The width at any temperature T and any pressure p can be calculated from the one at some standard values p_0, T_0 by [1]

$$\Delta\nu_{coll}(p, T) = \Delta\nu_{coll}(p_0, T_0)\frac{p}{p_0}\sqrt{\frac{T_0}{T}} \qquad (7.13)$$

Molecules in the condensed phase (in solution, or in a host matrix) will interact with their surroundings and 'collisions' will occur between the molecule and phonons in the medium, also leading to dephasing of the excited state. The homogeneous line width $\Delta\omega_{hom}$ of a molecular excited state is often expressed as

$$\Delta\omega_{hom} = \frac{1}{2\pi T_1} + \frac{1}{\pi T_2^*} \qquad (7.14)$$

where T_1 is the excited state lifetime and T_2^* is the pure dephasing rate. Note that T_1 is generally much larger than the natural lifetime, since in molecules many processes may contribute to the decay of the excited state. For a further discussion of the spectra of molecules in the condensed phase we refer to Section 7.6.3.

Inhomogeneous Broadening

The distribution of velocities of atoms/molecules in the gas phase leads to the so-called Doppler broadening, which is an example of inhomogeneous broadening, as different atoms/molecules contribute to slightly different parts of the spectrum. For instance, in emission the frequency of the emitted light suffers from a Doppler shift, which is due to the component u of the initial atomic/molecular velocity in the direction of the emitted photon's wave vector:

$$\omega \approx \omega_0\left(1 + \frac{u}{c}\right) \qquad (7.15)$$

A positive value $u > 0$ represents a particle moving towards the detector. Because in gases atoms and small molecules may approach high speeds, the whole Doppler-shifted range of frequencies will be detected in an absorption or emission spectrum. Since, according to the Maxwell distribution, the relative probability that a particle has a velocity component u in a particular direction has a Gaussian shape, the resulting Doppler broadened line shape will also be Gaussian. For a molecule/atom of mass M at temperature T the FWHM of this Gaussian line shape function is given by

$$\Delta v_D = 2v_0 \left(\frac{2kT \ln 2}{Mc^2} \right)^{1/2} \tag{7.16}$$

An illustrative application of eq. (7.16) is the measurement of the surface temperature of stars from the widths of particular emission lines in the solar spectrum. For instance, the sun emits a spectral line at 677.4 nm, which is due to a transition involving highly ionized ^{57}Fe. From the measured FWHM of 5.3×10^{-3} nm one would calculate, using eq. (7.16), the sun's temperature as 6800 K, which compares well with the temperature of 5800 K estimated from Wien's law.

For atoms or molecules in a non-gaseous environment, the interaction with the host will fine-tune the absorption frequencies. For a crystalline host this may give rise to a (limited) set of sharp, well-defined lines in low-temperature spectra. For molecules in solution or in a glassy environment (such as a protein or a membrane) a broad distribution of sites may exist, each with their own transition frequency, leading to the broad ($\Delta \omega \approx 200$ cm^{-1}) unstructured inhomogeneously broadened absorption bands observed for many molecules, amongst which are the natural pigments, such as chlorophyll and β-carotene. Specific techniques have to be applied to resolve the underlying homogeneous line widths (see Section 7.6.3).

Composite Line shapes

When two independent processes contribute to the line shape, the resulting line will be given by a so-called convolution of both line shapes:

$$L(\omega) = \int_{-\infty}^{\infty} L_1(\omega')L_2(\omega + \omega_0 - \omega')\,d\omega' \tag{7.17}$$

where ω_0 is the common central frequency of the two individual line shape functions.

7.2 ATOMIC SPECTRA

The spectra of atoms show transitions between electronic states $|i\rangle$ and $|f\rangle$, which are said to be allowed if the outcome of eq. (7.5) is sufficiently large. Atomic electronic states are denoted by the symbols $^{2S+1}L_J$, where $2S+1$ is the multiplicity with S the total spin quantum number, L the total orbital angular momentum quantum number and J the total angular momentum quantum number. We shall first discuss the spectra of 'one-electron' atoms and then the spectra of 'many-electron' atoms.

7.2.1 ONE-ELECTRON ATOMS

One-electron atoms, such as H, Li, K, Na,... have only one valence electron while all their remaining electrons are in closed shells. Consequently, the quantum numbers J, L and S only refer to the single valence electron. For one-electron atoms we have the selection rules

$$\Delta L = \Delta l = \pm 1 \tag{7.18}$$

$$\Delta J = \Delta j = 0, \pm 1 \tag{7.19}$$

in which l and j are the orbital angular momentum quantum number and the total angular momentum quantum number respectively of the single valence electron. These rules show immediately that in atomic hydrogen the optical $|1s\rangle \rightarrow |2p\rangle$ transition is allowed, while the transition $|1s\rangle \rightarrow |2s\rangle$ violates the selection rules.

As an example consider the sodium D-lines, observed in the emission spectrum of sodium excited by an electronic discharge. The yellow emission lines are observed at $16\,956.2\,\text{cm}^{-1}$ and at $16\,973.4\,\text{cm}^{-1}$ and are due to the transitions $^2S_{1/2} \leftarrow {}^2P_{1/2}$ and $^2S_{1/2} \leftarrow {}^2P_{3/2}$ respectively. The observed splitting of the sodium D-lines is called the fine structure and results from the interaction between the electron spin S and the orbital angular momentum L (also called spin–orbit interaction). The size of this splitting increases sharply with atomic number $(\sim Z^4)$.

7.2.2 MANY-ELECTRON ATOMS

For atoms with more than one electron outside a closed shell, the spectra rapidly increase in complexity. For the case that the spin–orbit interaction is relatively weak, the total spin quantum number S and orbital angular momentum quantum number L remain good quantum numbers. Then we obtain the correct atomic states by coupling these total momenta to give the total angular momentum J (Russell–Saunders coupling). For these many-electron atoms we have the following selection rules:

$\Delta S = 0$

$\Delta L = 0, \pm 1$ with $\Delta l = \pm 1$ (i.e. the orbital angular momentum quantum number of an individual electron must change, but whether or not this affects the total orbital angular momentum depends on the coupling)

$\Delta J = 0, \pm 1$ with $J = 0 \rightarrow J = 0$ forbidden.

Figure 7.2 shows all possible dipole allowed transitions between a p^2 and an sp configuration.

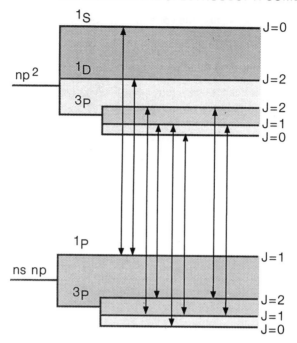

Figure 7.2 Schematic representation of all allowed dipole transitions between a p^2 and an sp configuration.

With increasing spin–orbit interaction (heavy atoms) the Russell–Saunders coupling fails and then individual spin and orbital angular momenta have to be coupled first into individual j angular momenta, which then are combined into one large J (jj coupling). In that case the selection rules given above break down and transitions between states of different multiplicity are observed. The most prominent example of this phenomenon is the intense line observed in the emission spectrum of a high-pressure mercury lamp at $\lambda = 253.7$ nm. This line corresponds to a transition between the singlet ($^1S_0$) and one of the triplet ($^3P_1$) levels of mercury.

7.3 MOLECULAR SPECTRA

Molecular spectra are generally much more complex than atomic spectra. Their complexity arises from the large number of states, rotational, vibrational and electronic, between which transitions may be observed. Moreover, the motion of the nuclei may be coupled to the electronic transitions, further adding to the complexity. Nevertheless, the molecular spectra constitute a rich source of information about molecular structure and molecular dynamics and we shall illustrate a few of their important features in the following sections.

7.3.1 ROTATIONAL TRANSITIONS

Rotational transitions can occur between the rotational energy levels of a particular electronic state, in general the electronic ground state, only if that state possesses a permanent dipole moment. The rotational energy levels will be calculated from the general expression of a free rotating body for a few limiting cases.

Spherical Rotors

Spherical rotors are molecules with all three moments of inertia equal ($I_{xx} = I_{yy} = I_{zz} = I$). An example is CH_4. The energy of a spherical rotor is given by

$$E_J = \frac{J(J+1)\hbar^2}{2I}; \qquad J = 0, 1, 2, \ldots \tag{7.20}$$

where J is the quantum number for rotational motion.

The separation between two neighbouring rotational levels is

$$E_J - E_{J-1} = 2hcBJ \tag{7.21}$$

with $B = \hbar/(4\pi cI)$ the rotational constant, by convention expressed in cm^{-1}. Typical values of B for small molecules are in the range 1–$10\,cm^{-1}$. Note that $E_J - E_{J-1}$ decreases as I increases and, consequently, large molecules have closely spaced rotational energy levels.

Symmetric Rotors

Symmetric rotors have two equal moments of inertia, with for example $I_{xx} = I_{yy} = I_\perp$ and $I_z = I_\parallel$. Examples of such molecules are NH_3 and C_6H_6. The energy of such a rotor can be expressed as

$$E_{JK} = hcBJ(J+1) + hc(A-B)K^2; \qquad J = 0, 1, 2, \ldots; \quad K = 0, \pm 1, \pm, \ldots, \pm J \tag{7.22}$$

with $A = \hbar/(4\pi cI_\parallel)$ and $B = \hbar/(4\pi cI_\perp)$. In eq. (7.22) K represents the component of J along the molecular symmetry axis. With $K = 0$, there is no rotational angular momentum component along the symmetry axis; with $K = J$ almost all the rotational angular momentum is along the symmetry axis.

Linear Rotors

In a linear rotor (such as CO_2 and HCl) the rotation occurs only about an axis perpendicular to the line connecting the atoms. This corresponds to the situation that there is no rotational angular momentum along the connecting line, and we can use eq. (7.22) with $K = 0$. We then obtain for the energy levels of a linear rotor

$$E_J = hcBJ(J + 1); \qquad J = 0, 1, 2, \ldots \qquad (7.23)$$

Note that, although $K = 0$, each J level still has $2J + 1$ components on some laboratory axis. This degeneracy may be removed, for instance, by an applied electric field (Stark effect).

Selection Rules

The selection rules for rotational transitions are again determined by an analysis of eq. (7.5). For a molecule to have a pure rotational spectrum it must possess a permanent dipole moment. Thus, homonuclear diatomics (O_2, N_2) and symmetric linear molecules are rotationally inactive. Spherical molecules can only have a rotational spectrum if they are distorted by centrifugal forces, but in general they are rotationally inactive.

For a linear molecule the selection rule for J is

$$\Delta J = \pm 1 \qquad (7.24)$$

From equation (7.23) it follows that a pure rotational spectrum consists of a set of equidistant lines, separated by $2B$ (in cm^{-1}). The relative intensities of each rotational line increase with increasing J, pass through a maximum and eventually tail off as J becomes large. This pattern is an immediate consequence of the Boltzmann population of the various rotational energy levels at room temperature, which for a linear molecule are determined by

$$N(E_J) \sim (2J + 1)e^{-E_J/kT} \qquad (7.25)$$

7.3.2 VIBRATIONAL TRANSITIONS

Vibrational transitions can occur between the vibrational states of an electronic state, which is under normal conditions the ground state. For vibrational transitions the molecule must undergo a change in dipole strength or dipole direction, due to the vibration. Within the framework of the Born–Oppenheimer approximation the vibrational energies and wave functions are calculated using the electronic energy $E_{el}(R)$, which is a function of all the nuclear coordinates symbolized by R and which may be interpreted as a potential for the motion of the atomic nuclei.

Diatomic Molecules

A diatomic molecule possesses one vibrational coordinate, the interatomic distance R. In regions close to the potential minimum, the molecular potential energy can be approximated by a parabola

$$V = \tfrac{1}{2}K(R - R_e)^2 \tag{7.26}$$

with K the force constant and R_e the equilibrium distance between the two atoms. Solving the Schrödinger equation by using the harmonic oscillator potential (7.26) shows that the reduced mass μ of the two atoms undergoes harmonic motion. Therefore the permitted energy levels are given by

$$E_v = (v + \tfrac{1}{2})\hbar\omega; \qquad \omega = \sqrt{\frac{K}{\mu}}; \qquad v = 0, 1, 2, \ldots \tag{7.27}$$

in which v is the vibrational quantum number. The corresponding wave functions that describe the harmonic motion can be found in any elementary textbook on quantum mechanics. Note that the steepness of the potential function, characterized by the force constant K, determines the energies of vibrational transitions.

Anharmonicity of the vibrational motion can be introduced by using a more complex form of V that resembles the true potential more closely. The most frequently applied form is given by the Morse potential energy

$$V = D_e(1 - e^{-a(R - R_e)})^2 \tag{7.28}$$

where D_e now represents the depth of the potential minimum and $a = (\mu/2D_e)^{1/2}\omega$.

Now the permitted energy levels are given by

$$E_v = (v + \tfrac{1}{2})\hbar\omega - (v + \tfrac{1}{2})^2 x_e \hbar\omega; \qquad x_e = \frac{a^2 \hbar}{2\mu\omega} \tag{7.29}$$

where x_e is called the anharmonicity constant. In practice even higher order terms in $(v + \tfrac{1}{2})$ are added to eq. (7.29) and the resulting expression may be fitted to the experimental data.

The selection rule for transitions between vibrational states is again derived from eq. (7.5). If we denote our total vibronic wavefunction as $\Psi_{el}(r, R)\phi_v(R)$, where r represents all electron coordinates, then a vibrational transition requires that

$$\langle \psi_{el}\phi_v | \mu | \psi_{el}\phi_{v'} \rangle \neq 0, \qquad \text{with } v \neq v' \tag{7.30}$$

which is only the case if μ is a function of the internuclear coordinates R. Thus expanding μ around the equilibrium position R_e gives

$$\mu = \mu_0(r, R_e) + \left[\frac{\partial\mu(r, R)}{\partial R}\right]_{R_e}(R - R_e) + \cdots \tag{7.31}$$

Substituting eq. (7.31) into eq. (7.30), the first term in the expansion yields zero, but the second term is of the form

$$\langle \psi_{el} | \left[\frac{\partial\mu(r, R)}{\partial R}\right]_{R_e} | \psi_{el} \rangle \langle \phi_v | R | \phi_{v'} \rangle \tag{7.32}$$

From the properties of the harmonic oscillator wave functions it follows within the harmonic approximation that eq. (7.32) is only non-zero for

$$\Delta v = \pm 1 \tag{7.33}$$

or

$$\Delta E = E_{v+1} - E_v = \hbar\omega \tag{7.34}$$

For example, HCl has a force constant $K = 526 \, \text{N m}^{-1}$. The reduced mass of HCl $(1.63 \times 10^{-27} \, \text{kg})$ is very close to the mass of the proton $(1.67 \times 10^{-27} \, \text{kg})$. Note that this implies that in HCl the Cl atom is hardly moving. From these values it follows that $\omega = 5.63 \times 10^{14} \, \text{s}^{-1}$, which implies that a transition is observed at $\lambda = 3.35 \, \mu\text{m}$ or $2990 \, \text{cm}^{-1}$, in the infrared region of the spectrum. Since at room temperature $kT \approx 200 \, \text{cm}^{-1}$ only the $v = 0$ level is populated, the vibrational absorption (and emission) spectrum of a diatomic molecule consists of a single line.

Additional lines in emission spectra may be observed if hot states are present, formed by chemical reactions. In addition, anharmonicity as expressed by the Morse potential leads to the transitions $2 \leftarrow 0$, $3 \leftarrow 0$, etc. From their appearance the dissociation energy of the molecule can be calculated, using the Birge–Sponer extrapolation (as described in Ref. 2).

As for rotational motion, molecules of the type O_2, N_2 (i.e. homonuclear molecules) do not exhibit electric-dipole vibrational spectra. However, quadrupole- and pressure-induced transitions of homonuclear molecules can be observed faintly.

Vibrational–Rotational Spectra

The rotation of a molecule is influenced by molecular vibration and therefore the two motions are not independent but coupled. A detailed analysis shows that the rotational quantum number J changes by ± 1 when a vibrational transition occurs (in exceptional cases $\Delta J = 0$ is also allowed). Thus a vibrational transition of a heteronuclear diatomic molecule is found to consist of a large number of closely spaced components and the observed spectrum may be understood in terms of the set of energy levels:

$$E_{v,J} = (v + \tfrac{1}{2})\hbar\omega + hcBJ(J+1), \qquad \Delta v = \pm 1; \quad \Delta J = \pm 1 \tag{7.35}$$

In a more detailed treatment the dependence of B on the vibrational quantum number is taken into account, but this is ignored here.

From eq. (7.35) one observes that the absorption spectrum consists of two branches. The P-branch has $\Delta J = -1$ and $v = v_0 - 2BJ$ (in cm^{-1}) and an intensity distribution following the Boltzmann population over the various J states. The R-branch has $\Delta J = +1$ and $v = v_0 + 2B(J+1)$ (in cm^{-1}). Sometimes the Q-branch ($\Delta J = 0$) is also visible. Figure 7.3 shows the high-resolution vibrational–rotational spectrum of gas-phase HCl, in which the $\Delta J = 0$ transition

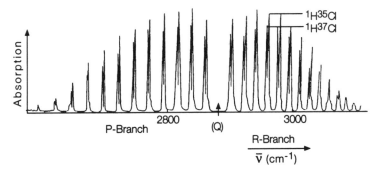

Figure 7.3 High-resolution vibrational–rotational spectrum of HCl. The lines appear in pairs because the spectrum reflects the presence of both $H^{35}Cl$ and $H^{37}Cl$ in their natural abundance ratio of 3:1. The $\Delta J = 0$ branch is absent in the spectrum of HCl. (Adapted and reproduced from P. W. Atkins, *Physical Chemistry*, 3rd edn, Oxford University Press, 1986, Fig. 18.14, p. 452)

is absent. We finally note that the distribution of the absorption intensities over the rotational states is a sensitive function of the temperature.

Polyatomic Molecules

In a molecule with N atoms there exist $3N - 6$ modes of vibration ($3N - 5$ for a linear molecule), for one has to subtract the three translational and three rotational (two for a linear molecule) degrees of freedom from the total $3N$. Thus, H_2O, a triatomic non-linear molecule, has three modes of vibration, while CO_2, which is linear, has four vibrational modes. These modes may be chosen such that each of them can be excited individually; they are called 'normal modes'. For instance, for the stretching modes of CO_2, one should not take the individual

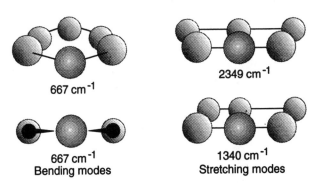

Figure 7.4 Representation of the two stretching modes and the two bending modes of CO_2.

stretching modes of the two C—O bonds, but the symmetric and antisymmetric linear combinations (see Fig. 7.4). Together with the two bending modes one obtains the four normal modes of CO_2.

For more complex molecules a general method of dealing with molecular vibrations is to use group theory to classify the vibrational modes according to their symmetries. In essence, each normal mode behaves like an individual harmonic oscillator with $E_Q = (n + \frac{1}{2})\hbar\omega_Q$, where ω_Q is the frequency of mode Q. This frequency depends on the force constant K_Q and the reduced mass μ_Q through $\omega_Q = (K_Q/\mu_Q)^{1/2}$.

The reduced mass μ_Q may be very different for the various normal modes, depending on which atoms are actually moving. For instance, in the symmetric stretch of CO_2 the C atom is static, while in the antisymmetric stretch both the C atom and the two O atoms are moving. Similarly, the value of K_Q is made up of a combination of the various stretches and bends that contribute to that particular mode.

The overall selection rule for infrared activity remains that the motion corresponding to the active normal mode should lead to a change in dipole moment. For CO_2 this can be evaluated by simple inspection. The symmetric stretch ($\nu_S = 1340\,\text{cm}^{-1}$) leaves the dipole moment of CO_2 unchanged at zero and this mode is infrared inactive. The antisymmetric stretch ($\nu_{AS} = 2349\,\text{cm}^{-1}$) leads to a change in dipole parallel to the line connecting the three atoms and consequently gives rise to a parallel infrared transition. Both bending modes of CO_2 ($\nu_B = 667\,\text{cm}^{-1}$) are infrared active and appear as bands polarized perpendicular to the long axis of CO_2 in the IR spectrum.

7.3.3 ELECTRONIC TRANSITIONS IN MOLECULES

Electronic transitions in molecules occur in the visible part and in the near- and far-UV parts of the spectrum. For instance, H_2O absorbs UV light below 180 nm, chlorophyll shows strong absorption bands in the red (680 nm) and blue (400 nm), etc. In principle the transitions between electronic states are governed by the same rules as those of atoms and the selection rules are again determined by the outcome of eq. (7.5). There are also some important differences with atoms, which complicate the electronic spectra of molecules but provide a wealth of information concerning the interaction of the molecule with its environment.

Molecular Orbitals and Transition Dipoles

In most molecules, the optical transition cannot be located on one or even a few atoms. Molecules are more or less complex arrangements of atoms, and their electronic wave functions are linear combinations of (all) atomic orbitals. Therefore, like the vibrational states discussed above, the electronic states of molecules are classified according to their symmetry properties and the outcome of the integral in eq. (7.5) is only non-zero if it contains a totally symmetric contribution under the symmetry operations of the molecular point group. These general statements directly lead to the important conclusion that often an absorption line in a molecular electronic absorption spectrum corresponds to only one of the spatial components of the transition dipole. Thus the absorption and emission corresponding to that transition may be strongly polarized.

For instance, in H_2O the molecular orbitals are linear combinations of the two H 1s and the oxygen 2s, $2p_x$, $2p_y$ and $2p_z$ atomic orbitals (the two 1s electrons of O are ignored, since they are too low in energy). The atomic wave functions are denoted by $(H 1s_a)$, $(H 1s_b)$, $(O 2s_a)$, $(O 2s_b)$, $(O 2p_x)$, $(O 2p_y)$ and $(O 2p_z)$. The z axis is chosen coincident with the C_2 symmetry axis of H_2O and the y axis is in the plane containing the O atom and the two H atoms. One uses group theory to classify the possible linear combinations of the atomic orbitals that form the bonding and antibonding molecular orbitals of H_2O. The symmetry group of H_2O is C_{2v} (one C_2 axis and two mirror planes (σ_v, σ'_v) containing the C_2 axis). Using the atomic orbitals given above, we can construct a symmetry adapted basis for the H_2O molecule which consists of the following set of wave functions:

$$a_1 = c_1(H 1s_a + H 1s_b) + c_2(O 2p_z) + c_3(O 2s) \text{ with } A_1 \text{ symmetry,}$$
i.e. totally symmetric for all symmetry operations

$$b_1 = (O 2p_x) \text{ with } B_1 \text{ symmetry,}$$
i.e. changes sign upon the operation of C_2 and σ'_v (7.36)

$$b_2 = c'_1(H 1s_a - H 1s_b) + c'_2(O 2p_y) \text{ with } B_2 \text{ symmetry,}$$
i.e. changes sign upon operation of C_2 and σ'_v

The coefficients can be found by the variation principle and this results in a set of three a_1 orbitals ($1a_1$, $2a_1$ and $3a_1$), one b_1 orbital (the original $(O 2p_x)$ orbital) and two b_2 orbitals ($1b_2$ and $2b_2$). The wavefunctions corresponding to the $1a_1$ molecular orbital (denoted by MO) show no nodes; the $1b_2$ MO has a nodal plane corresponding to the xz plane, while the $1b_1$ MO, the original $(O 2p_x)$ atomic orbital, is a non-bonding orbital. The $2a_1$ MO is mainly located on the O atom, while the $2b_2$ and $3a_1$ MOs contain more than one node in their wave functions, characteristic for antibonding MOs. The energetic order of the H_2O MOs is as follows:

$$E(1a_1) < E(1b_2) < E(1b_1) < E(2a_1) < E(2b_2) < E(3a_1) \quad (7.37)$$

Since we have to accommodate eight electrons (one from each H atom and six from the O atom) the ground state configuration of H_2O is $(1a_1)^2(1b_2)^2(1b_1)^2(2a_1)^2$. The lowest unoccupied molecular orbital has B_2 symmetry ($2b_2$) and it is not difficult to show that only the y polarized component of the electric dipole transition moment is allowed.

Similar considerations can be applied to all molecules and these methods allow us to predict ground and excited state configurations of molecules and the associated transition dipoles with great accuracy.

Sometimes the absorption of a photon can be traced back to the presence of a particular set of atoms. For instance, molecules containing the $—C=O$ (or carbonyl group) often show absorption around 290 nm. Similarly, molecules with many unsaturated $—C=C—$ bonds (benzene, retinal, chlorophyll) may show strong absorption in the near-UV or visible part of the spectrum.

In the case of a single $—C=C—$ double bond, the highest occupied molecular orbital (HOMO) is a π orbital; the lowest unoccupied molecular orbital (LUMO) is a π^* orbital. The π and π^* orbitals are the symmetric and antisymmetric combinations of $2p_z$ orbitals on the two C atoms, where the symbol π refers to the component of the total angular momentum along the axis connecting the two C atoms, which is one. For a single double bond the transition $\pi \rightarrow \pi^*$ occurs at about 180 nm. In the above-mentioned case of the carbonyl group, the oxygen atom possesses an additional non-bonding (n) electron pair. Although the $n \rightarrow \pi^*$ transition is in first order forbidden, distortion of the $—C=O$ chromophore, for instance due to an out-of-plane vibration, gives the $n \rightarrow \pi^*$ sufficient dipole strength to be observable and its absorption is found at about 230–290 nm (for instance in formaldehyde and in proteins).

For molecules with extended conjugated chains or conjugated ring systems (e.g. benzene, β-carotene, chlorophyll, the visual pigment retinal, the aromatic amino acid tryptophan) the $\pi \rightarrow \pi^*$ absorption is shifted to the near-UV and the visible regions. Often the optical transitions are strongly allowed and molecules like chlorophyll and β-carotene are among the strongest absorbing species in nature.

The Franck–Condon Principle

The excitation of an electron to a higher electronic state may be accompanied by the simultaneous excitation of a vibrational (or rotational) transition. Consequently, a progression of vibrational lines may be observed, or just a broad line, which is due to unresolved vibrational structure. These structures are explained by the Franck–Condon principle, which states that, because nuclei are so massive, they can be considered as static on the time scale of an electronic transition. Dependent on the displacement of the ground state potential energy curve relative to the excited state potential energy one or sometimes several progressions of absorption/emission lines may be observed. This is schematically indicated in Fig. 7.5, which illustrates that the vibrational wave function with the greatest overlap with the ground state vibrational wave function will dominate in the spectrum. Formally this is expressed as

$$\mu_{if} = \langle \psi_{el}^i \phi_v^i | \mu | \psi_{el}^f \phi_{v'}^f \rangle \approx \mu_{if} \langle \phi_v^i | \phi_{v'}^f \rangle = \mu_{if} S_{vv'} \qquad (7.38)$$

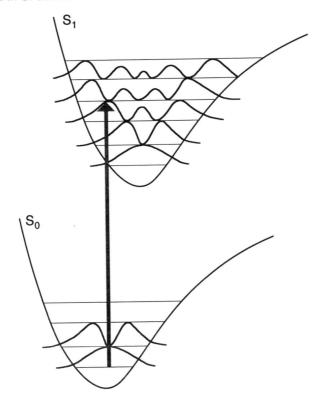

Figure 7.5 Illustration of the Franck–Condon principle in molecular spectroscopy. The process of absorption of a photon corresponds to a purely vertical transition in this diagram. The most intense vibronic transition occurs from the ground state to a vibrational level of the excited state for which the overlap of the vibrational wavefunctions is maximal

in which the $S_{vv'}$ are called the Franck–Condon factors. The Franck–Condon factors represent the overlap between the ground and excited state vibrational wave functions. Often several Franck–Condon factors may be sufficiently large for a progression of vibronic lines to be observed in the spectrum.

Decay of Excited States

Upon excitation to a highly excited vibronic state the molecule will relax due to interaction with its surroundings. In general the excess energy will be redistributed among other intramolecular vibrational and/or rotational states and eventually it will be dumped into vibration, rotation and translation of the surrounding matrix. Vibrational relaxation from a higher vibrational level to the $v = 0$ level of a particular electronic state may occur within 10^{-12} s (or less) in the condensed phase; for isolated small molecules these decay times may be much longer.

From the $v = 0$ level of a high electronic state the molecule will further relax to the lowest excited state, a process called internal conversion, which again occurs on a time scale as fast as 10^{-12}. If the molecule in its ground state was a singlet, these excited states will all be singlets, since the rate of interconversion

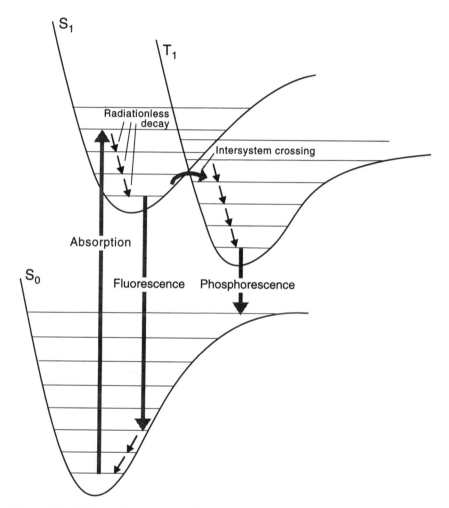

Figure 7.6 Schematic representation of the sequence of events leading to fluorescence, triplet formation and phosphorescence. Note that the maximum in the fluorescence spectrum again corresponds to the transition with the largest Franck–Condon factor

between singlet and triplet states is usually of the order of 10^8–$10^9\,\mathrm{s}^{-1}$ for these large molecules. Once in the lowest excited singlet state the molecule may decay through a variety of pathways. Radiationless decay to the ground state may occur, again through internal conversion. The rate of radiationless decay (k_{ic}) is highly variable and may range between $10^{12}\,\mathrm{s}^{-1}$ and $< 10^6\,\mathrm{s}^{-1}$. Alternatively the molecule may decay through spontaneous emission with a rate constant given by the relevant Einstein coefficient. The rate of radiative decay (k_{R}) is about 10^7–$10^8\,\mathrm{s}^{-1}$ for a strong optical transition. Note that since the emission occurs from the $v = 0$ level of the lowest excited state and since the Franck–Condon factors determine the relative intensities of the vibronic lines in the emission spectrum, most of the emission occurs at longer wavelengths than the absorption. Only the transitions between the $v = 0$ levels occur in both spectra. This phenomenon, called the Stokes' shift, is illustrated in Fig. 7.6.

An easy way to express the relative contribution of these different decay processes is by a quantity called the quantum yield. For instance, the quantum yield of fluorescence is given by

$$\phi_{\mathrm{F}} = \frac{k_{\mathrm{R}}}{\sum_i k_i} \tag{7.39}$$

where the summation is over all possible decay processes.

7.4 SCATTERING

When an atomic or molecular system is illuminated by light with a frequency which is not resonant with one of the available transitions, the light beam will not be absorbed, but may still be scattered. If the frequency of the scattered light is identical to that of the exciting beam, the scattering is called elastic or Rayleigh scattering. For particles with dimensions of the same order or larger than the wavelength of the scattered radiation, the process becomes a complicated function of particle size and shape and is called Mie scattering. Finally, scattered photons may emerge with lower or higher energy than the original illuminating beam. This is due to non-elastic or Raman scattering and the specific Raman lines which are observed are very informative about molecular structure and molecular dynamics.

7.4.1 RAMAN SCATTERING

The classical interpretation of the observation of several discrete frequencies in the scattered spectrum is that if a molecule vibrates with frequency ν_{vib} its polarizability may vary. Thus, the observed polarization and therefore the emerging (scattered) radiation field contains a term that oscillates with the incident electromagnetic field and corresponds to the Rayleigh scattering and two side bands: one at $v - v_{vib}$ which is called Stokes Raman scattering and one at $v + v_{vib}$ which is called anti-Stokes Raman scattering.

The intensity of these Raman lines may be between 10^{-3} and 10^{-6} of the intensity of the Rayleigh scattered light. Similarly, since the polarizability of a molecule may vary as it rotates, the rotational frequency v_{rot} appears in the Raman spectrum. In fact, twice the value of v_{rot} is observed, since the polarizability is the same for opposite orientations of the field relative to the molecule.

The specific selection rule for vibrational Raman spectra of diatomic molecules is the same as for the normal IR spectrum: $\Delta v = \pm 1$. Thus, both Stokes and anti-Stokes lines may be observed in a typical Raman spectrum, but the latter depends on the population of the $v = 1$ level, which may be relatively low. In addition to the selection rule for the vibrational quantum number it may be shown that a vibrational Raman transition can only occur if the polarizability changes as the molecule vibrates. This implies that both homonuclear and heteronuclear molecules, which swell and contract during a vibration, are Raman active. Superimposed on the $\Delta v = \pm 1$ transitions, one may observe a characteristic band structure arising from simultaneous rotational transitions for which the selection rule is: $\Delta J = 0, \pm 2$, as in pure rotational Raman scattering.

For polyatomic molecules normal vibrational modes are Raman active only if they are accompanied by a change in polarizability. For the CO_2 molecule this implies that the symmetric stretch, which was IR inactive, is Raman active. In general, the Raman activity of a particular normal vibrational mode is derived from group theory. One important rule is that for molecules with a centre of symmetry no vibrational modes exist that are both Raman and IR active.

For rotational Raman transitions the overall selection rule is that the molecule must have an anisotropic polarizability. All linear molecules, homonuclear or heteronuclear, show rotational Raman spectra. On the other hand, molecules such as CH_4 and SF_6, being spherical rotors, are rotationally Raman inactive. In addition, we have the Stokes rotational Raman lines for $\Delta J = +2$ and the anti-Stokes rotational Raman lines for $\Delta J = -2$. (This selection rule corresponds with the classical picture for rotational Raman scattering sketched above.) Note that at room temperature many rotational levels are occupied and in general both branches will be observed.

7.4.2 RAYLEIGH SCATTERING

Rayleigh scattering dominates elastic scattering of light for molecules whose size is much smaller than the wavelength of the scattered light. The intensity of Rayleigh scattering by a molecule with isotropic polarizability is given by

$$I = \frac{16\pi^4 c}{3\lambda^4} \alpha^2 E_0^2 \qquad (7.40)$$

where E_0^2 is the amplitude of the electric field vector and $\alpha = e^2/\hbar c$. Note the strong increase of the amount of scattered light with diminishing wave-length λ.

7.4.3 MIE SCATTERING

For molecules with a size comparable to the wavelength of the scattered light, interference effects between light scattered from different parts of the same particle complicate the description of the process. For Mie scattering the intensity of the scattered light increases towards shorter wavelengths with an approximate λ^{-2} dependence. In the atmosphere Mie scattering from particles is normally more prominent than Rayleigh scattering and determines the long distance visibility. The marked wavelength dependence of Mie and Rayleigh scattering is responsible for the blue colour of a clear sky and the red sunsets.

7.5 SPECTROSCOPY OF THE INNER ELECTRONS OF ATOMS AND MOLECULES

The inner electrons of most atoms and molecules are much stronger bound than the outer electrons and as a consequence transitions involving the levels that these inner electrons occupy require photons or particles with higher energy. In all cases, far-UV photons or X-rays are either emitted or absorbed. We shall briefly discuss the spectroscopy based on the emission or absorption of X-rays, photon electron spectroscopy and the Auger effect.

7.5.1 X-RAY EMISSION SPECTROSCOPY

Bombarding a material with electrons having an energy in the keV range gives rise to the emission of an X-ray spectrum ($\lambda \approx 0.01$–1 nm). The spectrum shows a quasi-continuous part, due to Bremsstrahlung of the decelerated electrons, and a set of lines characteristic for the atoms in the material. A typical example of an X-ray spectrum is shown in Fig. 7.7.

The Bremsstrahlung spectrum depends on the energy of the bombarding particles and arises from the deflection and slowing down of a charged particle in the field of a nucleus with charge Z. In principle the interaction of the electron with the nucleus may be viewed as a series of braking events and at each braking event a photon with energy corresponding to the difference in energy of the electron before and after the breaking event is emitted (see Fig. 7.8).

Since the initial and final states are not quantized, a continuum is emitted of which the intensity distribution is given by

$$I(v) \sim Z(v_{max} - v) \qquad (v < v_{max}) \tag{7.41}$$

Here Z is the atomic number and $v_{max} = E_0/h$ is the maximum frequency that occurs in the Bremsstrahlung spectrum.

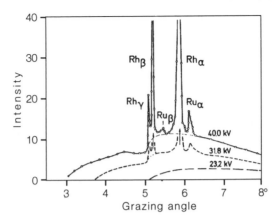

Figure 7.7 X-ray emission spectrum from a mixture of Rh and Ru, obtained for several acceleration voltages as indicated. The spectra show the continuous Bremss-trahlung radiation as well as the lines characteristic for the specific elements. (Reproduced by permission of Springer-Verlag from A. Haken and H. C. Wolf, *Atomic and Quantum Physics*, 2nd edn, Springer-Verlag, Berlin, 1987, Fig. 18.3, p. 300)

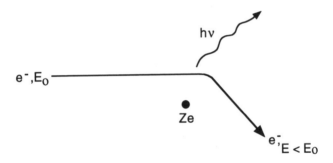

Figure 7.8 Origin of X-ray Bremsstrahlung

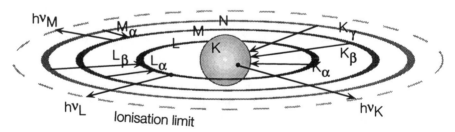

Figure 7.9 Schematic representation of the origin of the K, L and M series in a typical X-ray spectrum. After the ejection of an electron via excitation of the atom by the high-energy particle beam, the hole may be filled by an electron from a shell farther out. The corresponding energy difference is emitted as an X-ray photon

Superimposed on the quasi-continuous Bremsstrahlung spectrum a set of lines is observed, characteristic for the elements in the target. These lines arise from transitions between discrete atomic states. The X-ray emission spectra consist of a relatively limited number of lines which can be grouped in a few series of which the frequencies bear a clear relation to the nuclear charge Z. The explanation for these line spectra is schematically indicated in Fig. 7.9.

The kinetic energy of the incoming electron may be sufficient to excite inner electrons of the atoms in the material to high excited states that lie above the ionization limit. The vacancy which is produced is rapidly filled by an electron from one of the outer electron shells and the corresponding difference in energy between the states involved is emitted as one or more photons.

Figure 7.9 shows that if the vacancy was produced in the K shell, a series of lines is observed, denoted as K_α, K_β, K_γ, etc., corresponding to the transition of an electron from the L, M, N, etc., shell respectively, and the emission of the corresponding photon. Similarly, a set L_α, L_β, L_γ is produced upon removal of an electron from the L shell and a set M_α, M_β, M_γ upon removal of an electron from the M shell.

The position of the K_α line as a function of the atomic number Z is, to a good approximation, given by the expression

$$\gamma_{K_\alpha} = \tfrac{3}{4} R (Z-1)^2 \qquad \mathrm{cm}^{-1} \tag{7.42}$$

and the position of the L_α line by

$$\gamma_{L_\alpha} = \tfrac{5}{36} R (Z-7.4)^2 \qquad \mathrm{cm}^{-1} \tag{7.43}$$

where R is the Rydberg constant given in Appendix C. Note that for the K_α line the charge of the nucleus is screened by one electron (the remaining electron in the K shell) while for the L_α line the nuclear charge is effectively screened by more than 7 units.

Upon close inspection one may observe that the X-ray spectra show fine structure. These detailed properties of X-ray spectra can be understood once it is realized that a missing electron in an otherwise filled shell is equivalent to a single electron in an otherwise empty shell. Consequently, X-ray emission spectra can be understood on a basis similar to that explained in Section 7.2.1 for the spectra of alkali atoms and they are characterized by one-electron quantum numbers. This leads us directly to the term diagram 7.10. In these energy diagrams the l degeneracy is removed due to the different degree of screening of the nucleus for electrons with different l quantum number. In addition, the j degeneracy has been removed as a consequence of spin–orbit coupling.

Figure 7.10 shows that the transitions between the levels obey the same selection rules as those of the optical transition of the one-electron atoms:

$$\Delta l = \pm 1; \qquad \Delta j = 0, \pm 1 \tag{7.44}$$

For instance, for the transitions to the K shell only the transitions corresponding

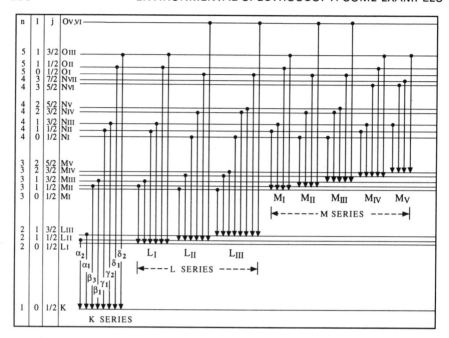

Figure 7.10 Fine structure diagram for the X-ray spectrum of a platinum anode with $Z = 78$. The notations for the series, the lines and the corresponding quantum numbers are indicated. Arrows pointing upwards correspond to absorption, those pointing downward refer to emission. The spacings between the L subshells L_I, L_{II} and L_{III} and the M subshells M_I–M_V are not shown to scale. For a given l value they arise from the normal doublet splitting; they are produced by different screenings of the nuclear charge and are therefore not all equal. (Reproduced by permission from V. Valcovic, *Trace Element Analysis*, Taylor and Francis, 1987)

to $^2P_{1/2} \rightarrow {}^2S_{1/2}$ and $^2P_{3/2} \rightarrow {}^2S_{1/2}$ are observed. Selection rules (7.44) allow us to understand the fine structure of the X-ray emission spectra of all atoms. The major sets of lines corresponding to transitions between shells with different major quantum number n are now split in subshells, characterized by the set of quantum numbers n, l, j. For instance, for the L set we obtain L_I, L_{II} and L_{III}.

We finally note that X-ray emission can also be induced by X-ray absorption and, in analogy with optical experiments, the emission is called X-ray fluorescence. X-ray fluorescence spectroscopy is a sensitive technique, which allows the detection of elements in a sample to an accuracy of about 10 ppm. Alternatively, accelerated heavy particles such as protons can be used to excite an X-ray emission spectrum. For more details concerning this PIXE particle-induced X-ray emission method we refer to Section 7.6.4.

7.5.2 X-RAY ABSORPTION SPECTRA

A typical X-ray absorption spectrum is shown in Fig. 7.11. The X-ray absorption spectrum consists of sets of steep increases or 'absorption edges' superimposed on a continuously decreasing value of the absorption coefficient with increasing frequency. The fall-off at higher frequencies arises from the v^{-3} dependence of the absorption coefficient. The frequency of each absorption edge corresponds to the series limit for the K, L_I, L_{II}, L_{III}, etc., series. For instance, the position of the absorption edge for the K series of lead is at 88 keV. Since the ionization energy for a K electron of lead is given by $Z_{eff}^2(13.6)$ eV, it directly follows that $Z_{eff}^2 = 80.4$. Thus, the screening of the $Z = 82$ nucleus for the inner electrons is about 1.6 units of charge.

In the X-ray absorption spectrum only the ionization transitions are observed (all shells are occupied) and the absorption spectra thus show one K edge, three L edges, five M edges, etc. Following the steep absorption edge, characteristic for a specific element, a fine structure is observed on the X-ray absorption spectrum, which reflects the chemical surrounding of the element. The fine structure arises from X-rays scattered off certain atoms near to the element under

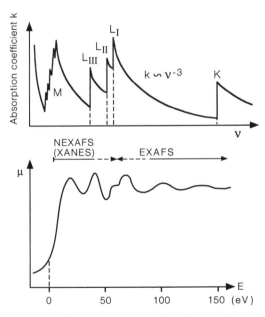

Figure 7.11   Top: schematic representation of an X-ray absorption spectrum with absorption edges. Bottom: fine structure at a K absorption edge at high resolution (EXAFS spectrum). (Reproduced by permission of Springer-Verlag from S. Svanberg, *Atomic and Molecular Spectroscopy*, 2nd edn, Springer-Verlag, Berlin, 1992, Fig. 5.9, p. 74)

consideration. Experiments in which these edge structures are accurately measured have become possible using synchrotron radiation. These experiments are generally referred to as EXAFS (extended X-ray absorption fine structure) spectroscopy.

7.5.3 THE AUGER EFFECT

For relatively light nuclei, the energy which is liberated upon the transition of an electron from an outer shell to a vacancy in one of the inner shells can be used to emit an electron from one of the outer shells. This effect is called the Auger effect and is illustrated in Fig. 7.12.

The Auger electron can be detected and its kinetic energy can be measured. The kinetic energy is a direct measure for the energies of the states involved and for the case illustrated in Fig. 7.12 is given by

$$E_{kin} = E_{K_\alpha} - E_B \tag{7.45}$$

where E_{K_α} is the energy of the quantum that normally would have been emitted and E_B is the binding energy of the Auger electron.

7.5.4 ∧ X-ray PHOTOELECTRON SPECTROSCOPY (XPS)

In case photons of sufficient high energy are absorbed by atoms of molecules, the excited electrons may be emitted. In photoelectron spectroscopy (XPS), the kinetic energy of the emitted electrons is accurately measured and used as a precise measure of the energy levels of the atom or molecule. The process is schematically illustrated in Fig. 7.13.

For atoms the binding energy E_B of a particular electronic level is measured by

$$E_{kin} = h\nu - E_B \tag{7.46}$$

in which E_{kin} is the measured kinetic energy and $h\nu$ the energy of the incoming X-ray or UV photon. Molecules may end up in a higher vibrational state after the emission of the electron and then the binding energy is calculated from

$$E_{kin} = h\nu - E_B - E_{vib} \tag{7.47}$$

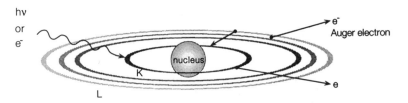

Figure 7.12 Illustration of the Auger effect. Auger electron emission completes
with X-ray emission

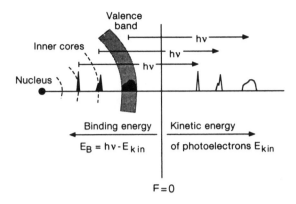

Figure 7.13 Illustration of photoelectron spectroscopy. The kinetic energy of the photo electrons is the difference between the quantum energy $h\nu$ of the exciting photons and the binding energy of the electrons in the atom or solid. The dashed lines represent the orbitals of the electrons. Note that here the binding energies are indicated and not the distances from the nucleus. (Reproduced by permission of Springer-Verlag from A. Haken and H. C. Wolf, *Atomic and Quantum Physics*, 2nd edn, Springer-Verlag, Berlin, 1987, Fig. 18.14, p. 310)

where E_{vib} is the amount of energy used to excite the vibrations in the ion. XPS-spectra of molecules typically show long progressions in particular vibrations since the equilibrium nuclear geometry of the neutral and the ionized molecule may be very different.

XPS also measures the energies of the core electrons of atoms and molecules and this is the basis for a chemical analysis technique called ESCA (electron spectroscopy for chemical analysis). The technique mainly measures the surface properties of the material, not because the X-rays do not penetrate, but mainly because the liberated electrons cannot escape. Fine structure and line shifts in ESCA spectra allow the sensitive detection of molecular structures.

7.6 SPECTROSCOPIC TECHNIQUES FOR ANALYSIS OF THE ENVIRONMENT

In this section we will discuss a few specific techniques that are currently employed to analyse the environment. Of course, this section is not complete; we have selected examples, that illustrate the principles outlined above and play a major role in environmental analysis. We will discuss two examples of remote sensing. In Section 7.6.1 satellite monitoring of the upper atmosphere is described, followed by a discussion of the LIDAR technique in Section 7.6.2.*

---

*Sections 7.6.1 and 7.6.2 are based on a draft by thesis student Matthieu Visser.

In Section 7.6.3 high-resolution spectroscopy is discussed as a tool to detect organic compounds with ultra-high sensitivity, while finally in Section 7.6.4 PIXE is described*.

7.6.1 A GLOBAL APPROACH TO MONITORING POLLUTION OF THE UPPER ATMOSPHERE: A BRIEF DESCRIPTION OF NASA's UARS SATELLITE

The destruction of the ozone filter was discussed in Section 2.3.3. It is therefore important to measure concentrations of ozone and of the gases destroying it. This is the aim of NASA's Upper Atmosphere Research Satellite, abbreviated UARS. This satellite was launched by the space shuttle *Discovery* on 12 September 1991. The satellite is carrying nine different instruments, to obtain the temperature in the atmosphere and the densities of sixteen key components of atmospheric chemistry. These parameters are measured daily for altitudes from 10 to 130 km altitude and for a large part of the global surface. Horizontal winds at altitudes from 10 to 45 km and from 55 to 110 km altitude can be measured as well. This is very useful for the interpretation of changes in atmospheric composition. This section serves as an introduction to the almost incredible possibilities that modern satellite observation offers atmospheric research.[†] First we will briefly describe satellite orbits, then we will summarize the various instruments that are carried and finally we will describe the information that the satellite provides as a whole. We will also mention the organization and accessibility of the measurements. Finally one of the instruments is discussed in more detail, following a simple measurement scheme which nicely illustrates the applicability of spectroscopy to atmospheric pollution detection.

UARS Mission

To obtain global coverage and rapid repetition of the measurements, the satellite's orbit needs to be carefully chosen and a low altitude of 585 km was selected. This results in a period of 97.0 minutes, at a speed of 7.6 km s^{-1}. The orbit is inclined 57° to the equator. Detectors mounted perpendicular to the spacecraft's velocity vector can thus see to 80° at one latitude and 34° at the other. The orbital precession rate is such that, after 36 days, the same part of the atmosphere is monitored at the same local time, which is believed to be fast enough to allow resolution of diurnal atmospheric effects in a time that is reasonably short compared to seasonal effects. One should realize, however, that a polar orbit

---

[*] For detailed descriptions of other techniques we refer to the books of Johansson and Campbell [3], Measures [4], Scharda [5] and Schulman [6].
[†] *The Journal of Geophysical Research*, 20 June 1993, contained a separate section on this mission. Objectives, organization and technical realization were described, as well as the first results.

would be more favourable: the whole earth can be viewed as it passes by. Satellites launched by space shuttle from Florida cannot reach polar orbits, which is the more practical reason for the UARS orbit. Besides the limited view, the UARS orbit has another drawback: the satellite needs to be turned 180° every 36 days, to meet thermal and viewing requirements of the instruments, as well as to keep the solar arrays exposed to the sun, thereby providing the power for the instruments to run. This has become a problem in the operation of this first satellite: turning was not perfect and due to misalignment of the solar arrays, insufficient power was generated to run all instruments after some months of flawless operation. Future satellites are planned to fly in polar orbits, and will not need reorientation in orbit. Before the mission started, baseline requirements for all measurements were defined: a maximal spatial resolution of 3 km in vertical direction and 5° in latitude, corresponding to about 500 km at the equator. Given the spacecraft's velocity of $7.6 \, \mathrm{km \, s^{-1}}$, this meant that in one minute or less the detectors should be able to make vertical density profile measurements of all the atmospheric constituents and the other measured parameters. Figure 7.14 shows the atmospheric constituents measured by the various instruments. As can be seen, most of the components can be monitored

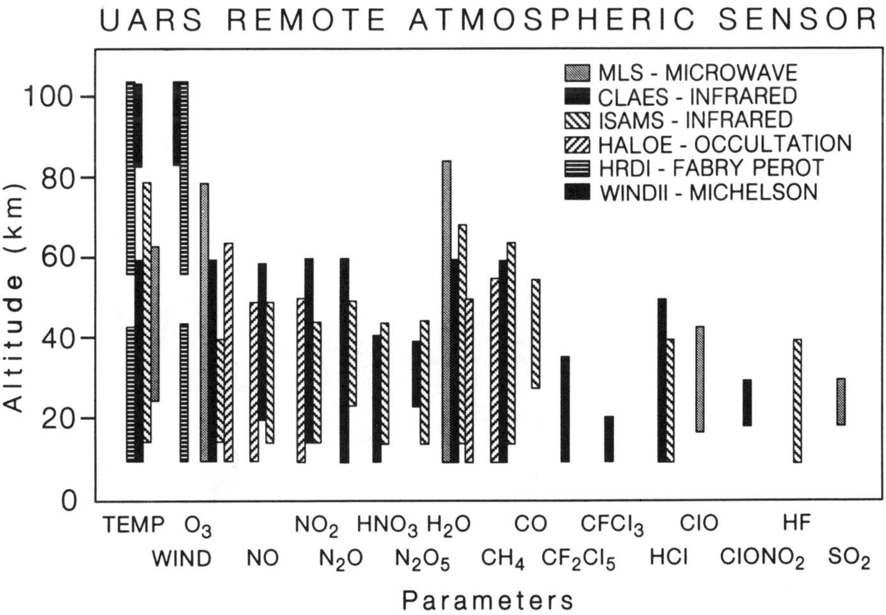

Figure 7.14 UARS measurements on wind, temperature and atmospheric composition. Shown is the altitude range that the instruments reach for the measured parameters. (From C. A. Reber *et al.*, Upper atmospheric research satellite, *Journal of Geophysical Research*, **98** (D6) (1993) 10643–7, Fig. 1, p. 10644)

by a variety of instruments. This facilitates data validation through intercomparison.

As a first approximation a spatial resolution of 500 km in the horizontal direction and 3 km in the vertical direction is sufficient to obtain a reliable view of the stratospheric chemical changes, with the given time resolution, for, as was noted in Table 5.4, mixing in the stratosphere is slow. In modern numerical models of the stratosphere (cf. Section 3.3), grid sizes of 100–500 km in the horizontal direction, 0.1–3.0 km in the vertical direction and time intervals of 1–24 h are used, close to UARS resolution. In these models one accounts for ≈ 50 molecular species, giving rise to ≈ 150 reactions.

Figure 7.15 shows the instruments on board. The three instruments listed under SSPP detect solar radiation input in the atmosphere. The Particle Environment Monitor (PEM/ZEPS) detects X-rays, protons and electrons. Four instruments measure atmospheric trace gases and the atmospheric temperature distribution, mainly by the detection of emitted or absorbed electromagnetic radiation. For example, CLAES uses IR emission from molecules, whereas HALOE uses the absorption of sunlight by molecules between the detector and the sun. Notice that CLAES is the instrument that is able to measure the highest number of different molecular species. The lifetime of this instrument, however, is limited to approximately 20 months, because of the limited amount of cryogens

UARS OBSERVATORY

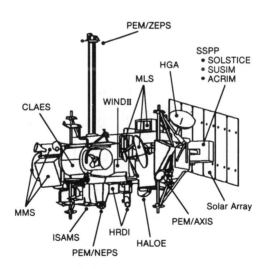

Figure 7.15 Instruments on board UARS. (From C. A. Reber *et al.*, Upper atmospheric research satellite, *Journal of Geophysical Research*, **98** (D6) (1993) 10643–7, Fig. 2, p. 10645)

(solid Ne and CO_2) to cool the detector. HRDI and WINDII are used to detect winds, important to understand detected transport phenomena.

The data obtained by the UARS observatory are telemetered down to the Goddard Space Flight Centre (Maryland, USA), where some rough data cleaning is done. The resulting 'level-0 data' are then distributed by a dedicated electronic communication system among remote analysis computers, for further data processing. Then the desired geophysical parameters can be read out and analysis can be done both at the investigators' sites and at the GSFC where the processed data are collected and stored. The whole process, from obtaining the data to processing and extracting interpretable geophysical parameters, takes some 3 days. Notice that elaborate data processing is necessary: deconvolution of instrument responses and correction of the distortion from geometrical projection. All the software necessary to do this was developed and tested extensively before the launch of the satellite.

The geophysical data obtained should be interpreted in a coherent way; i.e. results from various instruments on board UARS and from other sources have to be combined. The results then need to be interpreted within the framework of a model for the atmosphere's chemistry, using the obtained parameters (solar energy and particle input, winds at several altitudes and atmospheric composition). Therefore ten 'PIs' (principal investigators), leading scientific groups specialized in radiative transfer, atmospheric dynamics and photochemistry, were associated with the mission.

The Halogen Occultation Experiment: A Description of One of the UARS Instruments and Its Measurements

We will now discuss one of the instruments on UARS performing the halogen occultation experiment, abbreviated HALOE. It is based on the IR absorption into the vibrational levels of O_3, HCl, HF, CH_4, H_2O, NO, NO_2 and CO_2 by molecules located between the UARS detector and the sun. These gases are key components in the chemistry of the middle atmospheres, as illustrated in Fig. 7.16 and mentioned in Section 2.3.3. As can be seen, many components play a role and HALOE detects some of its key components.

Figure 7.17 shows the experiment's geometry. This measurement can therefore only be made at sunrise and sunset. Because of the UARS orbit, this means that fifteen sunrise and fifteen sunset measurements can be made every day, so that it takes HALOE some 25 days to collect data for a global distribution of a particular species. This is a drawback of this method with respect to the other UARS instruments. However, advantages of (horizontal) limb viewing are a higher sensitivity as 30–60 times more absorber is observed than in vertical viewing and also a high vertical resolution because of the geometry and the exponential decrease of density with altitude. Since the solar occultation method is relative (each mixing ratio is determined by comparing the solar intensity

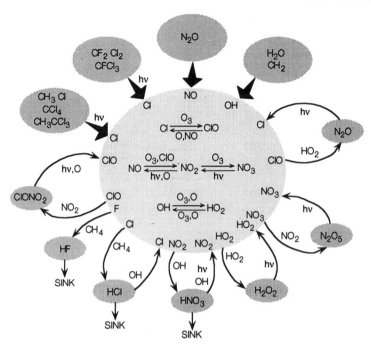

Figure 7.16 A simplified scheme of the middle atmosphere's chemistry. (From J. M. Russell III *et al.*, The halogen occultation experiment, *Journal of Geophysical Research*, **98** (D6), (1993) 10777–97, Fig. 1, p. 10778)

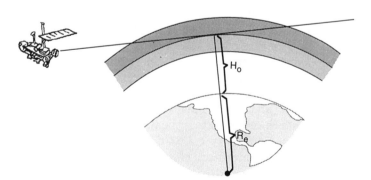

Figure 7.17 Limb-viewing solar occultation geometry. HALOE uses light from the sun, but some instruments on other satellites can use the light from bright stars as well. (From J. M. Russell III *et al.*, The halogen occultation experiment, *Journal of Geophysical Research*, **98** (D6), (1993) 10777–97, Fig. 2, p. 10778)

reduced by atmospheric absorption with the unattenuated solar intensity outside the atmosphere), it is essentially self-calibrating. This is very useful if long-term trends are to be observed. The method requires a good set of laboratory spectra of all the gases that one would like to observe. Both NASA and ESA therefore use extensive spectroscopic databases, which are accessible for the remote sensing community.

To detect trace gases like HCl or HF in limb viewing is not easy, since the IR absorption of the molecules is weak compared to the absorption of the much more abundant CH_4 (≈ 1000 times higher concentration) that is absorbing in the same spectral region. The case for a 30 km altitude HCl channel (at 3.401 μm) is shown in Fig. 7.18.

At first glance it seems impossible to detect HCl at all. However, using a simple and elegant trick yields great results. Besides a broadband filter ($\approx 80\,cm^{-1}$ FWHM), to select for the proper spectral region one uses a second filter, which consists of a gas cell containing the gas (in this example HCl) to be detected. This concept is illustrated in Fig. 7.19.

The incoming light is split into two parts by the beamsplitter (BS). The light that follows the lower path is sent through vacuum to detector D2. The light following the upper path is sent through a gas cell containing HCl to detector D1. The signal of D1 can be adjusted by means of an adjustable amplifier, G. This signal and the signal from D2 are fed to the positive and negative inputs of a differential amplifier. The gain V is adjusted so that when the detector is viewing the sun outside the atmosphere, the output of the differential amplifier is zero.

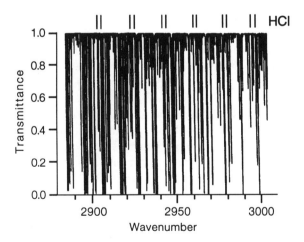

Figure 7.18 Calculated 30 km absorption spectrum in HALOE's HCl channel. The tickmarks at the top show the position of HCl lines. The other lines are mainly due to CH_4. (From J. M. Russell III *et al.*, The halogen occultation experiment, *Journal of Geophysical Research*, **98** (D6), (1993) 10777–97, Fig. 4, p. 10780)

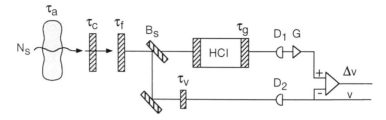

Figure 7.19 HALOE's gas filter radiometer concept. (From J. M. Russell III *et al.*, The halogen occultation experiment, *Journal of Geophysical Research*, **98** (D6), (1993) 10777–97, Fig. 5, p. 10780)

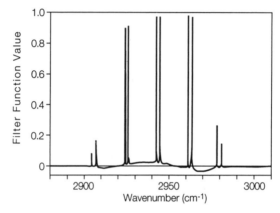

Figure 7.20 HALOE's HCl narrow-band spectral function to selectively detect HCl. (From J. M. Russell III *et al.*, The halogen occultation experiment, *Journal of Geophysical Research*, **98** (D6), (1993) 10777–97, Fig. 6, p. 10781)

In this way one obtains a detector that is sensitive to one specific gas only. Figure 7.20 shows the HALOE narrow-band spectral function.

From Fig. 7.20 it is clear how one manages to detect HCl even in the presence of much higher CH_4 concentrations present. For proper functioning of the apparatus, it is not important to know the exact concentration of the gas in the cell (of course the order of magnitude should be properly chosen for an optimal signal-to-noise ratio), but the amount of gas in the cell, as well as its spectrum, should remain constant. Therefore, the cell needs to be well sealed, made of material that cannot be oxidized by the contained gas (gold with sapphire windows is used for the highly corrosive HCl and HF gases) and kept at constant temperature ($\pm 0.05°$). As always in optical designs, sensitivity is optimized by using big-diameter input optics and sensitive detectors. HALOE is equipped with a 16 cm diameter primary mirror with 96 cm focal length. The field of view is 2 arc minutes (the sun covers 30 arc minutes seen from the earth). For the four radiometer channels, broadband detection of the spectrally well-resolvable higher abundancy gases H_2O, O_3, CO_2 and NO_2 uncooled bolometer detectors are employed. For the HCl, HF, NO and CH_4 channels, thermoelectrically cooled indium arsenide detectors are used and photovoltaic mercury–cadmium–telluride is used for NO detection. These materials are chosen for high sensitivity at the respective detection wavelengths, indicated in the HALOE optical diagram shown in Fig. 7.21.

Before a scientific space operation of any space instrument is done, extensive ground testing is performed. HALOE ground tests were done with gas mixtures resembling the atmospheric composition at various altitudes, as measured in previous missions. Only after these experiments yielded satisfactory results, within 1–4% repeatability when viewing the same artificial atmosphere, was the instrument installed on UARS.

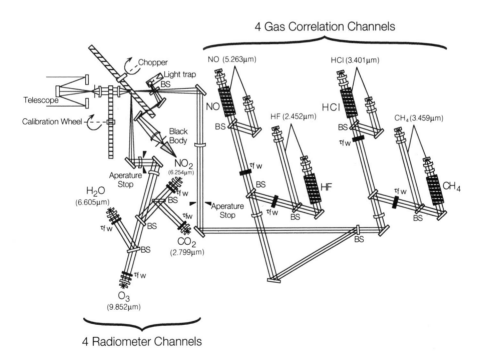

Figure 7.21 HALOE optical diagram. The different gases are detected at different wavelengths. By splitting the spectrum by means of dichroic beamsplitters, parallel detection of all eight gases is possible. (From J. M. Russell III *et al.*, The halogen occultation experiment, *Journal of Geophysical Research*, **98** (D6), (1993) 10777–97, Fig. 12, p. 10784)

We conclude with some (preliminary) examples of in-flight data obtained by the HALOE instrument. Figure 7.22 shows a comparison of data obtained by the ATMOS satellite on 5 May 1985 at 29° N (zonal mean, data over all longitudes but with the same latitude averaged) and zonal mean data obtained by HALOE on 5 May (top left figure) and 7 May (other figures) 1992. Because of the large differences in time, as well as the small difference in latitude, quantitative agreement is not expected. Still, the profiles have similar shape and values, except for HF, which has increased by $\approx 50\%$ in 7 years, in agreement with expectations, based on a 5–6% annual increase. HF is not of natural origin and its strong increase in the upper atmosphere is an indication of the presence of man-made CFCs at this high altitude. The HCl concentration, of vital interest because of its role in catalytic O_3 destruction, has not increased as much as expected.

To conclude, the UARS data when validated with other measurements will form a reliable and significant database for scientific discussions on the ozone layer and the greenhouse effect for the years to come. It will not be the only source though: the ENVISAT, planned for launch to a polar orbit by ESA in 1998, will

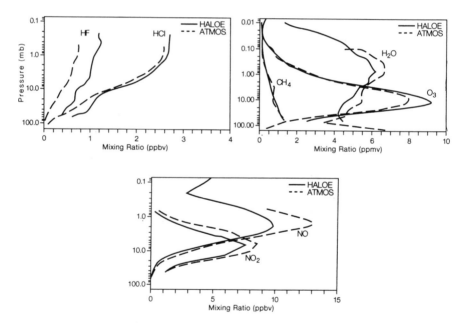

Figure 7.22 HALOE HCl, HF, CH_4, O_3, H_2O, NO_2 and NO zonal mean mixing ratios obtained on 5 May (HCl and HF) and 7 May (the other gases) 1992 at 34° N compared to these same mixing ratios obtained on 5 May 1985 at 29° N by the ATMOS satellite. Note the different axes in the different figures! (From J. M. Russell III *et al.*, The halogen occultation experiment, *Journal of Geophysical Research*, **98** (D6), (1993) 10777–97, Fig. 20, p. 10791)

carry four instruments to detect global changes in the atmosphere. NASA plans to launch its NASA polar orbital platform in the year 2000. These satellites will apply basic spectroscopic principles in combination with extensive data processing at the ground, just like that sketched for the UARS mission. All this information will be important for the successful and realistic modelling of atmospheric chemistry and may contribute to the formulation of a global environmental protection policy.

7.6.2 UV LIDAR: A TECHNIQUE TO MEASURE THE STRATOSPHERIC OZONE DISTRIBUTION AT A FIXED POSITION

In this section we will discuss a specific example of active remote sensing: the UV-LIDAR. In the specific case discussed here the UV-LIDAR is used to measure the ozone distribution profile in the stratosphere (from 20 to 50 km altitude). Similar methods are used to obtain density profiles of N_2, O_2 and H_2O, or to characterize atmosphere temperature as a function of altitude. Like the example discussed in Section 7.6.1, the UV-LIDAR technique is complicated and expensive and therefore one has to examine critically in what cases this method is superior to other methods, like using balloon or rocket-borne sensors. Intercomparison between different LIDAR systems and other methods is needed to estimate the accuracy of the measurements. Since local ozone concentrations are of rather limited interest, one has to integrate ground-based measurements in a network, so that calibration, information access and exchange is facilitated. The basis for such a network was created during a meeting in Boulder, Colorado, in 1986, held shortly after the first publication on the Antarctic 'ozone hole' in 1985 [7]. We discuss the physical and technical aspects of the LIDAR system that is used at the Jet Propulsion Laboratory, part of the Californian Institute of Technology.* Then we will show some results from an intercomparison campaign, held at the same location. In this campaign, independent ('blind') ozone measurements from different methods were critically compared. Finally, we will briefly discuss the operation of the Network for Detection of Stratospheric Change (NDSC), which organized this campaign. This institution will be involved in database and information exchange activities and operate as a forum for all activities in this area in the years to come.

LIDAR: Physical and Technical Aspects

LIDAR methods are used since the 1960s [4]. The most-often used method is called DIAL: Differential Absorption LIDAR, which is used since 1966 [8]. This method employs pulsed lasers, aimed vertically upwards. The light pulses, when

---

*The work was done under a contract with NASA.

travelling upwards, are gradually absorbed or scattered. The amount of absorption and scattering depends on the number and type of molecules that are encountered and their absorption and scattering cross-sections at the wavelength of the laser. The number of backscattered photons is detected as a function of the delay time between the firing of the laser pulse and the arrival time. To select for specific molecules in the atmosphere, two wavelengths are applied in the DIAL method: the first wavelength is absorbed and scattered by the molecule that one wants to detect, whereas the second wavelength is only scattered. These wavelengths are called 'on' and 'off' wavelengths respectively, as shown in Fig. 7.23.

We will now quantify these considerations [9]. Suppose $S(t)$ is the signal that monitors the amount of backscattered light at time t after the emission of the laser pulse at $t = 0$. A segment $S(t, \Delta t)$ of the signal is recorded during a time interval between t and $t + \Delta t$ and corresponds to the scattering by air molecules

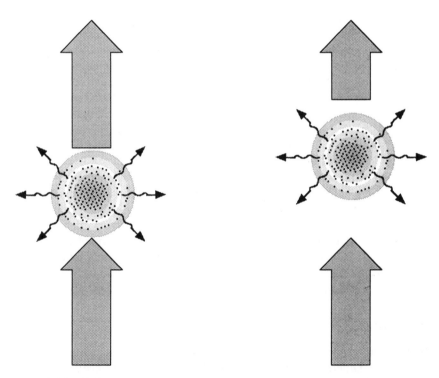

Figure 7.23 A simple representation of the optical DIAL process. The left graph shows the 'off wavelength': only scattering, no absorption. The graph on the right shows the 'on wavelength': the target molecules absorb and scatter radiation of this wavelength

in a volume extending from R to $R + \Delta R$ in the direction of the beam, with

$$R = c\frac{t}{2}, \qquad \Delta R = c\Delta\frac{t}{2} \tag{7.48}$$

Here c is the speed of light. It is assumed here that the laser pulse is short compared to the sampling interval, $\tau \ll \Delta t$. In almost all practical cases this is realized; one uses pulsed lasers with $\tau \approx 10\,\text{ns}$ pulse duration, corresponding to only 3 m extension in space. Normally, signals from a depth of at least 150 m (1 μs) are averaged in a channel.

When a laser pulse of energy E_λ and wavelength λ is emitted into the atmosphere, the backscattered light is collected by a receiver mirror of area A, and sampled by a photodetector of efficiency η_λ. The signal $S_\lambda(R, \Delta R)$ at wavelength λ and originating from molecules at an altitude between R and $R + \Delta R$ can be described by

$$S_\lambda(R, \Delta R) = E_\lambda \xi(R)\frac{A}{4\pi R^2}\eta_\lambda \beta_\lambda(R)\Delta R \exp\left\{ -2\int_0^R [\alpha_\lambda(r) + \sigma_\lambda N_{\text{abs}}(r)]\,dr \right\} \tag{7.49}$$

Here, $\xi(R)$ is a function describing the overlap of the emitted beam and the detection cone and B_λ is the backscattering coefficient at distance R. The attenuation integral, simply arising from Lambert–Beer's law (2.26), is divided in a part arising from the molecular species of interest $(\sigma_\lambda N_{\text{abs}}(r))$ and a part arising from all other particles $(\alpha_\lambda(r))$. Of course $N_{\text{abs}}(r)\,dr$ is the number of absorbing particles between r and $r + dr$. Equation (7.49) is called the 'LIDAR equation'.

To extract information on the distribution of the molecular species of interest, one uses two wavelengths, as mentioned before. From eq. (7.49) it can be seen why this is necessary: the extinction by the other particles $\alpha_\lambda(r)$ cannot be distinguished from the extinction $\sigma_\lambda N_{\text{abs}}(r)$ due to the molecules to be detected, the target molecules. Now if λ_1 is the 'on' wavelength (absorbed and scattered by the target molecules) and λ_2 is the 'off' wavelength (only scattered by the target molecules) then we can take the ratio $Q(R)$ of the two signals:

$$Q(R) = \frac{S_{\lambda 1}}{S_{\lambda 2}}(R) = \frac{E_{\lambda 1}}{E_{\lambda 2}}\frac{\eta_{\lambda 1}}{\eta_{\lambda 2}}\frac{\beta_{\lambda 1}(R)}{\beta_{\lambda 2}(R)}\exp\left\{ -2\int_0^R [\Delta_\lambda \alpha(r) + \Delta_\lambda \sigma N_{\text{abs}}(r)]\,dr \right\} \tag{7.50}$$

Note that the distance-dependent part of the detector efficiency $\xi(R)$ has dropped out and that only the wavelength-dependent part has remained. The exponent now contains two parts: the difference in extinction coefficient $\Delta_\lambda \alpha(r)$ and the difference in absorption cross-section $\Delta_\lambda \sigma N_{\text{abs}}(r)$. Ideally, the first part is much smaller than the second part. This is the case when λ_1 and λ_2 are close together, but λ_1 is in resonance with a transition in the target molecules whereas λ_2 is not.

To obtain accurate results, $\Delta_\lambda\sigma$ must be measured in the laboratory under temperature and pressure conditions equivalent to those in the atmosphere, and with identical laser line widths.

To obtain the density of the target molecules as a function of distance, one takes the logarithm of eq. (7.50) and differentiates with respect to R:

$$\frac{\mathrm{d}}{\mathrm{d}R}\ln Q(R) = \frac{\mathrm{d}}{\mathrm{d}R}\ln\left[\frac{\beta_{\lambda 1}(R)}{\beta_{\lambda 2}(R)}\right] - 2[\Delta_\lambda\alpha(R) + \Delta_\lambda\sigma N_{\mathrm{abs}}(R)] \qquad (7.51)$$

Note that in eq. (7.51) the pulse energies and spectral efficiencies no longer occur. The first term on the right side is usually zero (it can be non-zero when the wavelength-dependent backscattering varies with distance as a consequence of variation of droplet size of aerosols). Thus one obtains:

$$N_{\mathrm{abs}}(R) = \frac{-1}{2\Delta_\lambda\sigma}\frac{\mathrm{d}}{\mathrm{d}R}\ln Q(R) - \frac{\Delta_\lambda\alpha(R)}{\Delta_\lambda\sigma} \qquad (7.52)$$

As mentioned before, when λ_1 and λ_2 are sufficiently close together, the second term on the right side can be neglected. In the case of ozone measurements, the on and off wavelengths will be rather far apart, since the absorption band of ozone is rather wide, as is shown in Fig. 7.24. In the case of stratospheric ozone

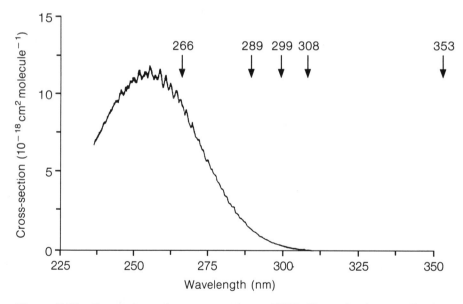

Figure 7.24 Ozone absorption cross-section at 195 K. Absorption is most effective at 255 nm. (Spectrum determined as described in Yoshino *et al.*, *Planet. Space Sci.*, **36** (1988) 395–8. Reproduced by permission of Dr. Yoshino)

measurements, however, this term is easily dealt with since it arises from Rayleigh scattering.

In conclusion, the DIAL technique derives its popularity from the fact that it is self-calibrating, for the detector characteristics do not influence the measured absolute concentration of target molecules. As long as the few approximations made are valid and the difference cross-section $\Delta_\lambda \sigma$ is known, measurements and their interpretation are relatively straightforward.

LIDAR in Practice

In the following section we will briefly discuss the LIDAR set-up that is used at the Jet Propulsion Laboratory (JPL) Table Mountain Facility, California, USA [10–12]. As follows from physical intuition, the amount of detected backscattered photons is very low (the reason simply being that the distance R is very large compared to the receiving mirror diameter, so that the probability of scattering photons into the receiver is very low). To optimize the system, the various parts have to obey the following specifications:

Laser:
(a) low beam divergence to have a good overlap between laser path and detection telescope;
(b) high pulse intensities at on and off wavelengths;
(c) short (ns) pulse durations to obtain sufficient range resolution;
(d) high repetition rates to obtain sufficient data to achieve acceptable statistics;
(e) narrow laser wavelength bandwidth (in this way the signal is restricted to a narrow wavelength region and narrow bandpass filters in the detection system can be used to reduce incoming scattered photons to an acceptably low level).

Detector:
(a) large area telescope to receive as much light as possible;
(b) stable positioning of detector with respect to laser to maintain constant overlap;
(c) high reflection mirror materials, low-loss beamsplitters and filters to have high detector efficiency;
(d) high photomultiplier efficiency;

For the JPL system, the beam divergence is $< 150 \,\mu$rad and the maximum total pulse energy is 500 mJ at 308 nm and 50–100 mJ at 353 nm. This off-wavelength is generated by focusing the 150–200 mJ 308 nm pulses from the excimer laser in a so-called Raman-shifting cell: a cell containing molecular hydrogen, H_2, at 35 bar pressure. Because of the high pressure and pulse intensities, the Raman scattering process reaches an efficiency of 50%. The pulse duration is 35 ns (≈ 5 m range) and the repetition rate is 150 Hz. The detector reaches an overall efficiency of 4% at 353 nm and 5% at 308 nm. The telescope has an aperture of 90 cm. To prevent saturation of the photomultipliers by the intense scattering from lower altitudes, the detector opening is closed for the first 100 μs after firing of the laser pulse, so that no scattering from the lower 15 km is detected. Bandwidth filters with an FWHM of 2 nm are placed in front of the photomultipliers to prevent too much scattered light from entering. The signal detected by the photomultipliers is divided into 4 μs channels, making the effective range resolution 600 m. Even though much scattered light is rejected, this system can only run at nighttime, because during the day too much scattered light from the sun enters the detector.

Typically, 1500 laser pulses are acquired (10 s at the 150 Hz repetition rate) before the data are transferred by IEEE connection to a VAX computer. The fact that all parameters mentioned are really critical to obtain a sufficient signal-to-noise ratio is illustrated by the fact that even with 10^{19} photons in the 500 mJ 'on' pulse at 308 nm, the average number of scattered photons from a layer of 600 m (4 μs) thick at 40 km altitude is only 2! Therefore, extensive averaging is necessary; one ozone profile takes 10^6 shots or in total two hours. This naturally implies that one can only detect concentration changes that take place over a time period of more than a few hours. Once a system like this has been installed, it can in principle be remotely operated, apart from service operations. This can be very convenient since some of the planned NDSC locations (Arctic, Antarctic) are rather inaccessible. To assess the accuracy of the LIDAR system, a comparison with several other systems has been carried out, including other LIDARs, balloon and rocket borne sensors, satellite measurements and microwave measurements. Figure 7.25 shows the result of the first 'Stratospheric Ozone Intercomparison Campaign'.* As one can see, the agreement between the various instruments is better than 5% over the range from 20 to 45 km.

Finally, we comment on the organizational, political and financial aspects of a world-wide, long-term measurement program like NDSC. First of all, organizing a long-term scientific commitment requires good management: political structure to guarantee funding, set priorities, evaluate campaigns, take care of information storage and exchange. Since NDSC's main value will derive from observing long-term trends, involving many identical (= tedious!) measurements for many years, it may be hard to keep scientists and politicians not only interested but

---

*The campaign was held during July and August 1989 and was organized by the NDSC. Data were gathered on 24 July 1989 by two LIDAR systems (JL and GL), a microwave instrument (MM), two balloon-borne electrochemical sondes (NS and WS), the SAGE II satellite (SA) and a rocket sonde (RO). A critical evaluation of the various methods can be found in 'Network for the detection of stratospheric change: a status and implementation report', issued by NASA's Upper Atmosphere Research Program and NOAA Climate and Global Change Program, edited by M. J. Kurglo, January 1990. Copies are available from: Code EEU, NASA Headquarters, 600 Independence Ave., S.W., Washington, DC 20546.

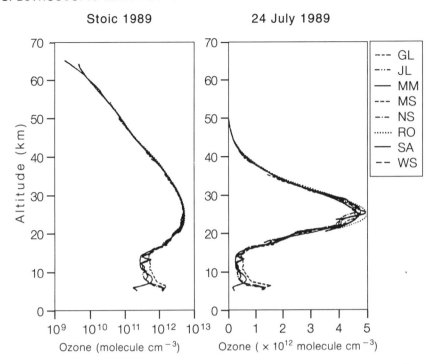

Figure 7.25   Ozone profiles as obtained during 'Stoic'. Results from eight different devices are shown. (From network defection change status report, NASA, Washington, January 1990, Fig. 1, p. 26)

also willing to contribute intellectually or financially. Another aspect is the data-distribution protocol: which scientist is allowed to publish measurements performed at NDSC locations. How much time should the NDSC keep its results for internal evaluation, to guarantee data quality? Decisions based on good management will prevent break-up of the network by rivalling groups or individual scientists.* One expects that the NDSC network will be of vital importance to calibrate satellite data, like those from UARS (Section 7.6.1), and to detect small global trends for the next 10–20 years.

7.6.3 ENERGY-SELECTIVE SPECTROSCOPY OF MOLECULES

In this section we illustrate the application of molecular spectroscopy in the analysis of the environment, for instance in the detection of polycyclic aromatic

---

*To give an impression of the size of NDSC: about 150 scientists are involved. Installation of an NDSC station equipped with several sensors was estimated at $2–3 million, with operating costs for all five planned NDSC stations at $1–2 million a year.

hydrocarbons. This section further illustrates some of the spectroscopic principles introduced in the preceding sections for the special case of molecules present as a guest in a quasi-crystalline or organic matrix. Molecules often exhibit broad spectra, due to inhomogeneous broadening, and little fine structure can be observed in the spectral properties. A typical value for the width of an electronic or vibronic transition is $100-500\,cm^{-1}$; this should be compared to about $1\,cm^{-1}$ for a molecule in a crystal. The major cause for the spread in transition energies causing the broad spectral features is that the molecule under study may occur in many sites of the surrounding matrix. There exist two independent strategies to avoid the loss of spectroscopic resolution due to inhomogeneous broadening. In the first strategy molecules are studied in a matrix that does not interact with the molecule and offers one or only a few relatively well-defined sites. In the second strategy a narrow-banded laser is employed for high-resolution spectroscopy. For the technique it is of particular importance that during the past few years many organic solvents and solvent mixtures that form glasses have been discovered which provide optically clear matrices for low-temperature high-resolution spectroscopy of a wide variety of compounds.

In this section we shall focus mainly on fluorescence techniques, and we will shortly mention hole burning. Molecular fluorescence spectroscopy is a well-known and frequently applied method for the analysis of organic compounds. It has the advantage of higher selectivity in comparison with absorption-based techniques (both excitation and detection wavelength can be varied independently) and the advantage of high sensitivity, since it is essentially a zero-background technique. Its disadvantage is the obvious limitation to fluorescent compounds. In principle, fluorescence (and other) energy-selective methods allow the identification and quantification of organic molecules in a complex mixture. However, this requires that the strong effects due to inhomogeneous broadening have successfully been eliminated. In the following we will discuss in some detail the physical origin of molecular spectra in glassy media and demonstrate how energy-selective spectroscopy can be applied to yield a dramatic increase in spectral resolution. Today energy-selective spectroscopy is a key technique in the quantitative analysis of the environment and we will demonstrate this by showing a few specific examples at the end of this section.

Molecular Spectra in a Low-Temperature Matrix

For energy-selective analytical spectroscopy the molecules are present as dilute guests in a matrix which can be a crystal, an amorphous glass or some other structure. In the following we will assume that (a) the guest molecules are not interacting with each other and (b) the electronic states of the host are not coupled to those of the guest molecules. In principle, since in a rigid matrix molecular rotations and translations are impossible, or at least strongly hindered, one may expect to observe sharp line spectra due to the molecular electronic and vibronic transitions. Under ideal conditions these sharp lines, or zero-phonon lines (ZPL), would only be lifetime broadened. Note that in general only the transition from the lowest excited state level is in the nanosecond time domain; all the higher energy levels have lifetimes in the picosecond range due to rapid internal conversion.

However, the fact that the molecules are present in a matrix has three distinguishable effects on the molecular spectra which we will discuss below: (a) homogeneous broadening of the purely electronic/vibronic transitions or zero-phonon lines, due to rapid fluctuations in the matrix; (b) the occurrence of a phonon wing (PW) or phonon side band due to excitation of one or more phonons during the absorption or emission process and (c) inhomogeneous broadening.

(a) *Broadening of the zero-phonon lines*
The physical process, called 'pure dephasing', responsible for the broadening of the ZPLs, arises from the interaction between the guest molecules and the matrix phonons. The fluctuating electric field, generated by the strongly temperature-dependent collective motion of the matrix atoms interacts with the electronic states of the guest molecules. Since (in general) the lattice motions occur on a time scale that is fast in comparison to the lifetime of the excited states, the interaction leads to a dephasing of the excited state and consequently a strongly temperature-dependent homogeneous broadening of all the energy levels. The phenomenon of pure dephasing shows a marked temperature dependence, in particular in amorphous glasses, that we will not further discuss here [13]. Below 30 K the contribution of pure dephasing to the ZPL width is at most a few cm^{-1}, and does not interfere with the resolution of the vibronic structure of molecular absorption/emission spectra, which is essential for the identification procedure.

(b) *The phonon wing or phonon side band*
The occurrence of a distinct phonon wing (PW) or phonon side band (PSB) in molecular absorption spectra arises from the interaction between the molecular electronic transition and the matrix phonons. The physical phenomenon is often referred to as 'electron–phonon coupling': a transition between two molecular electronic states may be accompanied by the creation or annihilation of one or more phonons in the lattice. Consequently, in the low-temperature absorption spectrum, the ZPL is associated to a phonon wing (PW) or phonon side band (PSB) at the high-energy side. Similarly, in the low-temperature emission spectrum of the guest molecule besides the ZPL a PW is observed at lower energies, for in emission the transition to the ground state leaves the system with an excited lattice phonon. Note that the ZPL corresponds to the transition in the absence of excitation of a phonon.

In principle, the relative probability of phonon transitions coupled to the electronic transition is determined by Franck–Condon factors introduced in

Section 7.3.3, but with the vibrational wave functions replaced by lattice phonon wave functions. Figure 7.26 illustrates the transition probabilities, both in absorption and emission, in this case associated to two phonon modes coupled to the electronic transition. Note that only in the ZPL the contributions from all the progressions add. In a real molecular system, there will be many phonons, and the summation over all states leads to a typical line shape as shown in Fig. 7.27, characterized by the sharp ZPL and a structureless PW.

The ratio of the total intensities of the sharp ZPL and the broad PW, which of course is a measure for the quality of the high-resolution spectrum, is expressed by the Debye–Waller factor:

$$\alpha = \frac{I_{\text{ZPL}}}{I_{\text{ZPL}} + I_{\text{PW}}} \tag{7.53}$$

The Debye–Waller factor α depends strongly on temperature and is determined by the strength of the electron–phonon coupling. For $T \rightarrow 0\,\text{K}$ the factor α is given by

$$\alpha = e^{-S} \tag{7.54}$$

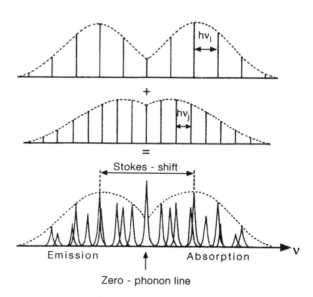

Figure 7.26 Transition probabilities for two phonon modes, *i* and *j*. Note that only in the zero phonon transition the contribution of the two phonon modes add. Upon the addition of a large number of phonons, a sharp zero-phonon line and a broad, structureless phonon wing is obtained. (Reproduced with permission from J. W. Hofstraat, C. Gooijer and N. H. Velthuis, in *Molecular Luminescence Spectroscopy: Methods and Application*, Part II (ed. S. J. Schulman), Wiley, Chichester, 1988, Fig. 4.7, p. 302)

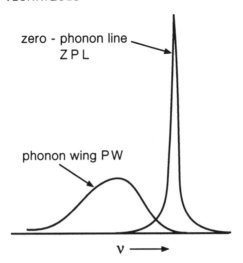

Figure 7.27 Schematic representation of the shape of a vibronic band in the emission spectrum of a guest molecule in a host matrix

where S is a dimensionless parameter indicating the strength of the electron–phonon coupling. For strong electron–phonon coupling ($S \gg 1$) the ZPL cannot be observed in the spectra, even at ultra-low temperatures. The parameter S is further connected to the experimentally observable Stokes' shift (see Fig. 7.26) by

$$\text{Stokes' shift} = 2S\omega_m \quad (7.55)$$

in which ω_m is the mean frequency of the phonons coupled to the electronic transition. In principle, knowing S and ω_m from experiments, and assuming some line shape for the ZPL and the phonon transitions, the single-site spectrum $L(\omega)$ of a guest molecule in a matrix is fully determined [14]. Note that S, the parameter that decides the quality of high-resolution spectra (and therefore the resolution), depends critically on the choice of the matrix.

(c) *Inhomogeneous broadening*
Inhomogeneous broadening of the spectra of molecules present as guests in a matrix is due to structural disorder. In general, a distribution of local interactions between the guest molecule and their environment gives a distribution of transition frequencies. The resulting absorption spectrum may be as broad as $500\,\text{cm}^{-1}$ for an amorphous glass. For a molecule with a single-site absorption spectrum $L(\omega - \omega_0)$ with the ZPL at ω_0, the total absorption spectrum is given by

$$A(\omega) = \int P(\omega_0 - \omega_m)L(\omega - \omega_0)\,d\omega_0 \quad (7.56)$$

in which $P(\omega_0 - \omega_m)$ is the inhomogeneous distribution function (IDF), which reflects the probability of finding the ZPL of a particular site at ω_0 relative to the maximum at ω_m. In general P is a Gaussian. In most of the relevant cases the widths observed in non-selectively obtained absorption or emission spectra are dominated by the IDF.

There are two essentially independent methods to reduce the effects due to inhomogeneous broadening. In the first one carefully selects a matrix, in which the guest molecules occupy essentially similar sites. Although there exists a variety of methods to reduce the IDF by selecting the matrix [15] we shall briefly discuss Shpol'skii spectroscopy [16]. In the second method the sample is studied with a laser, with a bandwidth much smaller than the width of the IDF. Only those molecules whose transition energy coincides with the laser frequency are studied. At low temperature the energy selection is maintained during the lifetime of the excited state and consequently one will observe narrow-line spectra. We will also discuss the fluorescence line narrowing (FLN) technique and its application in environmental analytical spectroscopy.

Shpol'skii Spectroscopy

When aromatic molecules, such as polycyclic aromatic hydrocarbons (PAHs), polyenes, big biomolecules like porphyrins or phtallocyanines, are dissolved in n-alkanes and the sample is cooled to 77 K or lower, fluorescence spectra consisting of sets of narrow lines are observed. The effect is illustrated in Fig. 7.28 for the emission spectrum of benzo (k) fluoranthene where it is clear that the narrow lines correspond to electronic and vibronic transitions to higher excited states. These spectra are named Shpol'skii spectra after their discoverer.

It is generally believed that the dramatic line-narrowing effect is a matrix-induced effect. The n-alkanes are known to freeze to a polycrystalline state. Apparently, the inhomogeneity in the environment of the guest molecules is strongly reduced. It is generally agreed that the guest molecules are trapped in the n-alkane matrix in such a way that they occupy substitutional sites, in which they replace several n-alkane molecules in the host matrix. For a certain PAH the quality of the Shpol'skii spectra depends critically on the choice of the n-alkane. Best spectra are obtained if there is a close match between the longest dimensions of the guest and the host. Although this 'lock and key' principle is generally valid, many exceptions exist, in particular for the larger aromatic molecules. The major *advantages* of Shpol'skii spectroscopy may be summarized as follows:

(a) The choice of the n-alkane is essential, but requirements of the fit of the guest in the n-alkane crystal are not as stringent as in single crystals.
(b) Laser excitation is not required to obtain vibrationally resolved spectra; nevertheless the selectivity of Shpol'skii spectroscopy can be amplified dramatically by laser excitation.

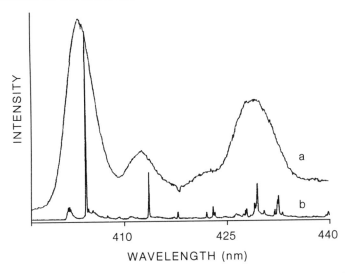

Figure 7.28 Emission spectra of benzo [k] fluorethane in n-octane. (a) Room temperature emission spectrum, concentration $= 10^{-4}$ M. (b) Shpol'skii spectrum at 26 K, concentration $= 10^{-6}$ M. Spectra a and b were recorded using the same experimental settings. The recorded intensities are plotted on the same scale. (Adapted and reproduced by permission from F. Ariese, Shpol's skii spectroscopy and synchronous fluorescence spectroscopy, bio(monitoring) of polycyclic aromatic hydrocarbons and their metabolites, thesis, Free University, Amsterdam, 1993, Fig. 3, p. 30)

(c) Also the $S_0 \rightarrow S_1$ absorption spectrum is narrow banded. Note that Shpol'skii spectra are still inhomogeneously broadened; typical widths range between 1 and $10 \, cm^{-1}$.

(d) Although electron–phonon coupling is not absent in the n-alkane matrix it is generally weak. This allows the detection of ZPLs up to relatively high temperatures $(T \leqslant 77 \, K)$.

(e) Phosphorescence spectra are also vibrationally resolved, even upon $S_0 \rightarrow S_1$ excitation.

(f) Shpol'skii spectroscopy can relatively easily be combined with other analytical techniques, for instance the chromatographic fractionation of a sample.

Major *complications* of Shpol'skii spectroscopy are:

(a) The multiplet structure of the quasi-line spectra. Multiplet spectra are due to the fact that the guest molecules may be inserted into structurally different substitutional sites, thus experiencing essentially different microenvironments. Of course, site selection is possible using laser excitation. Multiplet structure depends on the choice of the n-alkane and the thermal history of the sample. For instance slow cooling or annealing (i.e. heating the sample close to the melting temperature, followed by recooling) may reduce the number of available sites [15].

(b) The spectra often show both broad bands and narrow lines. The broad bands arise from a segregation of the guest molecules and the formation of aggregates. The latter occur in a glassy environment and do not exhibit quasi-line spectra. Again careful choice of the n-alkane, the concentration ($< 10^{-6}$ M) and the cooling procedure may shift the balance in favour of obtaining quasi-line spectra.

From the position of the lines in a Shpol'skii spectrum, the composition of a complex mixture of PAHs can be established. A quantitative analysis of Shpol'skii spectra depends on the availability of an internal reference standard. Deuterated analogues of the PAHs to be studied are extremely useful, since they interact in the same way with the n-alkane matrix as the original PAH, while their spectra are virtually identical, apart from a small spectral shift.

In the following we shall illustrate the application of Shpol'skii spectroscopy by the quantitative determination of polycyclic aromatic hydrocarbons in sediments from the Rotterdam harbour. Polycyclic hydrocarbons (PAHs) form a group of chemicals consisting of two or more fused benzoid rings, composed of only carbon and hydrogen. If aromatic compounds that contain heterocycles with nitrogen, oxygen or sulphur are also included the term polyclic aromatic compounds (PACs) is used. Often PAHs and PACs have similar physicochemical and ecotoxicological properties.

PAHs are natural constituents of oil and occur in many petrochemical products. PAHs are further produced during the burning of fossil fuels or other organic matter. In the latter case the dominant PAH fraction contains four or more fused benzoid rings.

PAHs have been recognized as potent carcinogenic compounds. Early in the twentieth century PAHs isolated from coal tar were found to induce tumours in rabbits. One of the compounds that was identified is benzo (a) pyrene (BaP), a five-ring aromatic hydrocarbon, the structure of which is shown in Fig. 7.29. Although BaP is by far the best-studied PAH as far as carcinogenicity is concerned, many PAHs or PAH derivatives are now listed as proven or suspected carcinogens [17–19]. For instance, in automobile exhaust over 80% of the carcinogenic activity was due to the PAH fraction, of which BaP was only a

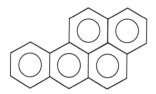

Figure 7.29   The structure of benzo[a]pyrene (BaP)

minor constituent. Nevertheless, the BaP concentration in a sample is often used as an indicator of the total carcinogenic potency. Note that BaP itself is a relatively inert molecule; only after metabolization by a living organism are reactive products formed, which eventually may form a complex with DNA. The presence of the DNA adducts in the genome may interfere with the correct reading of the genetic code, or induce mutations in the DNA due to mistakes in the repair process. Thus, tumour formation may be initiated.*

The quantitative detection of PAHs occurring on the Priority Pollutant List of the US Environmental Protection Agency is therefore of crucial importance in environmental analysis. Shpol'skii spectroscopy turns out to be extremely useful for this purpose. Figure 7.30 shows a typical Shpol'skii emission spectrum of a mixture of PAHs extracted from a sample of sediment from the Rotterdam harbour, obtained by Hofstraat *et al.* [15]. In the complex mixture the fluorescence from a variety of PAHs can easily be distinguished. Several intense vibronic bands from pyrene are indicated. The concentrations of the various PAHs in the n-octane solution range from 6×10^{-8} M to 3×10^{-7} M. From this experiment a detection limit of about 10^{-10} M can be estimated. From a

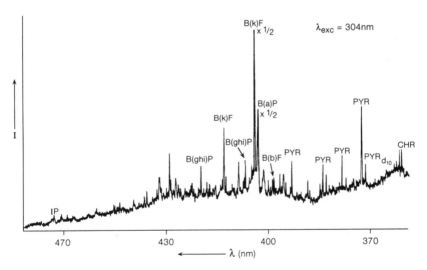

Figure 7.30 Fluorescence spectrum of a harbour sediment PAH extract in n-octane. The sample was excited at 304 nm. The abbreviations of the main PAHs are given in Table 7.1. The spectrum was recorded at 20 K. (Reproduced with permission from J. W. Hofstraat, C. Gooijer and N. H. Velthuis, in *Molecular Luminescence Spectroscopy: Methods and Application*, Part II (ed. S. J. Schulman), Wiley, Chichester, 1988, Fig. 4.7, p. 302)

---

*For a detailed review see Ref. 20.

calibration with reference compounds the amount of each PAH per kg harbour sediment could be established and these are listed in Table 7.1. Note the accuracy of the calculated amounts in comparison with the estimates obtained with a standard but low-selectivity HPLC method employing fluorimetric detection.

The internal standards used to obtain the quantitative amounts were pyrene deuterated at the C_{10} position (pyr-d_{10}) and perylene deuterated at the C_{12} position (per.d_{12}).

Fluorescence Line Narrowing Spectroscopy

Fluorescence line narrowing spectroscopy (FLNS) is a technique that diminishes the effects of inhomogeneous broadening on fluorescence spectra. The major strength of the technique arises from the fact that it yields high-resolution spectra in a variety of guest–host systems; in particular in amorphous glasses that make optically clear samples, FLNS is a powerful and simple analytical technique.

In FLNS the sample is excited with a monochromatic light-beam, usually produced by a tunable laser. The laser only excites those molecules whose energy difference perfectly matches the energy of the photon. As long as the distribution of excited molecules persists on the time scale of the fluorescence, a narrow-line fluorescence spectrum may be observed. At this point it should be noted that the laser only selects molecules with the same optical transition energy and not necessarily molecules in the same microenvironment. Consequently, the energy selection is often lost when levels are studied other than those involved in the transition selected by the laser (e.g. $S_0 \rightarrow S_2$ laser excitation does not necessarily result in line-narrowed fluorescence from $S_1 \rightarrow S_0$).

Typical problems associated with FLNS are the following:

(a) Electron–phonon coupling should not be too large. For small values of the Debye–Waller factor the ZPL is weak and consequently high-resolution spectra cannot be observed. By a judicious choice of the matrix material a strong interaction between guest and host can be avoided.

(b) Local heating. Molecules excited by the laser in high vibronic states will release a large amount of local heat during relaxation. Molecules with a low fluorescence quantum yield will similarly release heat upon radiationless decay to the ground state.

(c) Concentration of the guest. Even at intermediate concentrations of the guest, Förster energy transfer may occur between different guest molecules. Energy transfer will destroy the energy selection.

(d) Specific properties of glasses. Since FLNS is often applied to guest molecules in amorphous glasses, a few remarks about the properties of the glass are relevant. The energetic structure of a low-temperature glass is regarded as a statistical collection of double-well potentials or two-level systems (TLS). The TLS model was originally introduced to explain the low-temperature

Table 7.1 Results of the determination of PAHs in harbour sediment by Shpol'skii fluorimetry and fluorescence detected HPLC. (Adapted with permission from J. W. Hofstraat, C. Gooijer and N. H. Velthuis, in *Molecular Luminescence Spectroscopy: Methods and Application*, Part II (ed. S. J. Schulman), Wiley, Chichester, 1988, pp. 283–398)

Compound	Abbreviation	λ_{exc} (nm)	λ_{em} (nm)	Amount determined with pyr.d_{10} as standard (mg/kg sediment)	Amount determined with per.d_{12} as standard (mg/kg sediment)	Amount determined with HPLC (mg/kg sediment)
Chrysene	CHR	272	360.6	1.23 ± 0.13	1.20 ± 0.07	—
Pyrene	PYR	339	392.6	2.73 ± 0.24	2.64 ± 0.29	—
Benzo [*b*] fluoranthene	B(b)F	304	397.8	1.58 ± 0.19	1.50 ± 0.09	2.1
Benzo [*a*] pyrene	B(a)P	389	408.5	1.10 ± 0.17	1.12 ± 0.11	0.9
Benzo [*k*] fluoranthene	B(k)F	310	412.8	0.72 ± 0.09	0.72 ± 0.05	0.6
Benzo [*ghi*] perylene	B(ghi)F	304	406.3	0.86 ± 0.10	0.83 ± 0.06	0.5
Indeno (1,2,3-cd) pyrene	IP	380	462.7	0.75 ± 0.15	0.70 ± 0.10	1.3

behaviour of the heat capacity and thermal conductivity of amorphous glasses [13, 14]. The TLS also interact with the optical transition of the guest, giving rise to dephasing. Consequently the width of the ZPL is extremely sensitive to temperature and only at very low temperature ($T \leqslant 1$ K) does the ZPL of a guest in an amorphous glass show a width comparable to that in crystalline samples.

FLNS shows a remarkable wavelength dependence. The shape and bandwidths of the line-narrowed emission spectra are determined by the individual site spectra of the guest molecules, as discussed earlier: molecular spectra in a low-temperature matrix, by inhomogeneous broadening and by the vibronic structure of the guest spectra. We shall discuss the excitation wavelength dependence of FLNS for three different excitation regions.

(a) *Excitation in the 0–0 vibronic region*
Upon excitation in the 0–0 vibronic region two contributions to the line-narrowed fluorescence spectrum can be distinguished, which are illustrated in Fig. 7.31. The first arises from molecules resonantly excited in their ZPL at the

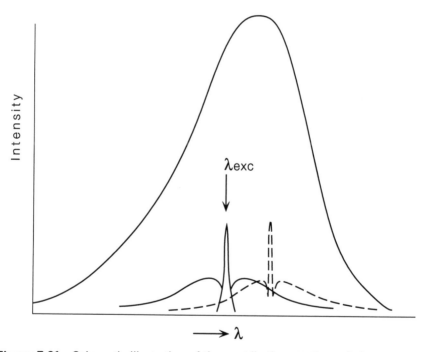

Figure 7.31 Schematic illustration of the contributions to the emission spectrum excited by a narrow-banded laser in the 0–0 vibronic region. Both resonantly and non-resonantly excited molecules are indicated

selected laser wavelength, λ_{exc}. Their emission spectrum similarly consists of a ZPL at λ_{exc} and a phonon wing (PW) at lower energy. The second contribution is due to guest molecules excited non-resonantly in their PW. Rapid relaxation to the corresponding zero-phonon excited electronic state occurs, followed by fluorescence. For such molecules all the emission is at the low-energy side of λ_{exc} and thus in the region of the PW of the resonantly excited guests. Thus the apparent Debye–Waller factor observed in FLNS, α_{obs}, is a result of both resonantly and non-resonantly excited molecules and is approximately given by

$$\alpha_{obs} \approx \alpha^2 \qquad (7.57)$$

where α is the single-site Debye–Waller factor as defined in eq. (7.53). The quadratic dependence of α_{obs} on α partly underlies the strong effect of temperature on FLNS. Finally, note that upon excitation in the 0–0 vibronic region, the ZPL coincides with the laser frequency. Since Rayleigh scattering from the glass is often severe, the ZPL is difficult to distinguish. Time-resolved techniques allow a separation of the long-lived fluorescence from the instantaneous scattered light.

(b) *Excitation into the region of the vibronic states (up to 2000 cm^{-1})*
This results in an increased complexity of the line-narrowed fluorescence spectra. In general, in complex molecules the inhomogeneous distribution functions of a large number of vibronic transitions overlap and as a consequence the laser excitation simultaneously probes several vibronic subbands. This is schematically depicted in Fig. 7.32.

Each vibrationally excited level relaxes to the corresponding ground state. From Fig. 7.32 it is immediately clear that a set of energy-selected excited states is created. As a result a multiplet structure is observed in the spectra. When the observed fluorescence emission is plotted versus $\lambda_{em} - \lambda_{exc}$, sharp ZPLs in the emission spectrum are observed, when the frequency difference corresponds to a vibrational quantum. Recently, in our own laboratory a line-narrowed fluorescence spectrum of chlorophyll *a* in a detergent micelle was measured. The detergent micelle was selected to model the biological membrane. The resulting spectrum is shown in Fig. 7.33. Note that a large number of vibronic transitions can clearly be distinguished in the FLN spectrum.

(c) *Excitation in the region of the higher vibronic states*
This generally results in a loss of the line-narrowing effect. In these regions the density of states is so high that many overlapping vibronic states are excited simultaneously, either through their ZPLs or by the PW, and only broad-banded emission spectra are obtained.

A wide variety of molecules including PAHs and molecules of biological interest have been studied by FLNS. The analytical potential of FLNS resides mainly in its versatility in solvent choice. Line-narrowed fluorescence spectra can be obtained in polar, apolar and even very complex biological matrices. For

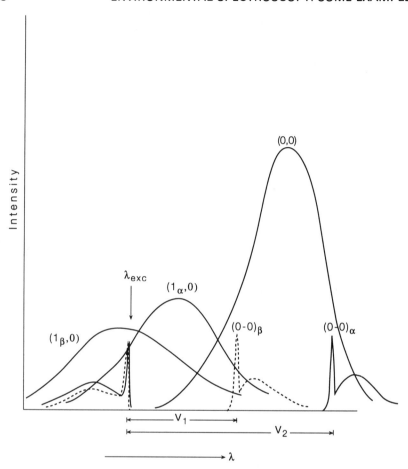

Figure 7.32 Schematic illustration of the contributions to the emission spectrum excited by a narrow-banded laser in the region of the vibronic states. The resonant excitation of two vibronic states (v_1 and v_2) is indicated

instance FLNS was used to quantify the amount and type of benzo (a) pyrene adduct in DNA. Most complicating effects can be avoided by carefully selecting the concentration, the host and the temperature.

The application of FLNS for the analysis of environmental samples is illustrated in Fig. 7.34. FLN spectra are shown for a variety of PAHs in a solvent refined coal sample. Note that in this case the sample preparation consisted solely of a 1:10 000 dilution of the coal sample with the glass-forming solvent.

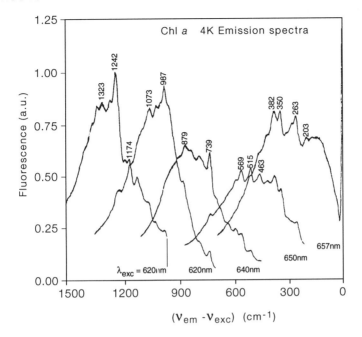

Figure 7.33 Fluorescence emission spectra of Chl *a* (CP47/TX100) at 4 K excited with a laser at 620, 630, 640, 650 and 657 nm. Here $(\nu_{em} - \nu_{exc})$ is the relative emission frequency calculated from the absolute emission frequency ν_{em} and the fixed laser excitation ν_{exc}. The spectra, all measured on the same sample, were normalized to the same incident laser. (Reproduced from *Photochemistry and Biology* Volume 59 (page 226) by permission of the American Society of Photobiology)

7.6.4 PARTICLE-INDUCED X-RAY EMISSION (PIXE)

Particle-induced X-ray emission (PIXE) is a spectroscopic technique for the analysis of the elemental composition of (dry) samples. PIXE is based on the detection of X-ray spectra characteristic for specific elements. The X-rays are emitted after the bombardment of the sample with accelerated particles, usually protons. PIXE is one of the many techniques used in environmental chemical analysis and quantitatively detects trace amounts of elements in soil, water residues, collected aerosol particles and biological material such as leaves, bark or tissue. The major advantages of PIXE are the high sensitively for multi-elemental detection (from Na $(Z = 11)$ to U $(Z = 92)$), short measurement time (1–10 min) and the small amounts of sample that are required. Under optimal conditions in a complex mixture a particular element with a relative concentration of 0.1 μg/g can be detected, even in samples of which only sub μg amounts are

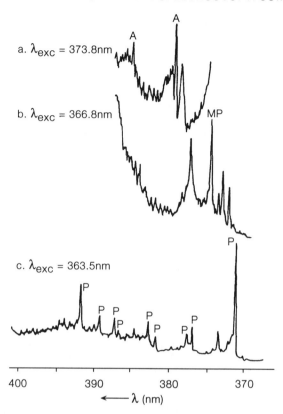

Figure 7.34 FLN spectra for (a) anthracene (A), (b) 1-methylpyrene (MP) and (c) pyrene (P) in a solvent-refined coal sample at 4.2 K. Excitation wavelengths were optimized for each PAH. (Reproduced by permission from J. C. Brown, J. A. Dunscanson Jr and G. J. Small, *Anal. Chem.*, **52**, 1980, p. 1711)

available. This corresponds to an absolute detection limit of 10^{-15} g (1 fg). In special μ-PIXE machines the proton beam can be strongly focused to a spot size of $< 1 \, \mu m^2$, which allows a measurement of the spatial distribution of selected elements, for instance in biological material. This also enables the measurement of single particles of μm size.

X-Ray Spectroscopy

The principles underlying the structure of X-ray line spectra observed upon the excitation of elements by high-energy radiation or high-energy particles have been explained in Section 7.5. In summary, after the bombardment of a sample with an electron or proton beam, the emitted spectrum consists of a continuous part, due to Bremsstrahlung of the electron beam or of the secondary electrons produced by the proton beam, with a line spectrum superimposed due to the emission of X-ray photons by the relaxing atoms. Figure 7.35 shows the energies of K, L, M (X-ray) lines as a function of the atom number Z (see eqs. (7.42) and (7.43)). Figure 7.35 emphasizes the fact that each element is characterized by its own set of X-ray emission lines, and this explains why PIXE spectra of complex samples allow a quantitative analysis of their composition.

In practice, the available equipment to detect X-ray spectra restricts the useful energy region to the interval between 4 and ≈ 30 keV. These limits are indicated in Fig. 7.35 by the two dashed lines. The major cause for this limitation originates from the sensitivity range of the commonly used X-ray detectors in PIXE: lithium drifted silicon detectors. These detectors consist of a semiconductor Si crystal doped with lithium. The charge produced in the crystal after an X-ray photon has entered generates a current pulse, the amplitude of which is proportional to the energy of the incident photon. The peak of each current pulse is stored and the X-ray spectrum is generated after the detection of a large number of single X-ray photons.

From Fig. 7.35 it follows that for elements with $Z = 20$ to $Z = 50$ the K lines can be detected. For the heavy elements with $Z > 50$ the L lines occur within the available energy window. The figure indicates the difficulty of recording the elements in the $Z = 45$ to $Z = 55$ region (Cd, Sb, Cs) and illustrates the minor significance of the M lines for PIXE.

Figure 7.36 shows the X-ray spectra detected with a lithium drifted silicon detector for Fe ($Z = 26$), Ag ($Z = 47$) and Pb ($Z = 82$). The K X-ray spectrum of

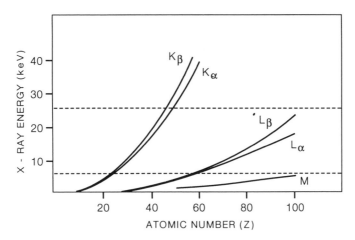

Figure 7.35 Level diagram showing the main K and L X-ray transitions as a function of the atom number Z. (Reproduced by permission from V. Valcovic, *Trace Element Analysis*, Taylor and Francis, 1987)

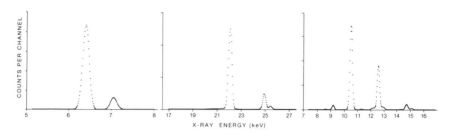

Figure 7.36   K X-ray spectra of iron (left), silver (middle) and the L X-ray spectrum of lead (right). All spectra were recorded with a Si(Li) detector. (Reproduced with permission from S. A. E. Johansson and J. L. Campbell, *PIXE, A Novel Technique for Elemental Analysis*, Wiley, Chichester, 1988, Figs. 1.5 and 1.6, pp. 17–18)

Fe is the simplest with the K_α line at 6.4 keV and the K_β line at 7.1 keV. For Ag the K_α transition is at 22 keV, while the K_β line is split. With increasing Z the spectra tend to become more complex. The effect is particularly pronounced for L X-ray spectra of the heavier elements. The X-ray emission spectrum of Pb exhibits peaks at 10.5, 12.6 and 14.7 keV, marked L_α, L_β and L_γ respectively. The additional small peak is at 9.2 keV from $L_3–M_1$ transitions. The L_α peak is due to $L_3–M_{4,5}$ transitions, while the other peaks, L_β and L_γ, represent complex multiples.

For the analysis of PIXE spectra of unknown samples, spectra like the ones shown in Fig. 7.36 serve as elemental standards. The absolute intensities and intensity ratios of the peaks that occur in the spectra are characteristic for each element and have been extensively tabulated.[*] These spectra provide a basis for the quantitative analysis of complex environmental samples.

PIXE in Practice

Several factors have promoted the development of PIXE as a sensitive technique for the analysis of the elemental composition of environmental samples. The first is the advent of the above-mentioned lithium drifted silicon detector, which allowed the energy-selective detection of X-ray photons with high efficiency. The second is that PIXE is a quantitative multielemental detection technique, which requires only small amounts of sample ($< 1\,\mu g$) while maintaining a high dynamic range (< 1 ppm).

The detection time is short (1–10 min) and the technique is non-destructive. One of the major advantages is that PIXE can easily be combined with other

---

[*] See Ref. 3 and references cited therein.

ion-bombardment techniques, such as particle-induced gamma-ray emission (PIGE), particle elastic scattering analysis (PESA) and nuclear reaction analysis (NRA). This allows the combined detection and identification of the light elements (Li, B, C, O and F). Finally we wish to mention the μ-PIXE variant, suitable for the measurement of spatially resolved ($\approx 1\ \mu m^2$) PIXE spectra.

PIXE spectroscopy requires a particle accelerator, usually a 1–4 MeV proton beam from a Van der Graaff accelerator. The major advantage of using a proton beam resides in the relatively low intensity of the Bremsstrahlung. The Bremsstrahlung intensity from the beam itself is proportional to $1/m^2$, where m is the mass of the incoming particle, which implies a reduction with a factor of $(1836)^2$ for a proton beam relative to an electron beam. The proton beam, however, produces secondary electrons, which in turn produce Bremsstrahlung. Due to kinematics this background is limited in energy by the energy transferred to an electron by a proton–electron collision. We finally note that selectivity for specific elements can be enhanced by changing the energy of the protons and by application of (combinations of) filters in the X-ray detection path.

Figure 7.37 shows a PIXE spectrum of the solid residue of a rainwater sample. Most of the observed lines originate from K X-ray spectra and one easily

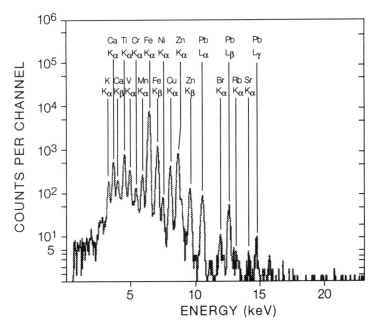

Figure 7.37 PIXE spectrum of a rainwater sample. (Reproduced by permission of Elsevier from S. A. E. Johansson and T. B. Johansson, *Nucl. Instr. Meth.*, **137** (1976) 473, Fig. 26, p. 505)

distinguishes the contributions from Fe, Cu and Zn. Also the L lines of are clearly present.

Aerosols

One of the major applications of PIXE is in atmospheric research and here we will specifically discuss the characterization of aerosols. Aerosols are dust particles of various origins, introduced in the atmosphere by, for instance, traffic or industrial activity (cf. Section 5.8). Dependent on the weather conditions, these aerosols may be transported over very large distances and monitoring their concentration in relation to their origin is an important form of air quality control.

A variety of complex filtering techniques exists that separate the aerosols of various sizes from the atmosphere. The aerosols are often collected on plates coated with mylar, which can be placed directly in a proton beam. Thus a large number of samples can be analysed allowing for high time resolution.

Figure 7.38 shows a typical aerosol spectrum recorded through a filter that suppresses the low-intensity region. The various elements present in the aerosol sample have been indicated. Note the presence of a typical Bremsstrahlung background which arises from the high-energy electrons ejected out of an atomic inner shell after excitation by the proton beam (secondary Bremsstrahlung).

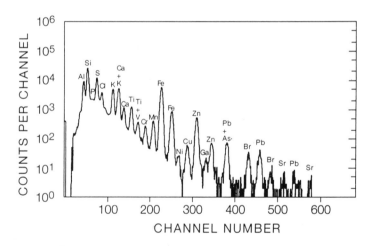

Figure 7.38 PIXE spectrum of an urban aerosol recorded using a 'funny filter'. Such a filter consists of a conventional plastic that blocks the K rays of all the elements below calcium. A hole is pierced in its centre corresponding to 5–25% of the detector solid angle. In this way only 5–25% of the intense low-energy portion of the spectrum is transmitted, allowing a more accurate detection of the heavier trace elements that are present in relatively low concentrations. (Reproduced by permission of Elsevier from J. L. Campbell *et al.*, *Nucl. Instr. Meth.*, **B14** (1986) 204, Fig. 4, p. 206)

The analysis of aerosol composition and origin requires an efficient sampling strategy by a network of stations with continuously operating samplers. Such measuring programmes have been initiated at several places in the world. In one relatively promising approach the size distribution of aerosols is correlated with their composition and source. Usually a few principal components are specified (cf. Fig. 5.23): industrial pollution, automobile pollution, sea-spray and soil-derived dust. Under extreme conditions one of these components is by far dominant and can be identified. For instance, heavy storms from the sea will bring mainly pure marine aerosol. On other occasions predominantly air that passes over highly polluted industrial areas is detected. Based on these back-trajectories aerosol composition and origin are initially classified. By further treating the data using statistical methods, based on a combination of pattern recognition and principal component analysis, the origin of the aerosols can be determined from the elemental composition alone.

Figure 7.39 shows the elemental size distributions for some typical aerosols as detected with PIXE and how they occur in the various air-transport back-trajectories passing through a measuring station in south Sweden. Relevant other examples are the detection of aerosols in the Antarctic and Arctic atmospheres. The latter may contain significant amounts of pollutants, sometimes called the Arctic haze, originating from the Northern hemisphere.

Finally, μ-PIXE can be applied to study the history of pollution of a particular area. For instance, the distribution of pollutants over the annual rings of a tree allows a determination of the annual variation in the atmospheric concentration of the relevant elements. By studying trees at different distances from a polluting centre, the spatial distribution of the pollutants can be determined as a function of time.*

Exercises

7.1   Evaluate the integrals $\langle 1s|\mu|2s \rangle$ and $\langle 1s|\mu|2p \rangle$ using the appropriate expressions for the H atom wave functions.

7.2   Calculate the rotational energy levels of NH_3. The NH_3 molecule is a symmetric rotator, with bond length 101.2 pm and HNH bond angle 106.7°. The expression for the moments of inertia can be found in Ref. 2, p. 438. Repeat the calculation for $CClH_3$. Take the C—Cl bond length as 178 pm and the C—H bond length as 111 pm. The HCH angle is 110.5°.

7.3   Which of the following molecules may show a pure rotational spectrum in the microwave region: H_2, HCl, CH_4, CH_3Cl, H_2O and NH_3?

7.4   Which of the following molecules may show infrared absorption: H_2, HCl, CH_3CH_3, O_3 and H_2O?

---

*For more information concerning the PIXE technique and its application in environment
al analysis we refer to Ref. 3.

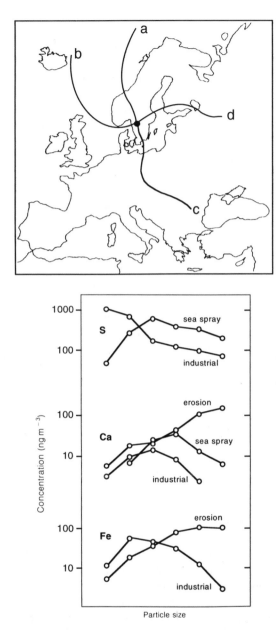

Figure 7.39 Elemental size distributions for some typical aerosols (top). Typical 48-hour back-trajectories going through a measuring station in Sweden (bottom). These trajectories correspond to the inflow of air masses characterized mainly by (a) pure Arctic air, (b) marine aerosol, (c) industrial pollution and (d) soil-derived dust. (Reproduced with permission from S. A. E. Johansson and J. L. Campbell, *PIXE, A Novel Technique for Elemental Analysis*, Wiley, Chichester 1988, Figs. 12.13 and 12.14, pp. 218–219)

7.5   Evaluate the Franck–Condon integrals (eq. 7.38)) for two harmonic poten-
      tials shifted relative to each other. Show that the resulting expressions
      correspond to a progression of vibrational transitions. (Harmonic oscillator
      wave functions are given in any quantum mechanics text.)

7.6   In a simple model for conjugated polyenes, such as the photosynthetic
      pigment β-carotene, the π-electrons are allowed to move freely along the
      chain of carbon atoms. In this model the 22 π-electrons of β-carotene (each
      carbon atom of the conjugated chain contributes one π-electron) are
      regarded as independent particles in a one-dimensional box and the MOs
      are to be the square well wave functions. With the dominant optical
      transition of β-carotene at 500 nm as the transition between the HOMO
      ($n = 11$) and the LUMO ($n = 12$), what would be the effective length of the
      β-carotene molecule? Compare this to the true length taking a C—C
      bondlength of 140 pm. Calculate the absolute value of the transition dipole
      moment using square well wave functions. Compare the result with the
      experimental value, using eq. (2.31) and taking an extinction coefficient in
      the maximum of $150\,000\,M^{-1}\,cm^{-1}$ and an FWHM of the β-carotene
      absorption spectrum of 40 nm.

7.7   Do you think a DIAL system can function on a rainy day? What about
      cloudy or foggy days? Try to find some data on visibility in several
      meteorological circumstances. (see, for example, Ref. 5.)

7.8   Discuss which improvements on lasers and detection would lead to shorter
      data collection times. Ask more experienced PhDs or senior scientists what
      they think about practical problems (if any!) regarding your suggestions.

7.9   Some (sometimes commerially available) DIAL systems for tropospheric
      measurements have data collection times of 15 minutes. These systems use
      a pathway that is nearly horizontal. Would you trust measurements on a
      windy day? Try to think about an upper limit for wind velocities for the
      data to be reliable in your opinion.

7.10 Evaluate the composition of the rainwater sample (Fig. 7.37).

References

1. I. I. Sobelmann, L. A. Vainshtein and E. A. Yukov, *Excitation of Atoms and Broadening
    of Spectral Lines*, Springer Series in Chemical Physics, Vol. 7, Springer, Berlin and
    Heidelberg, 1981.
2. P. W. Atkins, *Physical Chemistry*, 3rd edn, Oxford University Press, 1986. General
    textbook, with many relevant examples and good exercises.
3. S. A. E. Johansson and J. L. Campbell, *PIXE, A Novel Technique for Elemental
    Analysis*, Wiley, Chichester, 1988. A good introduction on the PIXE technique.
4. R. M. Measures, *Laser Remote Sensing: Fundamentals and Applications*, Wiley, New
    York, 1984. Best available review on laser remote sensing.
5. E. Scharda, *Physical Fundamentals of Remote Sensing*, Springer-Verlag, New York,
    Berlin and Heidelberg, 1986.

6. S. J. Schulman, *Molecular Luminescence Spectroscopy: Methods and Applications*, Wiley, New York, 1988.
7. J. C. Farman *et al., Nature*, 315 (1985) 207–210.
8. R. M. Schotland, Some observations of the vertical profile of water vapor by a laser optical radar, *Proceedings of the 4th Symposium on Remote Sensing of the Environment*, University of Michigan Press, Ann Arbor, 1966, p. 273.
9. U. Brinkman, Continuous monitoring of the atmosphere using DIAL, Lamba highlights (a publication from the Lamba Physik Laser Company), No. 30/31, 1991, pp. 1–8.
10. I. S. McDermid *et al.*, LIDAR measurements of stratospheric ozone and intercomparison, *Applied Optics*, **29** (1990) 4914–4923
11. I. S. McDermid *et al.*, Measurement of the JPL and GSFC stratospheric ozone LIDAR systems, *Applied Optics*, **29** (1990) 4671–4676.
12. I. S. McDermid *et al.*, Comparison of stratospheric ozone profiles and their seasonal variations as measured by the LIDAR and stratospheric aerosol and gas experiment during 1988, *Journal of Geophysical Research*, **95** (1990) 5605–5612.
13. S. Völker, *Spectral Hole-Burning in Crystalline and Amorphous Organic Solids; Optical Relaxation Processes in Molecular Excited States* (ed. J. Fünfschilling), Kluwer Academic, Dordrecht, The Netherlands, pp. 113–242.
14. S. G. Johnson, I.-J. Lee and G. J. Small, in *The Chlorophylls* (ed. H. Scheer), CRC Press, Boca Raton, Ann Arbor, Boston and London, pp. 739–768.
15. J. W. Hofstraat, C. Gooijer and N. H. Velthuis, in *Molecular Luminescence Spectroscopy: Methods and Application*, Part II (ed. S. J. Schulman), Wiley, Chichester, 1988, pp. 283–398.
16. R. N. Nurmurkhametov, *Russ. Chem. Rev.*, **38** (1969) 180.
17. W. Karcher, R. J. Fordham, J. J. Dubois, P. G. J. M. Glaude and J. A. M. Ligthart (eds.), *Spectral Atlas of Polycyclic Aromatic Compounds*, Vol. 1, Reidel/Kluwer, Dordrecht, The Netherlands, 1983.
18. W. Karcher, S. Ellison, M. Ewald, P. Garrigues, E. Gevers and J. Jacob (eds.), *Spectral Atlas of Polycyclic Aromatic Compounds*, Vol. 2, Reidel/Kluwer, Dordrecht, The Netherlands, 1988.
19. W. Karcher, J. Devillers, P. Garrigues and J. Jacob (eds.), *Spectral Atlas of Polycyclic Aromatic Compounds*, Vol. 3, Reidel/Kluwer, Dordrecht, The Netherlands, 1991.
20. M. Hall and P. L. Grover, *Polycyclic Aromatic Hydrocarbons: Metabolism, Activation and Tumour Initiation* (eds. C. S. Cooper and P. L. Groover), *Chemical Carcinogenesis and Mutagenesis*, Vol. I, Springer, Berlin, 1990.
21. R. Jankowiak and G. J. Small, *Chem. Res. Toxicol.*, **4** (1991) 256.

Bibliography

Banwell, C. E., *Fundamentals of Molecular Spectroscopy*, McGraw-Hill, London, 1983. General textbook on molecular spectroscopy.
Steinfeld, J. I., *Molecules and Radiation*, 2nd edn, MIT Press, Cambridge, Mass., 1985.
Svanberg, S., *Atomic and Molecular Spectroscopy, Basic Aspects and Practical Applications*, 2nd edn, Springer-Verlag, New York, Berlin, Heidelberg, 1992. Overview of atomic and molecular spectroscopy with many interesting applications of laser and optical spectroscopy.
van der Hulst, H. C., *Multiple Light Scattering*, Vols. 1 and 2, Academic Press, Orlando, 1980.
van der Hulst, H. C., *Light Scattering by Small Particles*, John Wiley, New York, 1957. Extensive texts on Rayleigh and Mie scattering.

8

The Context of Society

In Chapter 1 we summarized in Fig. 1.1 the activities of an industrialized, techno-logical society. In such a society many decisions have to be made in which costs and benefits are weighed. Costs may comprise environmental damage, harm to individuals and groups and the direct financial cost of building or reconstructing of an installation. Benefits are usually seen as financial profits, or more generally in terms of increased consumption for an expanding population.

Professional physicists have to deal with these matters if they work for govern-ment and protect the public interest by designing regulations and ways to enforce them. Alternatively they may work for private enterprises where the direct profit motive will be stronger, but again there is growing awareness that it is in the long term interest of a company to take environmental aspects into account.

Finally, physical scientists may be active in public life, in professional societies or political parties and may participate in the making of public policy. It will be understood that defining and implementing public policy in a modern society is a matter of discussing reports, writing comments, of consulting and media presentations. Here, on many levels physicists may advise or even take public office.

In this final chapter we therefore touch on a few points of which we believe that physical scientists should be aware. We start in Section 8.1 with discussing the concept of *risk* and how one is trying to quantify the probability of disasters. These concepts apply, for instance, to big technological installations such as nuclear power plants.

On a somewhat higher level of abstraction, policy on energy and environment are formulated. On that level one has to consider the resources and the consumption of energy and the environmental consequences of the choices to be made. This is the subject matter of Section 8.2. We will illustrate that there are vast differences between countries in the amount of energy required to produce a certain wealth. We will argue that all countries should adopt a Japanese type of industry, which would imply a tremendous shift in money from individual consumption to investments in environmentally benign production structures.

Such a harsh and unpopular measure will not easily be made. There will always be opponents who argue that the capabilities of nature to adapt itself are much bigger than imagined and that there are many more resources than we present in our tables. We admit that we cannot prove that a catastrophe is due. The question then is how to deal with uncertainties. Therefore, in Section 8.3 we discuss the role of policy analysis in policy making under such conditions. As a possible global catastrophe we will keep in mind the global warming, discussed in Chapter 3, and to be more specific we look at the emissions connected with several means of passenger and freight transport. In this area of economic activities there are many different choices possible.

One of the essential elements in sound decision making is that all points of view receive due attention. Many of them originate from the differing interests of the economic groups in modern society. We will discuss them briefly in Section 8.4. Others arise from a differing look at nature. We will discuss them in a little more detail and point out that for a long time the dominant Western view on nature has been to see it as an instrument only. Currently, the perspective is gaining strength that we can only avert an environmental catastrophe when we look at nature in a respectful way and not only as a means of production.

8.1 RISK ESTIMATION

Risk estimation is defined as 'the collection and examination of scientific and technical data to identify adverse effects and to measure their probability and severity'. Risk is a *number*, the probability that a specified event occurs during a stated period of time. We will use the words 'harm' or 'danger' when referring to adverse effects themselves without their probability.* Let us discuss some areas of risk estimation in order to explore the difficulties involved. We discuss man induced dangers like new chemicals and nuclear power stations and 'natural' dangers like the flooding of low countries.

Chemicals

When we study the dangers involved with certain chemicals, one has to look at their production, transport, use and ultimate end destination, usually as waste. The chemicals concerned are not only the principal desired products of the chemical industry; they may also be unwanted by-products, such as SO_2 and NO_x which result from the combustion of fossil fuels, or dioxins, which originate in low concentrations from some high-temperature chemical processes.

Not all man-made chemicals are dangerous nor are all natural chemicals harmless. Poisons exist in plants and animals, and we are educated not to eat

---

*We adopt the terms and definitions of the Royal Society Study Group. They are a little different from other conventions such as those from Ref. 1. In the latter what we call 'detriment' is called 'risk'.

certain mushrooms or plants from the nightshade family. In 1987 there were some 80 000 man-made chemicals in common use with 600–800 being added annually.* Not all of them are tested on their possible dangers, although industries are now making estimates of the effects they would have in the natural environment.[†]

In testing what harm a certain chemical can do, one distinguishes acute *toxicity* as the toxicological response of an organism following short-term exposure to a toxic substance, from 24 to 96 hours. This is easy to measure. In a laboratory one can perform animal tests and find a *lethal concentration LC50*, which is the concentration at which half the population will die. One usually assumes that 1% of this concentration will be harmless. More difficult to test is the chronic effect of a small dose over a long period of time (longer than 96 hours) which may diminish the overall strength of a population of organisms and make it an easy prey for its predators. These chronic effects are usually only discovered after the harm to the environment has been done.

For chemicals it is usually not clear whether there indeed exists a threshold concentration below which the situation can be considered harmless or at least acceptable. If this were the case then *dilution is the solution to pollution*. If there is no threshold, the consequences may be extinction and loss of species. It is the object of biological studies to analyse these consequences.

There are physical–chemical aspects of toxicity as well. For aqueous organisms certain organic chemicals may be concentrated in their bodies by diffusion through membranes against a concentration gradient. This will continue until the concentration inside the body is so large that the organic chemicals will be removed at the same rate as they are taken in. In this equilibrium the concentration inside may be 10^5 times as large as in the water outside (DDT in a fish called the fathead minnow). For inorganic chemicals the process goes more slowly and in many cases there is no time to establish an equilibrium during the lifetime of the organism.

Although a lot of information about possibly harmful effects of chemicals is available and much more will be known in the future, it is difficult to put risks down to a single number on the basis of testing. This is often only possible in retrospect for events where adverse effects can be clearly defined.

Floods

As an example which shows the difficulties inherent in any risk estimation, even when ample data seem available, we take the flood catastrophes in the

---

*UN Environmental Programme, quoted in Ref. 2, pp. 148 and 392.
[†]An overview of the recent state of affairs in the chemical industry is found in the papers collected by Cothern, Mehlman and Marcus [3]. It is interesting that the European Community is stimulating ECO audits in industry; they check the compliance to environmental rules and serve as a management instrument to estimate risks and to mitigate them. Ultimately firms may acquire a mark for sound environmental production. (From EC, Council of Ministers, 22 March 1993.)

Netherlands.* From the data for Hook of Holland on the North Sea coast one can make a diagram as in Fig. 8.1, where one notes on the vertical scale the high tide levels h and on the horizontal scale the probability m that a certain level (the height of a dike) is exceeded. On a logarithmic scale the data follow a straight line. Thus one may express

$$m = e^{-a(h-h_1)} \qquad (h > h_1) \tag{8.1}$$

where $h = h_1$ is the level that is exceeded on average once a year. Floods are due to a combination of high tides and heavy, long lasting North-Western gales on the North Sea, as the narrow Straight of Dover is not able to accommodate the water masses. The straight line in Fig. 8.1 suggests that the height of the tides can be described as a statistical effect. One may notice, however, that the highest point has been based on only one incident: the flood in 1953. This is therefore indicated by a horizontal error bar. Another very high data point could of course change the extrapolation. Therefore, even for a phenomenon that has been measured for centuries the uncertainties in the predictions of low-frequency effects may be unavoidably high, if the phenomenon is rare enough.

From eq. (8.1) it has been deduced how high the dikes should be to reduce the probability of a flooding of the dikes to 1 in 10 000 years. This number of 10^{-4} of course reflects a political decision. It may be based on the argument that 10^{-4} times the damage done should be comparable to the cost of the dikes (building and maintenance), calculated with the method of discounting. However, in the final analysis public discussion determines the adopted probability. The same is the case with nuclear power stations.

Nuclear Reactors

Nuclear power plants have an enormous inventory of radioactive materials (cf. Section 4.5). For decision making it is therefore necessary to obtain an estimate of the (low) probability that large amounts of radioactive materials escape as the result of an accident. The two accidents discussed in Section 4.5.3 do not provide enough statistics on which to base numerical values. Moreover, they do not take into account improvements in operation of the plants that have been effected since.

To analyse what would be the accident with the largest hazards for the population, we take as an example the reactor types where water acts both as a moderator and as a coolant (cf. Table 4.5). It appears that the loss of coolant by the breaking of a pipe is the most dangerous event. Although the chain reactions

---

*After the Middle Ages people have indicated on church towers the level to which the sea water had risen, after the dikes were broken. More systematic measurements have been performed since the eighteenth century, starting in Amsterdam. After closing the sea arms these data were no longer of use for forecasts.

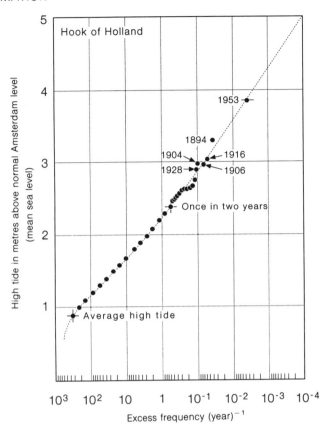

Figure 8.1 Summary of high tides at North Sea Coast, Hook of Holland. Vertically the water level is shown and horizontally the number of times it is exceeded in a year. The years indicate very high tides. The frequency is estimated by comparing with similar high tides. The 1953 flood was estimated in 1940 as having a probability of 3×10^{-3}. Since the flood many sea arms were closed, so the figure should be adapted for later use. The authors are indebted to Ir. J. van Malde for pointing out the literature. (Reproduced by permission of NV SDU Uitgeverij Semi-Officiale Publikaties from Rapport Deltacommissie, Beschouwingen over stormvloeden en getijdebeweging, Staatsdrukkerij en Uitgeversbedrijf, 's Gravenhage, 1961, Fig. 4, p. 61, 93)

would stop for lack of a moderator, the enormous decay heat of the many reaction products would cause a rise of temperature, which could lead to a core melt-down and a release of radioactivity through containment or the ground. To prevent this from happening an emergency core cooling system is present, using extra water pumped by an electric system. If they all fail, one has to rely on mechanisms to remove fission products and ultimately on the integrity of the containment.

The description just given may be summarized in an *event tree*, shown in Fig. 8.2. The failure of any part of the system may be given a probability P, so for any order of events one may calculate the total probability; the only point is whether one is allowed to see failures as unconnected so that their combined probability is the product of small numbers. Most probabilities for single events may be determined from experiences in industry, where failures of pipes or power are common. Most difficult is to estimate a failure probability of the emergency core cooling system, which cannot be imitated in practice.

Even if one doubts the numerical accuracy of the risk estimates by the event tree method, the method is valuable in a relative sense. By comparing the risks on release of radioactivity for several designs one may identify weak spots and choose the most reliable design.

The experience with commercial nuclear power stations may be expressed in reactor-years, which is the sum of years of operation for all reactors together. In 1993 this number was of the order of 10^4. This number is too small to deduce probabilities in the order of 10^{-5} or 10^{-6} (compare the example of floods discussed above). Also it does not seem appropriate to take all reactor types and designs together in a single estimate.

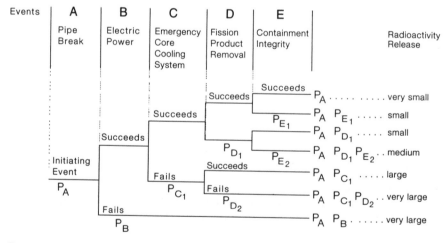

Figure 8.2 Simplified event tree, showing possible subsequent events in a nuclear 'loss of Coolant' accident. After a pipe break with probability P_A the bottom branches give the probability of failure P_B, etc., and the upper branches the probability of success in handling the problem. The latter probability $(1 - P_B)$, etc., is usually approximated as one. (Adapted from US Atomic Energy Commission, Reactor Safety Study, WASH-1400 (Rasmussen Report), 1975, Appendix I and Samuel Glasstone and Alexander Sesonske, *Nuclear Reactor Engineering*, 3rd edn, Van Nostrand, New York 1981, p 723 (cf. Chapter 4)

Detriment

For cost–benefit analyses connected with advanced technology one has to calculate costs in financial terms. This implies the evaluation of the effects of a certain accident. In the example of a nuclear power station, after estimating the probabilities of a chain of events one may calculate the composition and intensity of the radioactive materials released and use the Gaussian plume model of Section 5.6 to estimate the doses that members of the population would receive. With dose–effect relations discussed in Section 4.5.3 one may then estimate the number of resulting fatalities. Besides fatalities (sometimes expressed in money) one could calculate economic damage in terms of destroyed production power.

When comparing the risks connected with several options one often multiplies the probability of an adverse event with the harm done to obtain the detriment:

$$\text{Detriment} = \text{probability} \times \text{harm} \tag{8.2}$$

Eventually one could sum over partial detriments to find a total 'cost' for a certain installation.

A definition like (8.2) is used by insurance companies for calculations of their rates. It is a well-established method for 'ordinary' hazards. It is questionable whether the definition would hold as a suitable criterion to calculate the detriment for an event like the explosion of a nuclear power plant where the probability is low but the adverse consequences are very high. The same question could be asked to controlled genetic engineering of food where with a very small chance a disastrous reduction in resistance against diseases could occur by a combination of many very small effects. In such cases the product in eq. (8.2) would become similar to the limit of zero times infinity in mathematics where the result is undetermined.

A second limitation on the validity of eq. (8.2) is the following. Compare the 10 000 annual traffic deaths in medium-sized industrial countries with the single catastrophe of a chemical plant fire with the same number of fatalities. In the first case the tragedies are distributed over the country and society has by experience learned to cope with it. In the latter case the social structures in the region concerned are totally out of joint and society is prepared to go to large expense to prevent these accidents.*

The point made here is recognized in Table 8.1, where acceptable risk is defined in terms of the annual risk of death to members of the general public, singly or in groups. For a young male the annual probability of death is around 10^{-4}. A serious occupational hazard is the one run by coal miners. Their excess chance of death happens to be in the same order of 10^{-4}. Therefore, when the individual

---

*The 1953 flood in the Netherlands with 2000 deaths gave rise to a drastic delta plan of protection against the seas, even when the probability of another flood would be only once in 300 years. The annual 2000 traffic deaths are taken for granted.

Table 8.1 Acceptable risks for harm from a certain technological installation. (Data from the Netherlands Department of Health)

Number of deaths in single event	Acceptable risk per year (example)
1 death	10^{-6}
10 deaths	10^{-5}
100 deaths	10^{-7}
Individual death from all sources	10^{-5}

risk from a technological installation in the neighbourhood is significantly lower, e.g. 10^{-6}, one may see it as acceptable.

When there is a probability that in a single event several people are killed one notices from Table 8.1 that the acceptable risk for 10 deaths is put as 10^{-5} and for 100 deaths at 10^{-7}, although application of eq. (8.2) would give 10^{-6} for 100 deaths when the number for 10 deaths is accepted. The lower numbers for deaths of large groups reflect the public outrage after a big accident, although the hazard to a single individual is very low. There are no internationally accepted standards, so Table 8.1 should be interpreted as an example.*

8.2 'ENERGY AND ENVIRONMENT' POLICY

In this section we first give some data on energy consumption and resources and then define some problems in energy and environmental policy. Of course, we restrict ourselves to some major points and cannot discuss the differences between countries and regions in the world in any detail.

On the left-hand side of Fig. 8.3 the development of the energy consumption per capita (= per head of the population) is given for the world as a whole, for a typical established industrial country like the United States and a typical newly industrializing country like India. The scale on the left indicates as units both 10^9 J/yr and W. The energy consumption may be compared with the manual output of a human being, which is estimated as 60 W during a third of a day or 20 W averaged over the day. One observes that even in developing countries the equivalent of fifteen 'energy servants' have been available for a long time.[†] The average world inhabitant at present has some 90 energy 'servants' available.

In Fig. 8.3 one notices a steady increase in energy consumption. Although this seems to level off at the end, one may expect it to rise again as the industrialization of the world continues. One also notices the dramatic increase in world population (scale on the right), which is expected to double between 1980 and

---

*Data from the Netherlands Department of Health.
[†] Ferguson [4] estimates that the man-day is at most $553\,000$ kg m in 8 hours, which is 62.7 W.

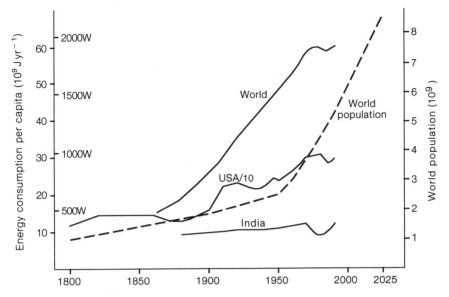

Figure 8.3 On the left scale the development of the energy demand per capita since the nineteenth century is shown for India and the World; for the United States the data should be multiplied by 10. On the right scale the increase in world population (dashed line) is given. (Energy data from Chauncey Starr, Energy and power, *Scientific American*, **244** (September 1971) 37 for the period up to 1970. Data for 1980–90 from UN *Energy Statistics Yearbook 1990*. Population data were taken from *World Resources 1986*, p. 10, and *World Resources 1990–1*, p. 50. The numbers for 2000 and 2025 are demographic predictions.)

2025. This raises the question of how long the energy resources will last and what the environmental constraints will be in harnessing them.

In Table 8.2 we show the consumption of fossil fuels and their resources for the world as a whole around 1990. The proven recoverable resources correspond with present prices. It should be realized that for a higher price the resources will increase.* The maximum may be a factor of two higher for oil and gas, a factor of ten higher for coal and a factor of twenty higher for oil shales and oil sands.† It is clear that at present levels of consumption the low-priced resources of oil and gas will run out within 50 years and for coal within a few hundred years (cf. Exercise 8.2).

---

*For a higher price the resources will increase for two reasons: first, poorer ores may be exploited; Second, it is worth while for exploration companies to look further—they may even find more cheap resources.
†Holdren in Ref. 5 gives somewhat higher values for the 'probable remaining recoverable resources'. Optimistic studies even point to resources of methane caught in ice crystals, twice as much as all earlier proven and estimated fossil fuel reserves together (cf. Ref. 6).

Table 8.2 World consumption (1990) of fossil fuels and proved commercially recoverable resources (1987). (Data from the United Nations *Energy Statistics Yearbook 1990*, Tables 4 and 38. *World Resources 1990–1* gives in Table 9.2 somewhat higher values for coal (22 000 × 10^{18} J for hard and soft coal together). CO_2 emissions are based on Shell data)

	Consumption (1990) (10^{18} J yr^{-1})	Resources (10^{18} J)	CO_2 emission (g/10^6 J)	Other emissions
Oil	117	5 000	74	High
Natural gas	72	4 000	56	Low
Coal	98	17 000	104	High
Oil shales/sands		2 000		

The fossil fuels show large differences in emissions of CO_2 and other materials. This is indicated in the column representing CO_2 emissions. In another column we give an indication of the other emissions.* It is clear that natural gas is the 'best' fossil fuel from an environmental point of view. A change from oil and coal to natural gas would therefore mitigate the environmental consequences of burning fossil fuels as a short-term solution.

After the Second World War nuclear power stations were built because they were available, provided a cheap source of electricity and reduced the dependence on oil-producing countries. Since then the price of nuclear electricity has gone up, mainly because of required safety measures. However, the argument is being made that this electricity does not pollute and would have only limited environmental consequences. Indeed, even when one takes into account the fact that fossil fuel energy is needed in the building of a nuclear power station, still the CO_2 emission per kW h would be less than 10% of that of a conventional power station.[†]

With the present-day technologies the easily available uranium resources are limited. In Table 8.3 we give the resources in GW_e years, which is the number of years a full-time operating power plant of 1 GW_e would have fuel. The assured resources would only last for some 45 years.

There is a difficulty in comparing uranium and fossil fuel resources. The first produces electric energy and the latter usually burns as heat. Fossil fuels may be converted into electricity with an efficiency of 38.5%. Therefore 1 electric joule

---

*From Culp (used in Chapter 4) [7] one may deduce for electric power plants the following generations: SO_2 coal and oil, respectively 0.5 and 1.4 g/10^6 J; NO_x coal and oil, respectively 0.13 and 0.10 g/10^6 J; dust from coal and oil, respectively 0.08 and 0.10 g/10^6 J. For coal 4.4 g ash/10^6 J has to be added. Note that these data will depend on the quality of the fuels; what is actually emitted depends on the precautions taken at the plant.

[†] From Ref. 8, p. 9, based on Japanese studies.

Table 8.3 Nuclear power installed in 1990 and resources for light water reactors. (Data on installed nuclear power and predictions for 2000 (based on building data) are from Ref. 8. Uranium resources are from the UN *Energy Statistics Yearbook 1990*, Table 38. With present light water reactors 130 tonnes of natural uranium (or 150 tonnes of oxide) may produce 1 GW$_e$ year of electricity. This gives the third column. The last column assumes fossil fuelled power plants with 38.5% efficiency of converting heat to electric power. It must be added that some reports give uranium reserves of 4.8 million tonnes and speculative reserves of 24 million tonnes; this confirms what was stated in the first footnote on p. 416. Holdren in Ref. 5 quotes even higher numbers)

| | Installed (GW$_e$) | Resources with present technology | | | |
			10^6 tonnes U	GW$_e$-yr	Equivalent fossil (10^{18} J)
1990	344	Assured	2.4	18 500	1500
2000	400	Estimated	1.4	11 000	900

would correspond to 2.60 fossil fuel joule. In this way an 'equivalent fossil' joule may be defined. This is used in the last column of Table 8.3. The number of 1500×10^{18} J is small compared with the fossil fuel reserves of Table 8.2. The estimated additional uranium resources are not large either, as is shown in the bottom line of the table. It is therefore essential to develop more advanced fuel cycles (cf. Sections 4.5.1 and 4.5.4) and nuclear fusion (cf. Section 4.5.2) if one wants to continue with nuclear power. It is not clear at what prices these options under investigation will produce electricity.

In discussing energy policy it may be helpful to distinguish between 'stock' resources like fossil fuels of which the reserves are limited and 'flow' resources originating from the sun that will last as long as or perhaps longer than human civilization—the renewable energy sources. Those resources are estimated according to the data given in Table 8.4. For devices that directly produce electric power we give the equivalent amount of fossil fuel required to produce the same power, which may be seen as the fossil fuel 'saved' by their implementation. One may compare and add the different tables. Table 8.5 summarizes our tables.

The resources in Table 8.4 look enormous, especially when compared with the annual consumption of 378×10^{18} J in 1990. However, as is noted in the table, the resources are not readily available. Expansion of hydropower, for example, would destroy a lot of animal life and vegetation. The expansion of biomass would compete with arable lands for food production necessary to accommodate the growing population of Fig. 8.3. Here the present conversion efficiency of biomass (1%) could be increased through better understanding of the photosynthetic process.

The hydrothermal resources are typically found in beautiful spots, which

Table 8.4 Consumption and resources of renewable energy with present technology. (Data and notes from Ref. 8. The UN *Statistical Yearbook* gives a lower number for biomass as 'traditional fuels'. It comprises forestry only, whereas in the number we give all kinds of agricultural wastes and dung are covered. The uncertainties in these estimates will be clear by noting that Holdren in Ref. 5 estimates the resources of biomass a factor of twelve higher, those of wind a factor of seven lower and of photovoltaic a factor of four lower).

	Installed (1989)		Estimated potential		
	GW_e	Equivalent fossil[a] $(10^{18} J yr^{-1})$	GW_e	Equivalent fossil[a] $(10^{18} J yr^{-1})$	Comments
Hydropower	240	16	1700	110	b
Biomass		52		75	c
Geo(hydro)thermal	6	0.4	4000	265	d
Wind	1.5	0.007		700	e
Photovoltaic	Small			21 000	f

[a] Equivalent fossil energy is the fossil fuel energy required to produce the same amount of electric energy, assuming an efficiency of 38.5%.
[b] Using all hydropower potential will have large environmental consequences and will meet huge capital costs.
[c] Limitations of potential arise because of lack of arable land and competition with food production.
[d] As most geothermal resources are in national parks only 1% of the number given is thought realistic. However, hot dry rocks and magma may offer extra resources.
[e] Energy storage is required to compensate for fluctuations in wind supply and electricity demand.
[f] Energy potential refers to desert areas. Transportation of energy (e.g. liquid H_2) will be expensive. For houses photovoltaics on roofs may provide a significant fraction of the required electricity. For apartment houses only hot water production on roofs may be economic. In a country such as Japan some 40% of household hot water production may thus be provided.

should not be destroyed; it is therefore realistic to take 1% of the potential resources given. For wind energy the price is still somewhat higher than for fossil fuel fired plants. The same holds for photovoltaic electricity. In both cases research and development may produce cheaper electricity. Both resources are largest far from population and industry. Thus one needs transportation of energy, sometimes over large distances, which makes these energies more expensive.

For these reasons we did not put a number on 'easy resources' of renewables in Table 8.5. Of course, governments could impose a tax on polluting activities like coal fired power plants, aeroplane use or car driving. Or they could grant tax reduction on clean activities like the production of power from the wind or the sun. Both would stimulate renewable energies and correct the fact that the

Table 8.5 Summary of energy data. (The number for 'renewables' is higher than in the UN *Statistics*, as all biomass is covered and electricity from hydro, nuclear and geothermal power is recalculated in terms of equivalent fossil energy).

	Consumption (1990) (equivalent) (10^{18} J yr^{-1})	Easy resources (equivalent) (10^{18} J)
Fossil	287	28 000
Nuclear	23	1 500
Renewables	68	
Total	378	
Solar energy (eq. (1.2))	All earth: 3.8 million $\times 10^{18}$ J yr^{-1}	

'free market economy' does not put a price on a clean environment. The measures would imply a shift from personal consumption to investments to ease the environmental burden of our economic system. For most countries we believe this to be necessary.

To this point we have discussed the energy supply and the options to use flow

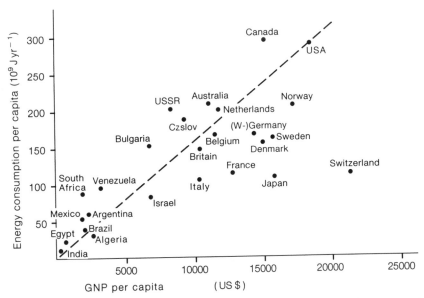

Figure 8.4 Energy consumption per capita and gross national product per capita for a variety of countries in 1987. (Energy data from UN *Energy Statistics Yearbook* and GNP data from *World Resources 1990–1*. Data are from 1987 but did not change much until 1992.)

resources and ease the environmental constraints of energy production. Another point is the consumption of energy and in particular the efficiency with which households, industry and services use this energy. This varies widely for different countries. In contrast to Table 8.5 where the world total was presented, we show in Fig. 8.4 for a number of countries the energy consumption and the gross national product (GNP), both per capita. The GNP measures the total economic activity in a country and indicates its total wealth. The ratio of both numbers is the amount of energy required to produce a certain wealth. The dashed line corresponds to the value for the United States. Countries above the dashed line use more energy per unit of wealth than the United States and countries below need less.

There may be special reasons why a country performs well in Fig. 8.4. For Switzerland, for example, one may presume that the importance of the banking and service sectors in the national economy reduce the energy efficiency of its economy. For larger economic blocs like the United States, Japan or the European Community these special factors will cancel. From Fig. 8.4 it is clear that Japan produces with a much smaller energy use than the other large

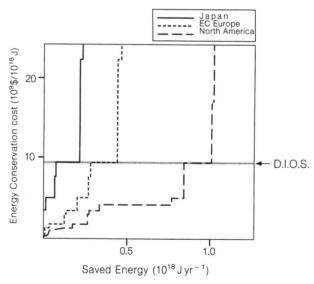

Figure 8.5 Energy conservation potential and its cost. The line marked DIOS indicates the introduction of the Direct Iron Ore Smelting process, saving much energy. (Reproduced by permission from Y. Kaya, Y. Fujii, R. Matsuhashi, K. Yamaji, Y. Shindo, H. Saiki, I. Furugaki and O. Kobayashi, *Assessment of Technological Options for Mitigating Global Warming*, Energy and Industry Subgroup, EG3 of the Intergovernmental Panel on Global Change, Geneva, 6–7 August 1991, Fig. 3.1.2, p. 6. This study refers to OECD Europe instead of the European Community. In practice the difference is negligible

economies. To achieve this Japan has invested considerably in energy conservation, stimulated by a high energy price.

This point is illustrated in Fig. 8.5, where the energy saving potential for the steel industry and its cost is displayed for the United States, Japan and the European Community. It is apparent that the inefficient economies can improve their energy efficiency for relatively low cost, whereas efficient economies have to make large investments to increase their efficiency even further. Figure 8.5 shows that Japan has to invest a lot to gain more efficiency and it obviously has the most efficient economy from an energy point of view. However, even Japan could operate with much higher energy efficiency with the right investments. The political question is who will pay for this, the poor or the wealthy?

The need for a world-wide policy on energy and environment is generally recognized. One often adopts the aim of *sustainable development*. This would mean an organization of the economic system that takes into account environmental constraints. Policy makers then have to accommodate the environmental arguments given here and elsewhere. The American economist Daly has concluded that a sustainable economy requires a *steady state economy*: one whose throughput remains constantly at a level that neither depletes the environment beyond its regenerative capacity nor pollutes it beyond its absorptive capacity [9]. At present we are far from this ideal.

8.3 AVERTING CATASTROPHE

Policy makers on energy and environment have to deal with all the factors discussed so far and perhaps many more. The points of the previous sections are summarized in Table 8.6. In the four columns of Table 8.6 we identify a problem, its effects or dangers, and solutions in the short, medium and long terms. We notice that virtually all measures bear an economic cost to society as a whole. Tax facilities for improving energy efficiency or for cleaning flue gases, for example, would imply a reallocation of public money or an increase in taxation elsewhere. It must be remarked, however, that improving energy efficiency on the demand side would slow down the increase in energy production and ease its environmental tensions. In this way, it would 'buy time' for further technical and political solutions.

To show that anything has its economic cost we remark that stabilizing the installed amount of nuclear power would have the positive effect of saving the limited resources of high-grade uranium and would limit the amount of high-level nuclear waste. However, the profits of the nuclear industry would be too low to pay for much research on 'better' or 'safer' use of resources. Consequently public money has to be spent on this subject, which is in fact what is done in many industrialized countries.

The only sociopolitical measure displayed in Table 8.6 is limiting population

Table 8.6   Problems in 'energy and environment' policy and their presumed solutions

Problem	Effects/dangers	Short/medium term solutions	Long term solutions
Limited resources of high-grade fossil fuels	(Armed) competition	Improving energy efficiency	Renewable energies
	Economic disruption	Limiting population growth Low-grade fuels	Nuclear power
Burning low-grade fossil fuels	Mining damage	Cleaning flue gases	Renewables
	High emissions High costs		Nuclear power
Fossil fuel CO_2 emissions	Climate change	Change to gas CO_2 sequestration Reforestation	Renewables Nuclear power
Limited resources high-grade uranium	Rising prices	No growth of nuclear power	Renewables
	Opposition to mining	Reprocessing	Breeding Fusion
Long lived nuclear waste	Environmental pollution	No growth of nuclear power Limited number of waste disposal sites	Renewables Fusion
High cost of renewables	Slow dispersion	Tax facilities	More research
Transportation solar hydrogen	Explosions	Keep out of populated areas	Absorption in metals
Transportation of oil	Oil spills	Safety on tankers Special sea routes	Pipelines Better tankers
Transportation of liquid gas	Explosions	Keep out of populated areas	

growth, which would reduce the strain on all resources. We showed in Fig. 8.3, however, that the demographic composition of the world population is such that growth will continue until after the year 2025, so limitation of population growth can have an effect in the long run only.

This discussion shows that many countries will meet difficulties in providing energy in the future. There are indeed solutions, but they all have their costs, question marks and sometimes dangers connected with them.* Often the dangers are well known, such as the explosion of tanks with liquid hydrogen. Society has had experience with tank explosions for some time and has by way of trial and error defined risks and imposed regulations that would put the risk to the general public as low as possible. In fact, Table 8.1 was originally produced to deal with the risks of transporting LPG (liquefied petroleum gas).

Given uncertainties concerning risks, costs and benefits, policy making necessarily takes the form of trial and error. The method of trial and error is only possible under the conditions that the consequences of errors are known and within acceptable limits, and that there is time to correct an error and improve technology and regulations. For several of the problems given in Table 8.6 this trial-and-error procedure is not possible.

Let us discuss as an example climate change from continued emissions of CO_2 and other greenhouse gases. In Section 3.3 we found that the resulting global warming is real, but we pointed out large uncertainties in the rate of the temperature rise as a function of time (cf. Fig. 3.17). It also appeared doubtful whether the uncertainties as to the rate of temperature rise will be reduced very soon. It is therefore clear that the method of trial and error will not work, for the effect of policy measures or neglecting to take them will only be clear after some decades. Correcting measures again will take a long time to have effect and in the meantime society will have to cope with consequences of—in hindsight—wrong decisions. These consequences in terms of local change of climate, rainfall, temperature, cloudiness, summer–winter difference, etc., only follow from model calculations, not from experience. It will therefore be difficult to take drastic policy measures, as experience with international climate agreements shows.

The social scientists Morone and Woodhouse recommend the following catastrophe-aversion system for cases like the greenhouse threat ([10], p. 135):

(a) Protect against the possible hazard. Do so conservatively.
(b) Reduce uncertainty. Do so through prioritized testing and prioritized monitoring of experience.
(c) As uncertainty is reduced and more is learned about the nature of the danger, revise original precautions. Strengthen them if new dangers are discovered or if the risks appear to be worse than originally feared; weaken them if the reverse proves true.

It is too early to discuss the third point. The second point is being executed already by climate research and climate modelling, but—as was said earlier—it may take decades before more satisfactory conclusions can be drawn. Main points of research are the influence of water vapour (cf. Table 3.3), the exchange

---

* In the second part of Section 8.3 we draw extensively on the book of Morone and Woodhouse [10].

between atmosphere and the oceans of heat and greenhouse gases (eq. 3.17) and the changes in albedo of the atmosphere and the earth's surface. Decisions on research priorities, however, are taken within a small circle of interested scientists with only marginal input from policy makers.

The first point of Morone and Woodhouse regards the mitigation of possible harmful effects. It is officially announced as a no-regret policy: take those measures that are least costly, have a value in themselves and slow down global warming. These measures were put down in Table 8.6. Besides the points mentioned before (transition to gas, renewables and perhaps nuclear power) one notices CO_2 sequestration (cf. Ref. 8, pp. 32–5) and reforestation. The first is the binding of CO_2 in flue gases by technical means and disposing of the gas in deep oceans or old gas wells. The second is the binding of CO_2 by biological means. This measure can only have short-term use as the land area that can be reforested is limited. One should also be aware that CO_2 disposal does not sequester the other greenhouse gases, which give half of the total effect of the man-made gases.

Reduction of Environmental Disruption

We should not only be aware of the limitations on easy energy resources and the danger of continuing CO_2 emissions to the climate, but also of the negative

Table 8.7 Energy consumption and emissions[a] in West German freight transport. It must be noted that within each means of transport larger units are more economical from an environmental point of view than smaller ones. Therefore trains with many wagons are 'better' than short trains and large trucks with one or two trailers are 'better' than small ones. There are many parameters that need to be optimized to find the 'best' transport system. (Reproduced by permission of Elsevier from Werner Rothengatter, Cost–Benefit-analyses for goods transport on roads, in *Freight Transport and the Environment* (eds. Martin Kroon, Ruthger Smit and Joop van Ham) Elsevier, Amsterdam, 1991, pp. 187–213, with some additions from Dr Rothengatter).

Emissions ($g\,ton^{-1}$-km^{-1})	Rail 1987	Rail 2005	Inland waterways 1987	Inland waterways 2005	Road 1987	Road 2005
CO_2	41	30	42	40	207	189
CH_4	0.06	0.04	0.06	0.05	0.30	0.30
$TVOC^b$	0.08	0.06	0.13	0.13	1.10	0.70
NO_x	0.20	0.07	0.50	0.50	3.60	2.40
CO	0.05	0.02	0.17	0.17	2.40	1.10
Equivalent energy ($kJ\,ton^{-1}$-km^{-1})	677	623	584	550	2889	2587

[a]Emissions during the process of constructing the vehicles and infrastructure have not been taken into account.
[b]TVOC = total volatile organic compound.

impact of other emissions into the environment. Let us look at the transport system, as there are many options for choice available. We discuss the emissions of C_xH_y, NO_x and SO_2 and first consider Table 8.7 where for West Germany the emissions for freight transport are given for the year 1987 and forecasted for the year 2005. They are compared for rail, inland water and road transport. For electric traction the fossil fuel energy consumed at the power stations and their emissions were taken into account. Although energy and emissions to build the vehicles have not been tabulated, it is expected that they do not change the overall picture.

We notice that energy use as well as most emissions are decreasing for all three means of transport, due to technical improvements. From an environmental point of view rail transport is the 'best' and inland water transport a 'second best'. If advantage was taken of this,* industries would need to concentrate along waterways with good rail connections. This will not happen without government regulations.

For passenger transport inland waterways are not very appropriate. However, from Fig. 8.6 one may draw similar conclusions as for freight transport. Emissions and energy use as well as accidents are given, all compared with rail transport over the same distance. It is again clear that a shift from the private car to rail transport would be the 'best' from an environmental point of view.

It is not easy for governments to push rail transport and discourage private transport. It would require a change of infrastructure, not building any more motorways but building and operating more rails, not only profitable intercity transport, but costly regional and local rail connections as well. Not many, if any, government will dare to do this for fear of losing public support.†

We doubt that governments will easily take steps to avert the threat of global warming. Besides the points already discussed, we add two. In the first place, it will not be attractive for one country to take expensive and perhaps unpopular measures when neighbouring countries do not comply. In the second place, developing countries with a low energy consumption per capita will want to catch up with the economic wealth of the richer countries and cannot bear the expense of complicated environmental measures. Ideally they should build up their industry and infrastructure to reach the low energy point of Japan in Fig. 8.4. This would require large scale aid from the richer countries, reducing the consumption there, and therefore would result in a reallocation of global wealth. At present such an approach seems not to be realistic.

---

*More sophisticated means of transport could be by means of pellets in underground pipelines. This technical solution requires high investment and strong backing.

†Some colleagues who read this believe that governments are taking steps, accepting declarations, etc., but that they are difficult to implement because of all kinds of vested interests. Others believe that increasing urbanization will stimulate rail transport naturally. However, the UN environmental agency expects a continuous increase in the number of private cars up to 10^9 by the year 2030 (cf. *Update, Science and Technology for Development*, 50/Summer 1992, p. 3).

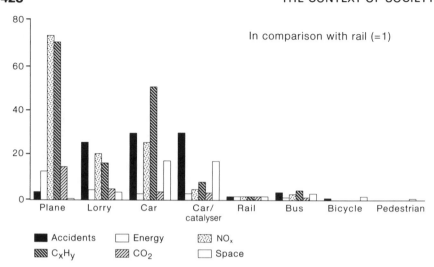

Figure 8.6 Environmental burden of passenger transport. (Reproduced by permission of Verkehrsclub Osterreich from Sante Mobilität: Strategien gegen den Verkehrsinfarkt, Verkehrsclub Osterreich, Vienna, 1991, p. 30. Based on West German data)

8.4 RENOVATING OUR THINKING

Why are drastic and—in our opinion—sound policy measures so difficult to take? There are several types of answers possible. Let us mention a few.

The most cynical explanation is that wealthy people and populations do not want to share with others and do not care about a future for their children and grandchildren. They just consume and follow the aristocrats on the eve of the French revolution with their slogan: *Apres nous le déluge* (After us, the flood). The economic powerful exploit these feelings and stimulate consumption by a never ending stream of new products, many of which are not basically different from existing ones. Companies have to change their machines at a vast pace to survive economic competition and old machines are discarded before the end of their technical life. The result is an increasing 'wealth' in terms of national product and an increasing amount of waste.

A somewhat kinder explanation may be that the need for drastic policy measures has not been proven beyond doubt. Policy makers know that they have to deal with a complicated society where in the course of many years a careful balance between differing interest groups has been established. They will follow the careful procedure discussed in Section 8.3. They hope that future technology will solve many of today's problems and allocate some money to research and development of renewable energies, nuclear fusion and generally of environmentally friendly production structures. This in fact corresponds with the behaviour of most governments.

As one cannot argue with selfishness, we hope that the truth is closer to the second argument. Then there are indeed policy changes possible following a well-informed public debate. We believe that it is the *responsibility of the scientist* to contribute to the dissemination of relevant knowledge. He or she should be aware of interest groups that tend to distribute partial (or even wrong) information to support their interests. The scientist could try to be as objective as possible but will realize that any reallocation of public resources is subject to political decisions in which views on man and nature play a role. As nature is the object of study in the physical sciences we elaborate a little on the argument that possibly our view on nature has to be questioned as well.

Our present way of thinking about nature can be traced back to the sixteenth century, the time of Descartes in France and of Francis Bacon in England. Nature was interpreted in religious terms as a gift of God which could and should be exploited for human purposes. Nature was regarded as something without a soul or a spirit. This approach was essential for the development of the modern scientific method with its systematic experiments.

Francis Bacon expressed it as follows: 'The passages and variations of nature cannot appear so fully in the liberty of nature, as in the trials and vexations of art'. Usually this is summarized by the somewhat stronger statement: 'Nature should be tortured to force her to witness her secrets' ([12], p. 5). Descartes supported in philosophical terms a dualism between spirit and matter where nature was modelled as a machine, that one could separate into its parts and put together again.

It must be added that early modern scientists like Newton and Kepler were very religious people. The latter wrote, for example, in the treatise where his third law was put forward: '...and finally deign graciously to effect that these demonstrations give way to Thy glory and the salvation of souls and nowhere be an obstacle to that' [13].

However, the modern scientific way of thinking eventually led to the general understanding that the laws of nature determine everything once its initial conditions are known. There is nothing mysterious, nothing spiritual in nature. For the general public this meant that one could handle nature as one wished.

The change in scientific thinking in the sixteenth century did not come alone. It was preceded by a revolutionary development in agriculture, where the peasants in North-West Europe were using an entirely new kind of plough with a vertical knife to cut the line of the furrow, a horizontal share to slice under the slod and a moldboard to turn it over. This more than anything else made possible the increase in food production necessary to feed the expanding populations. Since the Middle Ages nature has been exploited in many ways to feed the growing population shown in Fig. 8.3. Here it was aided in many ways by applied science and technology.

One may well question the view that science as such must lead to the understanding of nature as without purpose and without soul. The American social scientist Roszak claims that the agnostic philosophical program of the eithteenth century forced itself into the scientific way of thinking ([14], p. 115). The result is the underlying assumption that all things in nature can be and should be described without reference to premeditated design or purpose. It became a rule of the scientific method, but always was a metaphysical premise. Roszak may well be right. It is interesting that even developments in science put question marks or warning signs here.

The first is chaos, already mentioned in Section 3.3. As the laws of nature are non-linear one would need an impossibly precise knowledge of the initial conditions to be able to predict weather and climate precisely. This holds more generally for most natural phenomena.

The second question mark is posed by the experiments of the Frenchman Aspect, which show that even on a macroscopic scale of metres one may not consider parts of a physical problem separately. As this points closer to our argument, we show a very schematic representation of the Aspect experiment in Fig. 8.7.

It can be seen that a Ca source is excited into its first 0^+ excited state. It decays by an intermediate 1^- state into two photons. One is detected at the left, one at the right. The total angular momentum is zero, so they are circularly polarized, either both to the right or both to the left. Assume that it is both to the right. The detectors measure linearly polarized light only. Each photon has a 50% chance of going through; therefore there should be a 25% chance that both go through. The experiment shows, however, that when the left one goes through, the right one does the same and when the left one does not go through, neither does the right one.

A modern scientist will find this obvious as it follows from quantum mechanics that a single wave function describes the process. However, even without evoking quantum mechanics (which may not be the ultimate theory after all) it is clear that both photons are perfectly correlated. It means that one may not consider both detectors separately and conclude from their individual response to a single

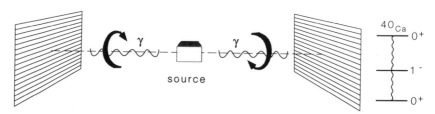

Figure 8.7 A Ca atom is excited in a 0^+ state releasing two polarized photons, one to the left and one to the right, which are 100% correlated. (Primas [12] gives a more elaborate description)

photon the total result. One may not even say that the whole is more than the sum of its parts as the parts may not be considered as separate entities.

In the opening article of a new Journal devoted to 'Ecological Perspectives in Science, Humanities and Economics', the Swiss scientist Primas concludes from these results that one may not think about nature in the old ways of Descartes and Bacon [12]. Perhaps we can say that nature is a unity in a deeper sense than scientists were aware of. Neither chaos theory nor the Aspect experiments prove that nature is inspired, nor that the agnostic programme of science leads to logical contradictions. They do suggest, however, that scientific knowledge is partial and that nature is more than just a few equations.

It is interesting that the same ideas come up in several parts of society. Architects are now talking about 'sacred design', referring to the old times where space, light, number and geometry were regarded as 'sacred', that is treated with due respect and awe. This would imply that a building, whether dwelling or office, should reflect nature as a microcosm and that the changes of natural light should connect man with his environment [15].

In the general public there is increasing interest in native cultures, for example of the American Indians. Well known is the story of the Pueblo Indians who believe that every human breath changes the world. The Indians, of course, did not anticipate chaos theory, but rather showed their feeling and understanding of the unity of nature. Another story tells how an American Indian in the middle of the nineteenth century refused to plough*, thus presenting a view of nature quite opposed to that of the North-West Europeans at the end of the Middle Ages:

You ask me to plough the ground.
Shall I take a knife
and tear my mother's bosom?

You ask me to dig for stones.
Shall I dig under her
skin for her bones?

You ask me to cut grass and make hay and sell it
and be rich like white men,
But how dare I cut off my
mother's hair.

With this way of thinking the earth could not support more than perhaps 100 million people and certainly not more than one billion. We do not advocate imitating the Indian approach in our society. Primas suggests thinking in terms of complementarity, a concept taken from quantum mechanics indicating two

---

*Smohalla, initiator of the dream culture, quoted in Ref. 16, p. 56. Reproduced by permission of Simon and Schuster.

Table 8.8 Two ways to look at the world. (Reproduced by permission of Shell International Petroleum Company Limited from Adam Kahane, global scenarios for the energy industry: challenge and response, Selected Papers, Shell Oil Company, January 1991, p. 9)

	Global mercantilism	Sustainable world
Challenge	Hegemonic decline and economic instability	Degradation of the environment (especially through global warming)
Response	Multipolar world and mercantilism	International cooperation and management
Implications for energy	New rules for business and reconfiguration of markets	New values for fuels and reconstruction of the energy industry

ways of thinking that both are required to understand nature. In a certain context one needs one approach, in another the other. In dealing with environment and nature we should find a way of incorporating nature in the economic system, but still treating it with due respect.

Perhaps we hear some echoes of this view in a study made by the Shell Oil Company. We reproduce a graph in Table 8.8. It says that with the same information one could look at the world from two points of view. The first, represented on the left in the table, observes the decline of the USA/Western hegemony. The answers then are given in terms of economic competition.

Alternatively one could view the degradation of the environment as the major point. The answer then is put in terms of a 'sustainable' world, respecting the limitations of nature and keeping enough of it for the developing regions of the world and for generations to come.

Needless to say we adopt the position of 'sustainable development'. The implications for science and society need to be worked out. Environmental scientists should have this in the back of their minds, while applying their professional skills.

Exercises

8.1 Find examples of hazards where the detriment definition (8.2) becomes questionable and indicate whether to your knowledge these have occurred. Discuss (a) the small probability of a catastrophic accident, (b) risks to the ecosystem. Are there risks that become greater for coming generations?

8.2 Assume that the energy consumption in each of the fossil fuels given in Table 8.2 increase by 3% annually. Calculate how long the proved commercially recoverable resources will last. Compare with the numbers without growth in energy consumption.

8.3   In finding the data of Table 8.3 we used the fact that 127 tons of natural uranium produce 1 GW_e year of electric energy. Use eq. (4.234) and the fact that three-quarters of the ^{235}U will fission to find its fission heat. Compare with the data given and conclude that roughly half of the heat results from ^{238}U burn-up.

8.4   Look for your own country in Fig. 8.4. If it is not there, find the data and put it in. Discuss the position of your country on the graph. Look at the energy price, the infrastructure, tax facilities and incentives to explain the data. Suggest how your country could improve its energy efficiency.

8.5   *Topic for class discussion.* What are the arguments in favour of or against an eco tax on polluting activities—the same for a CO_2 tax on CO_2 production or an energy tax on energy. If you are in favour of one or more of these taxes, how would you deal with developing countries not able to or willing to adopt such a tax?

8.6   *Topic for class discussion.* Would it be possible to define an optimum of energy consumption above which extra energy consumption gives only a small increase in well being? Look at Figs. 8.3 and 8.4.

8.7   *Topic for class discussion.* In densely populated countries like Germany one is considering a new measure for national well being, the gross ecological product (GEP) instead of the gross national product (GNP). The environmental costs of cleaning ground or water are taken as negative in the GEP and not positive as in the GNP. What problems do you see in this approach? Do you support it?

8.8   *Topic for class discussion.* Put yourself in the shoes of a politician. How would you argue in favour of investments in an energy infrastructure by which your country could reach the Japanese or Swiss data points in Fig. 8.4.

8.9   *Topic for class discussion.* At the beginning of this century NH_3 was used as a refrigerant in cooling devices. As was shown in Section 4.2.9 it has the right pressures and boiling point. Small leaks caused headaches; therefore a technical solution was found in CFCs. However, after leaking into the atmosphere they destroy the protecting ozone layer. Does this teach you to be careful with the claim that technical solutions work? Can you give more examples of technical solutions that gave unforeseen problems? Could you suggest alternatives to the use of CFCs in freezing and cooling that have less or no negative environmental consequences?

8.10 *Topic for class discussion.* In Table 8.7 and Fig. 8.6 the need for transport was not discussed. One could argue, however, that much freight transport could be avoided if people were prepared to lead a simper life, e.g. by using food grown in the neighbourhood. Also much passenger transport could be avoided if people could enjoy themselves nearer home. Do you believe such an approach to be realistic? Would it require another organization of society?

8.11 *Topic for class discussion.* In April 1977 the US president, Carter, gave a speech to the Americans, titled 'The Third Transition'. He gave ten principles of a national energy plan. Number 7 stated that the energy price should reflect the true replacement cost of energy. A litre of gasoline costs about $0.15 or ECU 0.12 off factory. A litre of alcohol fuel (renewable biomass) would cost $0.75 or ECU 0.60 off factory. So apparently President Carter would support alcohol as fuel instead of gasoline. What are the problems in putting President Carter's recommendation into effect? Consider taxation, reallocation of national income, competition with food production.

8.12 *Topic for class discussion.* In many countries kerosine as aircraft fuel is free of taxes and will cost less than $0.20 or ECU 0.16 a litre. Would you support taxation to make kerosine as expensive as fuel for rail or bus transport?

8.13 *Topic for class discussion.* In the first paragraphs of Section 8.4 reasons are given why policy measures in favour of an environmentally friendly system of production and consumption are so difficult to take. What would be your own position in the debate?

8.14 *Topic for class discussion.* Do the developments of chaos theory and the experiments of Aspect influence your way of thinking about nature, or are there other factors dominant in your thinking? If so, which?

8.15 *Topic for class discussion.* In looking at nature from an ethical–moral point of view one may distinguish three positions: a management ethic (one should protect the environment to guarantee production and quality of life, also in future), a kinship model (man, animals and plants are related; they all have rights to be protected) and an ethic of the environment (all living things have a moral status, to some extent even rivers and landscapes; man belongs to the ecosphere). What is your position? (See Ref. 17.)

References

1. Anne V. Whyte and Ian Burton (eds.), *Environmental Risk Assessment*, John Wiley, Chichester, 1980. May be used for Section 8.1; note, however that 'risk' here is defined as what we called detriment.
2. Al Gore, *Earth in the Balance, Ecology and the Human Spirit*, PLUME, Penguin, New York, 1993. A plea for a strong environmental policy, written by the later Vice-President of the United States of America.
3. C. Richard Cothern, Myron A. Mehlman and William L. Marcus, *Risk Assessment and Risk Management of Industrial and Environmental Chemicals*, Princeton Scientific Publishers, Princeton, New Jersey, 1988.
4. E. S. Ferguson, The measurement of the man day, *Scientific American*, **224** (October 1971) 96.
5. Lee Schipper and Stephen Meyers, with Richard B. Howarth and Ruth Steiner, *Energy Efficiency and Human Activity: Past Trends, Future Prospects*, Cambridge University Press, Cambridge, 1992. The book covers the subject matter of the present Sections 8.1, 8.2 and 8.3 in much more detail. It contains a prologue by John Holdren which makes the point that transition to a costlier form of energy is unavoidable.

6. T. Appenzeller, *Science*, **252** (June 1991) p. 1790.
7. Archie W. Culp Jr, *Principles of Energy Conversion*, McGraw-Hill, New York, 1991.
8. Y. Kaya, Y. Fujii, R. Matsuhashi, K. Yamaji, Y. Shindo, H. Saiki, I. Furugaki and O. Kobayashi, *Assessment of Technological Options for Mitigating Global Warming*, Energy and Industry Subgroup, WG 3 of the Intergovernmental Panel on Global Change, Geneva, 6–7 August 1991.
9. Herman E. Daly, *Steady-State Economics*, Freeman, San Francisco, 1977, and *GAIA*, **1**(6) (1992) 333–38.
10. Joseph G. Morone and Edward J. Woodhouse, *Averting Catastrophe, Strategies for Regulating Risky Technologies*, University of California Press, Berkeley, California, 1986.
11. Francis Bacon, *Advancement of Learning*, Book II, I, 6.
12. H. Primas, Umdenken in der Naturwissenschaft, *GAIA, Ecological Perspectives in Science, Humanities and Economics*, **1** (1992) 5–15. The Primas paper may be regarded as the trend setter for a new way of regarding nature, to be elaborated in this German language Journal. Use for Section 8.4.
13. J. Kepler, Harmonies of the world, in *Great Books of the Western World* (ed. Charles Glenn Wallis), Vol. 16, Chicago, 1952, p. 1080.
14. Theodore Roszak, *The Voice of the Earth*. Simon and Schuster, New York, 1992. Use for Section 8.4.
15. Ed Marzia, *Sacred by Design*, Progressive Architecture 03/91, pp. 74–5.
16. T. C. McLuhan, *Compilation, Touch the Earth, A Self-Portrait of Indian Existence*, Simon and Schuster, New York, 1971.
17. T. Regan, The nature and possibility of an environmental ethic, *Environmental Ethics*, **3** (1981) 19–34.

Bibliography

Conway, Richard A. (ed.), *Environmental Risk Analysis for Chemicals*, Van Nostrand Reinhold, New York, 1982. A main source as to the methods in use in connection with the chemical industry; use for Section 8.1.

Kletz, Trevor A., *Cheaper, Safer Plants or Wealth and Safety at Work*, Institute of Chemical Engineers, Rugby, Warwickshire, 1985. The report applies to chemical plants, and is of use for Section 8.1.

The Royal Society, Risk assessment, Report of a Royal Society Study Group, The Royal Society, London, January 1983. Concepts used in Section 8.1 are clearly defined here.

United Nations, *Energy Statistics Yearbook*, 1990, New York, 1992. Use for Section 8.2.

White Jr, Lynn, The historical roots of our ecological crisis, in *Dynamo and Virgin Reconsidered, Essays in the Dynamism of Western Culture*, IT Press, Cambridge, Mass., 1968; published earlier as *Machina ex Deo*. Gives background reading for Section 8.4.

World Resources 1986. Basic Books, New York. Use for Section 8.2.

World Resources 1990–1, Oxford University Press, New York, 1990. Use for Section 8.2.

Appendix A

Gauss, Delta and Error Functions

For many simple examples in mathematical physics the *Gauss function* $f(x)$ turns out to be a solution. In normalized form it is defined as

$$f(x) = \frac{1}{\sigma\sqrt{2\pi}} e^{-x^2/2\sigma^2} \tag{A.1}$$

with

$$\int_{-\infty}^{\infty} f(x)\,dx = 1 \tag{A.2}$$

It is shown in many texts that σ in eq. (A.1) equals the mean square distance to $x = 0$:

$$\sigma^2 = \int_{-\infty}^{\infty} x^2 f(x)\,dx \tag{A.3}$$

For a few values of σ the Gauss function is shown in Fig. A.1. One observes from Fig. A.1 that for decreasing values of σ the peak becomes narrower and higher while the integral remains constant because of eq. (A.2). For $\sigma \to 0$ one obtains a representation of the *delta function* $\delta(x)$ with the properties

$$\delta(x) = \delta(-x)$$
$$\int \delta(x)g(x)\,dx = g(0) \tag{A.4}$$

The last equality should hold for any physical function $g(x)$.

One often needs the integral of the Gauss function $f(x)$ up to a certain value β. This integral defines the *error function* erf (β) by

$$\text{erf}\,(\beta) = \frac{2}{\sqrt{\pi}} \int_{0}^{\beta} e^{-x^2}\,dx \tag{A.5}$$

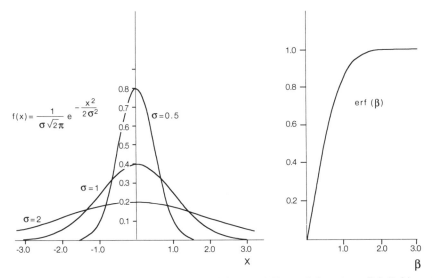

Figure A.1   The Gauss function eq. (A.1) for $\sigma = 0.5$, $\sigma = 1.0$ and $\sigma = 2.0$ (left) and the error function erf (β) (right)

It follows that

$$\text{erf}\,(0) = 0$$
$$\text{erf}\,(\infty) = 1$$
$$\text{erf}\,(\beta) = -\,\text{erf}\,(-\beta) \tag{A.6}$$

The error function is a monotonously increasing function of β, as displayed on the right-hand side of Fig. A.1. The complementary error function erfc (β) is given by

$$\text{erfc}\,(\beta) = \frac{2}{\sqrt{\pi}} \int_{\beta}^{\infty} e^{-x^2}\,dx \tag{A.7}$$

and therefore

$$\text{erf}\,(\beta) + \text{erfc}\,(\beta) = 1 \tag{A.8}$$

From Fig. A.1, or more accurately from tables of the error function, it may be found that 68% of the area of integral (A.2) is found in the region $-\sigma < x < \sigma$.

Appendix B

Some Vector Differentiations

Laplace Operator for Cartesian, Cylindrical and Polar Coordinates

The Laplace operator $\Delta = \nabla^2 = \text{div grad}$ is written in Cartesian coordinates as

$$\Delta = \frac{\partial^2}{\partial x^2} + \frac{\partial^2}{\partial y^2} + \frac{\partial^2}{\partial z^2} \tag{B.1}$$

In cylindrical coordinates (r, φ, z) in which z is the coordinate along the axis of the cylinder one may write the effect of Δ operating on a function $f(r, \varphi, z)$ as

$$\Delta f = \frac{1}{r}\frac{\partial}{\partial r}\left(r\frac{\partial f}{\partial r}\right) + \frac{1}{r^2}\frac{\partial^2 f}{\partial \varphi^2} + \frac{\partial^2 f}{\partial z^2} \tag{B.2}$$

For spherical polar coordinates (r, θ, φ) one writes

$$\Delta f = \frac{1}{r^2}\frac{\partial}{\partial r}\left(r^2\frac{\partial f}{\partial r}\right) + \frac{1}{r^2 \sin\theta}\frac{\partial}{\partial \theta}\left(\sin\theta\frac{\partial f}{\partial \theta}\right) + \frac{1}{r^2 \sin^2\theta}\frac{\partial^2 f}{\partial \varphi^2} \tag{B.3}$$

$$\Delta f = \frac{1}{r}\frac{\partial^2(fr)}{\partial r^2} + \frac{1}{r^3 \sin\theta}\left\{\frac{\partial}{\partial \theta}\left[\sin\theta\frac{\partial(fr)}{\partial \theta}\right]\right\} + \frac{1}{r^3 \sin^2\theta}\frac{\partial^2(fr)}{\partial \varphi^2} \tag{B.4}$$

The Gradient Operator ∇ in Spherical Polar Coordinates (r, θ, φ)

With e_r, e_θ and e_φ unit vectors in these three orthogonal directions one may write

$$\nabla\psi = e_r\frac{\partial \psi}{\partial r} + e_\theta\frac{1}{r}\frac{\partial \psi}{\partial \theta} + e_\varphi\frac{1}{r \sin\theta}\frac{\partial \psi}{\partial \varphi} \tag{B.5}$$

$$\nabla\cdot V = \frac{1}{r^2 \sin\theta}\left[\sin\theta\frac{\partial}{\partial r}(r^2 V_r) + r\frac{\partial}{\partial \theta}(\sin\theta V_\theta) + r\frac{\partial V_\varphi}{\partial \varphi}\right] \tag{B.6}$$

$$\nabla \times \mathbf{V} = \frac{1}{r^2 \sin \theta} \begin{vmatrix} \mathbf{e}_r & r\mathbf{e}_\theta & (r \sin \theta)\mathbf{e}_\varphi \\ \dfrac{\partial}{\partial r} & \dfrac{\partial}{\partial \theta} & \dfrac{\partial}{\partial \varphi} \\ V_r & rV_\theta & r \sin \theta V_\varphi \end{vmatrix} \tag{B.7}$$

with

$$\mathbf{V} = V_r \mathbf{e}_r + V_\theta \mathbf{e}_\theta + V_\varphi \mathbf{e}_\varphi \tag{B.8}$$

$$\nabla^2 \mathbf{V}|_r = \left(-\frac{2}{r^2} + \frac{2}{r}\frac{\partial}{\partial r} + \frac{\partial^2}{\partial r^2} + \frac{\cos \theta}{r^2 \sin \theta}\frac{\partial}{\partial \theta} + \frac{1}{r^2}\frac{\partial^2}{\partial \theta^2} + \frac{1}{r^2 \sin^2 \theta}\frac{\partial^2}{\partial \varphi^2} \right) V_r$$

$$+ \left(-\frac{2}{r^2}\frac{\partial}{\partial \theta} - \frac{2\cos \theta}{r^2 \sin \theta} \right) V_\theta + \left(\frac{-2}{r^2 \sin \theta}\frac{\partial}{\partial \varphi} \right) V_\varphi$$

$$= \nabla^2 V_r - \frac{2}{r^2} V_r - \frac{2}{r^2}\frac{\partial V_\theta}{\partial \theta} - \frac{2\cos \theta}{r^2 \sin \theta} V_\theta - \frac{2}{r^2 \sin \theta}\frac{\partial V_\varphi}{\partial \varphi} \tag{B.9}$$

$$\nabla^2 \mathbf{V}|_\theta = \nabla^2 V_\theta - \frac{1}{r^2 \sin^2 \theta} V_\theta + \frac{2}{r^2}\frac{\partial V_r}{\partial \theta} - \frac{2\cos \theta}{r^2 \sin^2 \theta}\frac{\partial V_\varphi}{\partial \varphi} \tag{B.10}$$

$$\nabla^2 \mathbf{V}|_\varphi = \nabla^2 V_\varphi - \frac{1}{r^2 \sin^2 \theta} V_\varphi + \frac{2}{r^2 \sin \theta}\frac{\partial V_r}{\partial \varphi} + \frac{2\cos \theta}{r^2 \sin^2 \theta}\frac{\partial V_\theta}{\partial \varphi} \tag{B.11}$$

The Gradient Operator ∇ in Cylindrical Coordinates (r, φ, z)

With $\mathbf{e}_r, \mathbf{e}_\varphi$ and $\mathbf{e}_z$ unit vectors in the three orthogonal directions one may write

$$\nabla \psi = \mathbf{e}_r \frac{\partial \psi}{\partial r} + \mathbf{e}_\varphi \frac{1}{r}\frac{\partial \psi}{\partial \varphi} + \mathbf{e}_z \frac{\partial \psi}{\partial z} \tag{B.12}$$

Appendix C

Physical and Numerical Constants

Planck's constant	$h = 6.625 \times 10^{-34}\,\text{J s}$
Planck's constant/(2π)	$h/(2\pi) = 1.0545 \times 10^{-34}\,\text{J s}$
Velocity of light	$c = 3.0 \times 10^8\,\text{m s}^{-1}$
Boltzmann's constant	$k = 1.38 \times 10^{-23}\,\text{J K}^{-1}$
Stefan–Boltzmann's constant	$\sigma = 5.672 \times 10^{-8}\,\text{W m}^{-2}\,\text{K}^{-4}$
Solar constant	$S = 1.353 \times 10^3\,\text{J s}^{-1}\,\text{m}^{-2}$
Universal gas constant	$R = 8.314\,\text{J K}^{-1}\,\text{mol}^{-1}$
Avogadro's number	$N_A = 6.022 \times 10^{23}\,\text{mol}^{-1}$
Rydberg constant	$R = 109\,678\,\text{cm}^{-1}$

Energy

1 EJ (exajoule)	$10^{18}\,\text{J}$
1 PJ (petajoule)	$10^{15}\,\text{J}$
1 TJ (terajoule)	$10^{12}\,\text{J}$
1 GJ (gigajoule)	$10^9\,\text{J}$
1 MJ (megajoule)	$10^6\,\text{J}$
1 eV	$1.602 \times 10^{-19}\,\text{J}$
1 ton coal equivalent	0.0293 TJ
1 ton oil equivalent	0.04187 TJ
1 m^3 natural gas	0.039021 GJ
1 kW h	$3.6 \times 10^6\,\text{J}$
1 barn	$10^{-28}\,\text{m}^2$

Air

Kinematic viscosity lower troposphere	$v = 14 \times 10^{-6}\,\text{m}^2\,\text{s}^{-1}$
Specific gas constant dry air	$R = 287\,\text{J K}^{-1}\,\text{kg}^{-1}$
Density (10 °C)	$\rho = 1.247\,\text{kg m}^{-3}$
Density (20 °C)	$\rho = 1.205\,\text{kg m}^{-3}$

   1 atm pressure $\qquad\qquad\qquad\qquad\qquad$ $p = 1.01325 \times 10^5 \, \text{Pa}$

   Fourier coefficient $\qquad\qquad\qquad$ $a = 22.5 \times 10^{-6} \, \text{m}^2 \, \text{s}^{-1}$

Earth

   Radius $\qquad\qquad\qquad\qquad\qquad\qquad$ $R = 6.37 \times 10^6 \, \text{m}$

   Angular velocity $\qquad\qquad\qquad\quad$ $\omega = 7.292 \times 10^{-5} \, \text{rad} \, \text{s}^{-1}$

Molecular weight M $\qquad\qquad\qquad\quad$ $1 \, \text{M} = M \times 10^{-3} \, \text{kg litre}^{-1}$

Index